Microorganisms for Sustainability

Volume 53

Series Editor
Naveen Kumar Arora, Environmental Microbiology
School for Environmental Science, Babasaheb Bhimrao Ambedkar University
Lucknow, India

Microorganisms have been in existence since the origin of life on earth and can survive the most extreme habitats or conditions on earth. Microorganisms are involved in regulating biogeochemical cycles, maintaining plant and animal health, and sustaining the global food chain. Moreover, they play crucial roles in addressing the challenges of climate change and achieving the targets of Sustainable Development Goals (SDGs).

This multidisciplinary book series captures the role of microbes towards building a sustainable world, while encompassing cutting-edge technologies and current needs across various fields such as agriculture sustainability, bioremediation, restoration of degraded habitats and wastelands, and food security. Additionally, this series explores microbial applications in industries, and building their utilization in clean and green energy solutions. Furthermore, themes like microbial secondary metabolites, extremophilic microbes and modern omics, including next generation sequencing and metagenomics, are also covered in this series.

With contributions from researchers across the globe, this series addresses the important call of 'One Planet-One Health-One Future'. It comprises a collection of diverse volumes that provides insights for scientists, young researchers, educators and decision-makers in the government, private sector, and non-governmental organizations, empowering their efforts to achieve the global goals.

The series invites, evaluates, and accepts book proposals to ensure a diverse, inclusive, and evolving program. The final decision regarding acceptance rests with the series editor.

Peer-review

This book series follows a stratified review process. Proposals for individual volumes are reviewed by the series editor and then the editorial board members. On a case-to-case basis, external reviewers are also invited for further evaluation of the book proposal. Review of the chapters is the responsibility of the volume editor(s). A manuscript submission platform has been recently made available to the authors, volume editors and the series editor.

Mukesh Meena • Andleeb Zehra
Prashant Swapnil
Editors

Multi-omics Approach to Investigate Endophyte Diversity

Springer

Editors
Mukesh Meena
Department of Botany
Mohanlal Sukhadia University
Udaipur, Rajasthan, India

Andleeb Zehra
Institute of Science, Department of Botany
Banaras Hindu University
Varanasi, Uttar Pradesh, India

Prashant Swapnil
Department of Botany, School of Basic Sciences
Central University of Punjab
Bathinda, Punjab, India

ISSN 2512-1901 ISSN 2512-1898 (electronic)
Microorganisms for Sustainability
ISBN 978-981-96-4099-7 ISBN 978-981-96-4100-0 (eBook)
https://doi.org/10.1007/978-981-96-4100-0

This Springer imprint is published by the registered company Springer Nature Singapore Pte Ltd.
The registered company address is: 152 Beach Road, #21-01/04 Gateway East, Singapore 189721, Singapore

Preface

Endophytes, the symbiotic microorganisms that reside within plant tissues without causing disease, have captivated the scientific community over the last few decades. Their ability to enhance plant growth, improve stress tolerance, and synthesize bioactive compounds places them at the forefront of research into sustainable agricultural and industrial practices. This emerging field of study presents a transformative opportunity for integrating microbial symbiosis into a broad spectrum of applications, ranging from agriculture to environmental management and biotechnology. However, the journey to fully understanding and utilizing the metabolic potential of endophytes is still in its nascent stages. With the global push toward sustainable practices, these microscopic allies may hold the key to addressing some of the most pressing challenges we face today—namely, food security, environmental sustainability, and the mitigation of climate change.

This book, *Metabolic Sustainability of Endophytes: Current Status, Challenges, and Their Potential*, seeks to present a holistic view of the role of endophytes in plant metabolism and sustainability, highlighting both their ecological importance and their potential in industrial applications. By focusing on the metabolic processes that endophytes influence within their plant hosts, this volume addresses several critical questions: How do endophytes modulate plant physiology to enhance resilience? What are the metabolic pathways involved in the biosynthesis of bioactive compounds? Can endophytes be engineered or applied on a large scale to meet the needs of modern agriculture?

Chapters will unfold the positions of fungal endophytes as reservoirs of bioactive metabolites with immense potential. The application of fungal endophytes in sustainable agriculture will emerge as a recurring theme, delving into their future role in plant health, anticipated abiotic stress management, and their contribution to the future production of extracellular enzymes. Moreover, the volume will explore the promising future applications of endophytes as biostimulant to offer sustainable solutions for anticipated crop production. This exploration will extend to their future in medicine and agriculture. Beyond agriculture, endophytes are also being explored for their ability to produce secondary metabolites with pharmaceutical value, such as antimicrobial, anticancer, and anti-inflammatory agents. This multifunctionality underlines the importance of deepening our understanding of these microorganisms and their metabolic capabilities.

This book brings together leading researchers and experts in the field to address key issues such as the variability of endophyte-plant interactions across different environments, the complexity of metabolic pathways, and the lack of standardized protocols for isolating and applying endophytes in real-world settings that pose significant hurdles. It provides a comprehensive overview of the current state of endophyte research while also exploring the potential of these microorganisms to contribute to sustainable practices. The chapters cover a wide range of topics, from the fundamental biology and diversity of endophytes to their metabolic interactions with plants, as well as their potential in biotechnological applications. Importantly, this volume also sheds light on the future directions of endophyte research. Advancements in genomics, metabolomics, and synthetic biology will undoubtedly accelerate the discovery of new endophyte strains with unique properties and the development of innovative applications for endophytes in agriculture, medicine, and beyond.

As editors, we are pleased to present *Metabolic Sustainability of Endophytes: Current Status, Challenges, and Their Potential* as a critical contribution to the field of endophyte research. We hope that this book serves as a valuable resource for researchers, academicians, and professionals alike, and that it will inspire future studies aimed at unlocking the full potential of endophytes for sustainable solutions in the face of global challenges. By deepening our understanding of the metabolic processes governed by endophytes, we can move closer to achieving a more sustainable and resilient world.

We extend our heartfelt thanks to all the contributing authors who have shared their expertise and insights, as well as to our reviewers for their critical evaluations. We are confident that this compilation of knowledge will be a timely and important addition to the scientific literature, inspiring further research and innovation in this exciting and rapidly evolving field.

Udaipur, Rajasthan, India — Mukesh Meena
Varanasi, Uttar Pradesh, India — Andleeb Zehra
Bathinda, Punjab, India — Prashant Swapnil

Acknowledgments

The successful completion of this book is the culmination of the collective efforts of many individuals and organizations. The editors extend their heartfelt appreciation to the contributing authors for their insightful chapters and unwavering dedication to this project. We are deeply grateful to the reviewers for their valuable feedback, which greatly enhanced the quality, coherence, and presentation of the chapters. Special thanks go to the members of the Editorial Advisory Board for their guidance and support throughout the compilation process. We also express our sincere gratitude to our respective institutions for their support and encouragement during this endeavor. In particular, we would like to thank the Vice-Chancellor, Director, Dean, and Head of the associated departments for their continuous backing and motivation. Finally, we are extremely appreciative of the publishing team at Springer Nature for their professional guidance and assistance in bringing this book to completion.

Mukesh Meena
Andleeb Zehra
Prashant Swapnil

Contents

About the Editors

Mukesh Meena completed his M.Sc. and Ph.D. in Botany from the Department of Botany, Banaras Hindu University, Varanasi, India. He is currently serving as an Assistant Professor in the Department of Botany at Mohanlal Sukhadia University, Udaipur, Rajasthan. Dr. Meena previously worked as a post-doctoral research fellow at the Department of Biotechnology, Faculty of Science, Jamia Hamdard, New Delhi, India. He is an active member of several prominent scientific societies, including the Mycological Society of India (MSI), the Indian Science Congress Association (ISCA), the Association of Microbiologists of India (AMI), and the Asian PGPR (plant growth-promoting rhizobacteria) Society for Sustainable Agriculture. Among his accolades, he received the Best Paper Presentation Award at the Fourth International Conference on Advances in Applied Science and Environmental Technology (ASET) held in Bangkok, Thailand, in 2016. He has participated in numerous international and national conferences and is well-versed in a range of analytical tools relevant to myco-pathological research. Dr. Meena has authored over 150 peer-reviewed publications, including research articles, review articles, and book chapters. Additionally, he has published six books with international publishers. He has been recognized in the "World Ranking of Top 2% Scientists" database, released by Stanford University, USA. This marks the third consecutive years—2022, 2023, and 2024—that he has been listed among the top 2% of scientists globally. He has also been ranked among the top 0.05% of scientists worldwide, according to the 2024 ScholarGPS® Rankings.

Andleeb Zehra has completed her Ph.D. from Centre of Advanced Study in Botany, Banaras Hindu University, Varanasi. She is a visiting lecturer in Mianz International College. She has extensive research experience in plant-microbe interaction, fungal biology, biological control, and plant growth promoting microbes. She is the recipient of one of the prestigious fellowships of UGC "Dr. D. S. Kothari Post-doctoral fellowship." She was awarded for her outstanding academic achievement by the governor of Uttar Pradesh, India. She has published more than 40 national and international articles and delivered numerous oral and poster presentations in several national and international conferences.

Prashant Swapnil graduated from Veer Bahadur Singh Purvanchal University and post-graduated from the Department of Botany, Banaras Hindu University, Varanasi, India. Dr. Swapnil did his Ph.D. from the Department of Botany, Banaras Hindu University, Varanasi. Currently, he is working as an Assistant Professor in the Department of Botany, Central University of Punjab, India. He has more than 20 peer-reviewed international publications and more than 12 book chapters published. He has participated in several International and National conferences. Dr. Swapnil has more than 7 years of teaching (from Banaras Hindu University, India, 2013–2015; Acharya Narendra Dev College, Delhi University, India, 2019–2020, Department of Botany, North Campus, Delhi University, India, 2020–2021) and 2 years of postdoctoral experience (from International Centre for Genetic Engineering and Biotechnology, New Delhi, India). Dr. Swapnil has expertise in Cyanobacterial Stress Physiology as well as Plant Pathogen Interaction.

Omics Approaches and Endophyte-Related Hormones for Promoting Host Plant Growth

1

Basanti Brar, Seema Joshi, Kiran Bala, Ravinder Kumar, Minakshi Prasad, Anju Manuja, Dilfuza P. Jabborova, Joginder Singh Duhan, and Baljeet Singh Saharan

Abstract

Endophytes are found within nearly all plants as part of diverse microbial communities. Thriving within their host plants, endophytes benefit from adaptation and protection, surpassing their counterparts in the rhizosphere. Like well-investigated rhizospheric plant growth-promoting bacteria (PGPB), endophytic PGPB employs various mechanisms to enhance productivity and plant growth. The current chapter examines these mechanisms and discusses their potential

B. Brar (✉)
Department of Biotechnology, School of Applied Sciences, Om Sterling Global University, Hisar, India
e-mail: basantibrar@osgu.ac.in

S. Joshi
Department of Environment Science, School of Applied Sciences, Om Sterling Global University, Hisar, India

K. Bala
Department of Biochemistry, School of Applied Sciences, Om Sterling Global University, Hisar, India

R. Kumar · J. S. Duhan (✉)
Department of Biotechnology, Chaudhary Devi Lal University, Sirsa, India

M. Prasad · A. Manuja
ICAR-NRCE, Hisar, India

D. P. Jabborova
Institute of Genetics and Plant Experimental Biology, Uzbekistan Academy of Science, Kibray, Uzbekistan

B. S. Saharan (✉)
Department of Microbiology, CCS, Haryana Agricultural University, Hisar, India
e-mail: baljeetsaharan@hau.ac.in

M. Meena et al. (eds.), *Multi-omics Approach to Investigate Endophyte Diversity*, Microorganisms for Sustainability 53, https://doi.org/10.1007/978-981-96-4100-0_1

applications. Endophytes offer environmentally friendly alternatives for agricultural practices, potentially replacing or supplementing chemicals like fertilizers, pesticides, and remediation agents. The interaction between endophytes and plant hormones, garnering attention due to scientific progress and the pursuit of sustainable agriculture, is explored here. Significant insights into the molecular mechanisms driving plant growth enhancement are gained by integrating omics approaches and studying endophyte-related hormones, such as auxins, cytokinins, gibberellins, and abscisic acid. This review aims to consolidate current knowledge and illuminate potential applications for sustainable agriculture. Technological advancements, a shift toward sustainable agriculture, the demand for stress-tolerant crops, and commercial opportunities drive the growing interest in omics approaches and endophyte-related hormones for promoting plant growth.

Keywords

Endophytes · Genomics · Plant growth · Proteomics · Transcriptomics

1.1 Introduction

Research into omics approaches and endophyte-related hormones to boost host plant growth is situated within the broader scope of sustainable agriculture. This field aims to develop eco-friendly practices to enhance crop yield and quality (Jin et al. 2024). The intricate relationship between plants and endophytes has long been acknowledged, and recent strides in omics technologies have illuminated the molecular mechanisms underlying these interactions, sparking a captivating area of study with profound implications for agricultural sustainability. Omics methodologies, spanning genomics, transcriptomics, proteomics, and metabolomics, provide potent tools to untangle the molecular complexities of plant-endophyte interactions (Kaki 2023). This chapter delves into the amalgamation of omics approaches and the influence of endophyte-related hormones on fostering host plant growth. Advancements in high-throughput sequencing and analytical techniques have significantly reduced costs and boosted the efficiency of acquiring omics data (Dai and Shen 2022). This has empowered researchers to conduct comprehensive investigations into plant-endophyte interactions across genetic, transcriptomic, proteomic, and metabolomic dimensions. Integrating omics data through system biology approaches yields a holistic grasp of the intricate networks governing these interactions. This enables analyses of the dynamic and interconnected molecular processes driving host plant growth (Suwinski et al. 2019). With a growing consciousness of environmental sustainability, there's an escalating emphasis on developing agricultural practices that minimize reliance on chemical inputs. Endophyte-mediated plant growth promotion offers a greener and more sustainable alternative, aligning with the global trend toward eco-friendly agriculture (Alori et al. 2017). As climate change presents new agricultural challenges, such as extreme weather and shifting disease patterns, a demand for stress-tolerant crops is heightened. Endophytes enhance plants' resilience to stressors by modulating hormones, making them invaluable assets in mitigating climate variability's impacts (Pandey et al. 2023).

With the world's population steadily rising, there's an increasing pressure on global food production. Finding sustainable methods to enhance crop yields is paramount for ensuring food security (Aryal et al. 2022). Endophytes' potential to bolster nutrient uptake and overall plant health positions them as promising contributors to meeting this demand. Insights gleaned from omics studies on endophyte-related hormones have practical applications in developing biofertilizers and biostimulants, offering natural and effective means of promoting plant growth while reducing reliance on synthetic fertilizers and chemical enhancers (Vejan et al. 2016; Chandran et al. 2021; Yadav and Meena 2022). Studying plant-endophyte interactions necessitates a multidisciplinary approach, drawing on expertise from microbiology, plant biology, genetics, and bioinformatics (Sharma et al. 2020). Enhanced collaboration across these domains has accelerated research progress and deepened our understanding of the underlying mechanisms. The prospect of commercial products derived from endophyte-related hormones has captured academic researchers' and the agribusiness sector's attention. Industry partnerships and investments in research and development further drive interest in harnessing the benefits of plant-endophyte interactions (Feng et al. 2017).

1.2 Omics Approaches in Understanding Plant-Endophyte Interactions

Omics approaches are pivotal in unraveling the intricate molecular mechanisms that govern plant-endophyte interactions (Shafi et al. 2021). These technologies, encompassing genomics, transcriptomics, proteomics, and metabolomics, offer comprehensive insights into the dynamic and complex nature of the symbiotic relationship between plants and endophytes (Fig. 1.1).

1.2.1 Genomics in Plant-Endophyte Interactions

1.2.1.1 Genomic Sequencing

Next-generation sequencing (NGS)-based technologies have transformed genomics, enabling fast and cost-effective complete genome sequencing. By applying genomic sequencing to both endophytes and their host plants, researchers can identify crucial genetic elements, including genes associated with plant growth promotion, nutrient uptake, and stress resilience (Toppo et al. 2023). Genomics, a field dedicated to studying genes and genomes, encompasses various aspects such as structure, function, evolution, mapping, epigenomics, mutagenomics, and genome editing. It plays a pivotal role in uncovering genetic variations that could enhance the efficiency of crop breeding, thereby facilitating genetic advancements in crop species (Mondal et al. 2023). Structural genomics delves into sequence modifications and chromosomal organization, aiding in the physical and genetic map

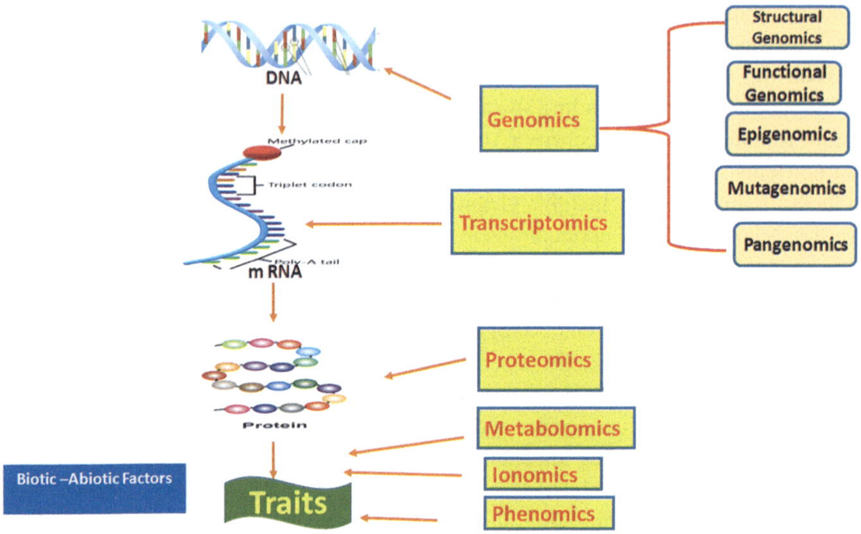

Fig. 1.1 Integration of multi-omics approaches in crop improvements

construction for identifying desirable traits in plants. In addition, functional genomics offers insights into gene functions concerning trait regulation (Gudi et al. 2024).

Epigenomics refers to epigenetic changes at the genomic level involving modifications like DNA or small RNA and histone modifications. Mutagenomics deals with mutational actions driving genetic modifications in mutant traits (Satam et al. 2023). Pangenomics which is defined as the sum of a core genome communal by all organisms and an expendable genome moderately shared or organism-specific, has emerged as a recent omics approach in crop (Zanini et al. 2022). These approaches focus on mutagenesis and the pangenome, respectively, offering new avenues for research and development in crop improvement.

1.2.1.2 Structural Genomics

Structural genomics relies on molecular markers, which are invaluable for identifying and mapping target genes, crucial for their integration into crop propagation programs. Marker-based techniques can be divided into PCR- and non-PCR-based (Amiteye 2021). Amplified fragment length polymorphism (AFLP), random amplified fragment polymorphism (RAPD), simple sequence repeats (SSRs), and single-nucleotide polymorphisms (SNPs) are PCR-based techniques (Betsy and Siva 2023), and restriction fragment length polymorphism (RFLP) is a non-PCR-based methods that investigate the DNA polymorphisms by DNA probe hybridization to a southern blot of restricted DNA fragment and then resulting in a disproportion DNA fragment report. RAPD markers amplify random DNA segments using single primers of arbitrary nucleotide sequences. AFLP conducts selective PCR amplification of restriction fragments from a total genomic DNA digest. SNPs are single-nucleotide changes in an organism's genome, detectable via genome sequencing PCR from diverse persons (Reshma and Das 2021). Diversity array technology (DArT) is a microarray-based high-throughput technique facilitating genotyping of several polymorphic loci throughout the genome.

The advent of NGS has facilitated the identification and utilization of SNPs. Approaches like genome-wide association studies (GWAS) and qualitative trait loci (QTL) mapping are employed to recognize numerous characteristics in crops (Yang et al. 2020). QTL mapping links complex genotypes with phenotypes, utilizing molecular markers like AFLPs and SNPs, correlating them with observed phenotypic data. GWAS identifies variants associated with traits by correlating genetic changes and phenotypes in residents depending on SNP sequence data. GWAS has been instrumental in enhancing crop tolerance to abiotic stressors, such as drought, by identifying relevant genomic regions and associated loci (Sukumaran et al. 2018). Structural variants (SVs) also play a crucial role in controlling agronomically important traits, as evidenced by GWAS in various crops like brassica, maize, and soybean.

Breeders utilize genotyping by sequencing and marker-assisted selection to enhance hybrid breeding, thereby increasing crop quality and yield. Multiparent mappings, such as nested association mapping (NAM) and multiparent advanced generation intercross (MAGIC), expose extensive phenotypic diversity achievable through experimental studies, aiding in breeding improvements (Susmitha et al. 2023). Analyses of genotype-phenotype relationships recognize QTLs that can be further validated via various functional approaches.

1.2.1.3 Functional Genomics and Multi-genomics

The wealth of information and resources provided by structural genomics will eventually be harnessed by functional genomics. It encompasses the advancement of comprehensive investigational approaches to evaluate gene function. Many biotechnological techniques have been devised to recognize and isolate target genes, clone and describe them, and conduct functional transgenic analyses by overexpressing or knocking out gene lines (George et al. 2020). Before the advent of genome sequencing approaches, identifying target genes involved arduous procedures, including expressed sequence tag (EST), suppression subtractive hybridization (SSH), and cDNA-AFLP sequencing. The introduction of NGS has streamlined these processes. Crop genome sequencing access has unveiled genes pivotal for resistance to disease and stress and yield determination (Zenda et al. 2021).

Additionally, genome engineering offers prospects for crop improvement through tools like the transcription activator-like effector nucleases (TALEN) and clustered regularly interspaced short palindromic repeats (CRISPR/Cas9 system). Genome editing tools, devoid of foreign DNA insertions, hold promise for enhancing yield by introducing pest and disease resistance in genetically modified crops (Singh et al. 2021). Using TALEN and CRISPR/Cas9 technologies, mutants like the bread wheat mildew resistance locus o (TaMlo) and SlMlo in tomato crops have been generated. The CRISPR/Cas9 system has been successfully employed for genome editing in various key crops, including soybean, rice, maize, and sorghum (Ansari et al. 2020).

Virus-induced gene silencing (VIGS) investigates gene functions in tomato crops in abiotic and biotic stress response. Comparative genomics identifies numerous mutants associated with crop development, growth, and stress tolerance in wheat, rice, and barley. A mutation in soybeans has been discovered to influence the transcriptome profiling of GmNARK (glycine max leucine-rich repeat receptor kinase), leading to nodulation independent of the jasmonate pathway (Karlo et al. 2020). Mutagenomics has emerged as a recent omics tool enabling the study of mutational changes, including genetic modifications in altered traits. High-throughput genomics technologies such as serial analysis of gene expression (SAGE), Targeting Induced Local Lesions IN Genomes (TILLING), high-resolution melt (HRM), and microarray investigation are utilized to illustrate these mutational events (Penna et al. 2023). TILLING, a functional genomics strategy initially, has evolved into a valuable technique for crop breeding and acts as an alternative approach to transgenics. Its feasibility has been demonstrated in essential crops like soybeans, rice, wheat, tomatoes, rapeseed (*Brassica napus*), and sunflowers, proving a possible technique for functional genetics and crop breeding enhancement (Yang et al. 2021). Mutants affecting seed oil synthesis have been screened through TILLING, exemplified by mutants wealthy in oils, isoflavones, and oleic acids (FAD2 and FAD2-1B) segregated in soybeans. Plowing has been broadly applied to identify yield changes, including rice, maize, wheat, grain, tomato, and soybeans.

Microarray investigations have uncovered that plant mutagenesis can induce more transcriptomic changes contrasted with those brought about by transgene inclusion. Mutagenesis is essential for distinguishing quality capabilities and fostering various agronomic characteristics with beneficial varieties (Raza et al. 2023). Roughly 3000 mutant assortments of different harvests have been grown universally, with 776 mutants guaranteeing dietary quality. Mutagenomics, through inverted hereditary methodologies, gives valuable chances to quiet and hinder applicant qualities for exploring quality capability (Narkhede and Kiranmayee 2023). Turnaround hereditary procedures like RNA interference (RNAi) and virus-induced gene silencing (VIGS) are used to screen/prompt crop changes when mutant alleles are inaccessible, permitting the examination of gene function. These reverse genetics methodologies have been used to separate transformations of rice, wheat, tomato, cotton, maize, chickpea, sunflower, soybean, and pea plants. Both utilitarian genomics and mutagenomics are instrumental in further developing crop yield, development, and resistance to stress (Gupta et al. 2023a, b, c).

1.2.1.4 Epigenomics

Epigenetics is the hereditary changes other than those in the DNA sequence and encompassing alterations brought about via post-transcriptional modifications of histones and DNA methylations. The convergence of epigenetics and genomics gives rise to epigenomics, a novel omics technique aimed at elucidating genetic regulation and its influence on the growth of cells and stress responses. Epigenomics can be affected by the environment's abiotic and biotic stress factors, but this is not the case with genomics. Genome-level studies can easily assess epigenetic events at various developmental stages or evaluate plant disease deformities (Chandana et al. 2022).

The technique (bisulfite sequencing) is employed to find the DNA methylation status of the genome, successfully validated in crops such as tomatoes, maize, and soybeans concerning DNA methylation and genes that regulate chromatin (Yang et al. 2021). DNA methylation quantification in the genome is commonly achieved via the methylation-sensitive amplified polymorphism (MSAP) method, demonstrated in crops like wheat and foxtail millet under salinity stress (Bhupenchandra et al. 2024). Furthermore, chromatin immunoprecipitation sequencing (ChIP-seq) presents a distinctive attitude for analyzing DNA methylation and histone proteins utilized in rice crops under drought stress. DNA methylation studies have explored epigenome modifications associated with tomato ripening and stable epigenome changes in tissue-cultured rice. Epigenomic investigations have proven beneficial in recognizing histone reforms linked with photosynthesis in maize (Perrone and Martinelli 2020). Nowadays, an epigenome-extensive research association investigated the MANTLED locus (hypomethylation) in oil palms (*Elaeis guineensis*) responsible for the mantled phenotype (hypomethylation). Karma (LINE) retrotransposon methylation was linked with normal fruit yield in clones compared to mantled clones. Whole-genome bisulfite sequencing (WGBS) identified ncRNAs in cotton crops under drought stress. Collectively, these findings highlight the potential of epigenomic applications in crop improvement in response to environmental stresses (Akhter et al. 2021).

1.2.1.5 Pangenomics

The pangenome concept encompasses a species' complete genomic repertoire, categorized into genes like core and dispensable. All individuals pool core genes; on the other hand, dispensable genes, also called accessory genes, are present in some individuals (individual specific) but not all (Liang and Lonardi 2021). Development of new genome analysis tools like various sequencing technology have facilitated the sequencing of multiple accessions of crop species. Pangenomic studies in crops such as rice, soybean, wheat, maize, *Brassica napus*, and *Brassica rapa* have unveiled the significant roles of dispensable genes in sustaining crop variety and enhancing quality. A pangenomic study involving nine morphologically diverse *Brassica oleracea* varieties and a wild relative revealed variable genes associated with major agronomic traits like vitamin biosynthesis, glucosinolate metabolism, and disease resistance (Mohd Saad et al. 2021). Various methodologies have been employed to construct pangenomes in crops and their wild relatives, including comparative de novo assembly, iterative assembly, and "map-to-pan" approaches (Khan et al. 2020). Comparative de novo approaches have been utilized with soybean, rice, and their wild relatives to analyze genetic variation and agronomic traits. Iterative assembly has been applied to species like *B. oleracea*, bread wheat, and *B. napus*, while a "map-to-pan" approach was adopted in rice.

Pangenomic studies have explored the diversity within genes of crop species, revealing that dispensable genes exhibit greater variability than core genes (Chapman et al. 2022). For instance, sets of dispensable genes in *Brachypodium distachyon*, rice, and soybean harbor higher densities of SNPs and insertions/deletions (InDels) compared to core gene sets. Dispensable genes are influenced by structural variation and have been reported to be enriched with genes related to disease resistance in crops like maize and rice and abiotic stress in barley and sorghum (Pagano et al. 2023).

Moreover, pangenomics is promising crop improvement by leveraging multiple crop wild relatives (CWRs) for their quantitative and adaptive traits (Bohra et al. 2022). These wild relatives have been instrumental in breeding programs, contributing genes for improving traits such as grain quality and stress such as biotic and abiotic resistance in crops like rice, sorghum, and wheat (Razzaq et al. 2021). Thus, pangenomic studies offer opportunities to identify elite genes in CWRs for crop improvement.

1.2.2 Transcriptomics and Gene Expression Profiling

High-Throughput RNA Sequencing (RNA-Seq) Transcriptomics, a field focusing on gene expression patterns using techniques like RNA-seq, enables researchers to delve into the entire transcriptome of plants and endophytes during their interaction (Xiao et al. 2024). The current approach recognizes upregulated or downregulated genes in response to endophyte colonization. Transcriptomics revolves around the transcriptome, encompassing the whole set of RNA transcripts formed by an organism's genome in a cell or tissue provided (Samuels et al. 2021). A special profiling technique has emerged as a promising gadget to check out quality expression for a long time, considering various upgrades. From the start, transcriptome components were explored using standard methodologies like cDNAs-AFLP, Differential Display Polymerase Chain Reaction PCR (DD-PCR), and SSH, yet these strategies offered limited objectives (Narkhede and Kiranmayee 2023). In this manner, more solid techniques, such as microarrays, high-level quality enunciation profiling, NGS, RNA-seq, and SAGE, changed RNA verbalization profiling. For example, microarray assessment has revealed differential quality enunciation on account of dry season strain in soybean and grain during different developmental and regenerative stages.

Moreover, quality explanation adjustments were found in soybeans without hydration stress using Affymetrix GeneChip arrays. The enunciation of characteristics has been associated with various record factors (TFs) in crops like *Arabidopsis*, soybean, and rice, considering abiotic stress. Novel TFs, such as Cys2/His-2-type zinc finger (C2H2-ZF) TF and drought and salt tolerance (DST), were recognized to oversee stomatal opening due to salt and drought tension in rice. Also, phytohormones, non-coding RNAs, and little peptides accept crucial parts in the quality explanation rule considering abiotic stress conditions in various yields (Jha et al. 2020). Transcriptomic studies have revealed differential gene expression patterns during growth and stress responses in crops like sorghum and rice. Recent progress in transcriptomics, for instance, in situ RNA-seq and spatially settled transcriptomics, has given encounters into tissue-unequivocal verbalization due to stresses in crops like foxtail millet and sweet potato (Parveen et al. 2023).

Moreover, RNA-seq has been instrumental in identifying drought stress-responsive genes in crops like rice, maize, and rapeseed oil. Comparative transcriptomics, including analyses across different crop species, has unveiled common genes responding to stress, paving the way for understanding regulatory networks involved in stress tolerance (Kumar et al. 2021).

Additionally, alternative splicing (AS) transcriptomics has emerged as a valuable approach to generate multiple transcripts in abiotic stress response conditions in crops like rice, maize, and sorghum (Rakkammal et al. 2022). These techniques collectively hold promise in regulating gene expression and improving crop species.

Temporal and Spatial Expression Patterns By examining temporal and spatial gene expression patterns, transcriptomic studies provide insights into the dynamic changes occurring at different stages of plant-endophyte interactions (Chen et al. 2021). This temporal resolution is crucial for understanding the progression of molecular events during the establishment of symbiosis.

1.2.2.1 Proteomics for Protein Profiling

Proteomics, a comprehensive method geared toward profiling the total expressed proteins inside an organism, is segmented into four wonderful aspects: sequence, structural, functional, and expression proteomics (Rajczewski et al. 2022). Proteomics unveils the amino acid sequences, normally discerned through excessive performance liquid chromatography. Structural proteomics delves into the architecture of proteins to understand their capabilities, utilizing techniques such as X-ray crystallography, nuclear magnetic resonance (NMR), crystallization and electron microscopy. Purposeful proteomics scrutinizes the functions of proteins through approaches like yeast-one or hybrids and protein microarray profiling (Sharma et al. 2022).

The progression in protein extraction and separation techniques has appreciably bolstered plant proteomic studies, each at pattern- and genome-wide scales. Conventional chromatography primarily based techniques include ion exchange chromatography (IEC), size exclusion chromatography (SEC), and affinity chromatography. But greater superior techniques like SDS-PAGE, two-dimensional gel electrophoresis (2-DE), and two-dimensional difference gel electrophoresis (2D-DIGE) have evolved for protein separation (Thikekar et al. 2023). Moreover, revolutionary techniques, including protein microarrays/chips, allow rapid evaluation of protein expression. The superior quantitative proteomic analysis employs methodologies like stable isotope labeling with amino acids in mobile lifestyle (SILAC), isotope-coded affinity tag (ICAT) labeling, and isobaric tag for relative and absolute quantitation (iTRAQ) (Yang et al. 2021). Functional proteomic investigations have exposed numerous pressure-responsive proteins in crops, including tomato, sunflower, wheat, and sugarcane. Quantitative proteomics using the iTRAQ approach has highlighted differential protein expression in potato and coconut plants below abiotic strain (Dikobe et al. 2023). Proteomic research in reaction to pathogens has been hooked up in *Vitis* species, while post-iTRAQ-based total analyses have unveiled translational modifications and protein-protein interactions within cell. Phosphoproteomics, geared toward reading protein phosphorylation, has been instrumental in identifying strain-related proteins in vegetation like wheat and grapevine (*Vitis vinifera* L.). Combining proteomics with metabolomics and purposeful genomics has supplied insights into strain biology and aided in molecular marker identification for clever legume breeding programs.

Comparative proteomic analyses have been conducted in numerous plants along with chickpea, *Medicago* spp., and common bean, uncovering differentially expressed proteins (DEPs) associated with abiotic stress responses. Procedures, such as MALDI-TOF, SDS-PAGE, LC-MS/MS, and iTRAQ, have been applied extensively in cereal vegetation like barley, maize, rice, and wheat below drought pressure, elucidating stress reaction pathways (Kuang et al. 2020). In essence, proteomics plays a pivotal position in unraveling practical mechanisms in crop technology in opposition to numerous stresses, ultimately improving crop yields.

1.2.3 Metabolomics for Comprehensive Metabolic Profiling

Metabolome refers to all the metabolites synthesized via metabolic pathways in a plant system. With the emergence of next-generation sequencing (NGS) technologies, researchers have received promising tools to understand the law of gene expression and the molecular foundation of cell responses in crops dealing with biotic and abiotic stresses (Ambrosino et al. 2020). Metabolomic interaction with NGS has furnished a foundation for predicting a preliminary metabolic community from the genome collection of an organ (Granata et al. 2021). By combining genome sequencing (NGS) with metabolite quantification (through mass spectrometry), researchers can increase techniques for crop development. Metabolites plays an important role in gene expression and also provides us the biochemical as well as phenotypic information about the organism. While proteomics focuses solely on gene merchandise, metabolomics allows the metabolic expression of proteins to be decided, identifying the biochemical processes important for gene functioning.

Given the numerous chemical and bodily houses of metabolites, separation and analytical techniques are essential to generate metabolic profiles from plant samples (Tsugawa et al. 2021). Several analytical techniques have been employed in plant systems to quantify metabolites, which include thin-layer chromatography (TLC), gas/liquid chromatography-mass spectrometry (GC/LC-MS), liquid chromatography-electrochemistry-mass spectrometry (LCEC-MS), nuclear magnetic resonance (NMR), direct infusion mass spectrometry (DIMS), Fourier transform infrared (FTIR) spectroscopy, and capillary electrophoresis-liquid chromatography-mass spectrometry (Dikobe et al. 2023). Among these, CE-MS, GC-MS, LC-MS, and NMR are the most normally used strategies in plant metabolomics, imparting unique benefits in selectivity, sensitivity, pace, and accuracy. NMR is thought for its rapidity and selectivity, even as mass spectrometry techniques (CE-MS, GC-MS, and LC-MS) provide high sensitivity and selectivity, albeit on the value of time.

Correct annotation and reporting of metabolomic information are crucial for identifying and studying metabolites well. Recently a new tool like liquid chromatography with tandem mass spectrometry (LC-MS/MS) has been developed for annotating and quantifying the metabolomic studies (Kitashova et al. 2023). Those recommendations embody diverse elements, including pattern instruction, replication, randomization, quantification, recuperation, recombination, ion suppression, and peak misidentification, serving as powerful gear for acquiring and reporting metabolite information. Moreover, sampling, quenching, metabolite extraction, and storage workflows have been elucidated. Adhering to those guidelines can prevent misinterpretation of metabolite statistics and ensure transparency in reporting LC/GC-MS-based metabolomics-derived statistics.

Flowers possess chemically complicated equipment as a main protection system toward abiotic strains and pathogens. The mechanisms of plant metabolic responses to pressure rely upon interactions between vegetation and stressors or pathogens (Dikobe et al. 2023). Metabolic profiling has identified the pivotal role of metabolites in cereal crops, along with rice, maize, and barley, in the presence of diverse biotic stressors. For instance, the metabolic profiles of rice types have revealed several metabolites or biomarkers in opposition to the gall midge biotype 1 (GMB1) pathogen using GC-MS (Sagare et al. 2021). Similar findings have been reported in response to pathogens like *Xanthomonas oryzae* pv. oryzae (Xoo) in rice plants and *Magnaporthe oryzae* in rice and barley plants. Phenylpropanoid and phenolic metabolites have been implicated in wheat vegetation in response to biotic strains.

Environmental metabolomics, focusing on plant responses to numerous abiotic stresses, holds promise in pressure physiology by elucidating metabolite modifications. Strategies like GC-TOF-MS have identified polyamine metabolites in rice vegetation below drought pressure conditions, and similar strategies have been employed to become aware of salt-tolerant cultivars in rice (Kaur et al. 2021). Metabolite analyses have also been carried out in wheat, maize, tomato, and soybean vegetation in reaction to drought, cold, and heat. Techniques along with LC/GC-MS, GC/EI-TOF-MS, HPLC, and NMR have been broadly applied in crop species like rice, tomato, maize, and soybean to research responses to abiotic (drought, salt, oxidative, and temperature) and biotic stress conditions. Manipulating plant metabolite pathways has shown promise in enhancing crop nutritional value and improving traits like anthocyanin production through metabolic engineering in tomato plants (Balestrini et al. 2021).

Consequently, in conjunction with other omics, including genomics, transcriptomics, and proteomics, metabolomics affords expertise in diverse features from the genome to the metabolome, including phenotypic characteristics. Sturdy correlations among incorporated omics have been determined in crop and plant responses to abiotic strains. In the end, metabolomics is critical in identifying and quantifying small molecules (metabolites) inside plant tissues and endophytes (Nagarajan et al. 2021). This approach affords precious insights into metabolic adaptations contributing to superior plant increase, nutrient metabolism, and stress responses. Integrating metabolomic statistics with other omics datasets enables pathway analysis, shedding light on interconnected metabolic pathways suffering from endophyte colonization and pressure.

1.2.4 Ionomics

Ionomics focuses on the ionome, which encompasses the whole nutrient content, including trace elemental composition, representing the cellular constituents of plant systems. It quantitatively measures an organism's elemental constituents and identifies mineral composition changes triggered by several physiological stimuli, developmental conditions, or genetic modifications (Niron et al. 2020). This dynamic approach can evaluate gene functions and networks characterizing the ionome in environmental and physiological stress responses.

High-throughput elemental analysis is performed in our ion SUMOylation facility using a variety of analytical tools, including inductively coupled plasma mass spectrometry/optical emission spectroscopy (ICP-MS/OES), neutron activation analysis (NAA), and X-ray crystallography. Induced plasma mass spectrometry is cost-effective and can analyze hundreds of samples daily with excellent sensitivity for content determination. However, despite its potential, ICP-OES may not be as sensitive as ICP-MS in some situations. X-ray fluorescence (XRF) allows for faster detection of elements but is less expensive than ICP-MS/OES. NAA is expensive and slow but can capture more than 30 elements simultaneously (Fadeel et al. 2017). Leaf ionome investigation has been instrumental in identifying plant ionome regulatory networks involved in phosphorus and iron homeostasis. ICP-MS was used to analyze grain and leaf ionomers to generate genetic maps, identify quantitative trait loci (QTLs), and detect the genetic diversity of mineral elements in rice crops (Campos et al. 2021).

Moreover, seed ionome analysis revealed differential gene expression, enhanced symbiotic response to mycorrhizal fungi, and altered growth phenotypes under phosphate deficiency in maize crops (Raza et al. 2021). Ionome profiling based on individual maize seeds can be influenced by environmental and genetic factors that affect seed ionome accumulation. During widespread elemental profiling, QTLs were identified as responsible for grain yield and mineral accumulation in maize crops. Additionally, ionomic studies revealed the relationships and responses of elements, minerals, and metabolites in barley under salt stress conditions. An ionic screen was performed for mutant lines of soybean crops that change the ionic composition of the seeds by measuring elemental concentrations by ICP-MS. Elemental profile analysis was also performed on tomato cultivars to monitor micro- and macronutrient concentrations under water stress conditions.

Similarly, ionomic profiling was performed to analyze the nutritional balance of different types of fruits, including orange, kiwi, apple, mango, and blueberry (Whitt et al. 2023). These ionomic studies suggest they play an important role in crop improvement and response to various abiotic and biotic stimuli. Integrating ionomics with other omics, such as genomics or metabolomics, can identify potential genes and their networks that improve crop resilience in response to environmental and physiological stresses.

1.2.5 Phenomics

Phenomics involves characterizing phenotypes by collecting extensive phenotypic data on a large scale. The phenome, on the other hand, encompasses the entire phenotype, influenced by interactions among genotype (G), environment (E), and management (M) factors, also known as genotype-phenotype-envirotype (G-P-E) interactions (Callister et al. 2024). Accurate phenotyping is essential for gene and QTL mapping of specific traits to improve genetic traits in crop plants. Automated high-throughput imaging technologies facilitate rapid phenotyping of traits using noninvasive methods such as color imaging of biomass, far-infrared canopies, LiDAR for measuring growth parameters, and magnetic resonance imaging for analyzing crop root systems. Furthermore, several basic imaging techniques like electrical resistivity, electrical capacitance, X-ray, computed, and positron emission tomography enable visualization of root systems without harming plant samples, especially in laboratory and greenhouse environments (Clark et al. 2020).

Red, green, and blue (RGB) imaging using visible light is utilized to assess canopy and root systems, with research employing RGB setups to determine crop phenotyping parameters. Infrared thermography has validated the significance of stomatal conductance under salinity stress in young barley and wheat seedlings. At the same time, chlorophyll fluorescence imaging has been applied to screen for abiotic stress responses in tobacco, canola, and cotton crops (Abu-Elsaoud et al. 2022). Digital imaging techniques have also measured boron toxicity under abiotic stress and identified disease symptoms caused by pathogens in barley crops, highlighting the crucial role of phenomic applications in evaluating phenotypic parameters in crops under biotic and abiotic stress conditions. Phenotyping methods are particularly essential for field analysis of crops, with hyperspectral and multi-technologies used to study different agronomic features (Ganeva et al. 2022). Among these, the laser-induced fluorescence transient (LIFT) technique is notable for studying the photosynthetic effectiveness of field crops.

Omics approaches have transformed our comprehension of plant-endophyte interactions by providing insights into molecular intricacies. Integrating genomics, transcriptomics, proteomics, and metabolomics offers a comprehensive view of symbiosis mechanisms (Khatabi et al. 2019). As technology progresses, omics approaches will continue to play a crucial role in advancing our understanding of these interactions and their potential applications in sustainable agriculture.

1.3 Integration of Multi-omics and System Biology Approaches for Crop Improvement

To comprehend the intricate workings of biological systems and their cellular components, integrating various omics techniques is imperative to understand how organisms respond in specific circumstances. Previously, combining metabolomics with functional genomics, proteomics, and transcriptomics allowed for a better understanding of the function of the genome from phenotype to environmental influences (Naik et al. 2023). This integration has been used in *Arabidopsis* and potato tubers to study metabolic and transcriptome profiles, revealing its usefulness in recognizing probable candidate genes for its functional analysis.

Progress in computational tools and omics approaches has led to the implementation of integrative omics approaches in plant sciences. For instance, epigenetic-based associations of multi-omics unveiled the regulation of the biosynthesis of lipids during the development of cotton fiber. Integrating GWAS and metabolic profiling has proven to be a potent method to analyze genetic and biochemical processes in target crop species such as maize, rice, and tomato. Importantly, integrating omics approaches has resulted in the development of abiotic stress-tolerant crop phenotypes (Gupta et al. 2023a, b, c; Chandran et al. 2020). Metagenomics and functional genomics have identified several mutants with précised variations related to development, growth, and stress tolerance in crops like maize, wheat, rice, and barley. Combined GWAS and HRPF approaches elucidated agronomic characteristics accountable for increased biomass, growth, and yield in rice plants. Similarly, QTL mapping combined with agronomic-associated gene performance identified numerous QTLs in maize crops, replacing traditional phenomics with a robust technique for breeding sciences and crop genetics (Ijaz et al. 2024).

Moreover, multi-omics methods can complement each other in reading organic techniques. This has been hooked up through the differential law of metabolites, proteins, and ions associated with salinity pressure in halophytes. Metabolomics serves as a bridge between genotypes and phenotypes (Katam et al. 2022). Combining ionome and metabolome strategies diagnosed the suppression of photosynthesis and boom costs in maize vegetation under alkaline situations. Similarly, leaf and grain ionome assessment discovered mineral detail genetic diversification in rice flora through genetic mapping and QTL identification.

Therefore, integrating multi-omics from genotype to phenotype might provide insights into the practical mechanisms of genes and their networks to enhance crop technology, genetics, yield increase, and resistance in opposition to physiological and environmental strain. System biology, aiming to apprehend whole organic structures through modeling, has furnished a foundation to combine multi-omics for holistic information of organisms concerning variation and development. These tactics have introduced molecular regulator networks' identity for salt stress tolerance in grapevine plants and proposed models for abiotic strain responses in crop vegetation (Mehari et al. 2021). In this context, a model integrating system biology with multi-omics, based on phenotype-to-genotype and genotype-to-phenotype approaches, may advance crop breeding and genetic manipulation (Fig. 1.1).

1.4 Endophyte-Related Hormones and Their Impact on Plant Growth

Microbes play vital roles in the lives of plants, ranging from useful to neutral to harmful. Among these, PGPBs are pivotal, offering various benefits to plants and fostering their growth through different mechanisms (Fig. 1.2). These mechanisms

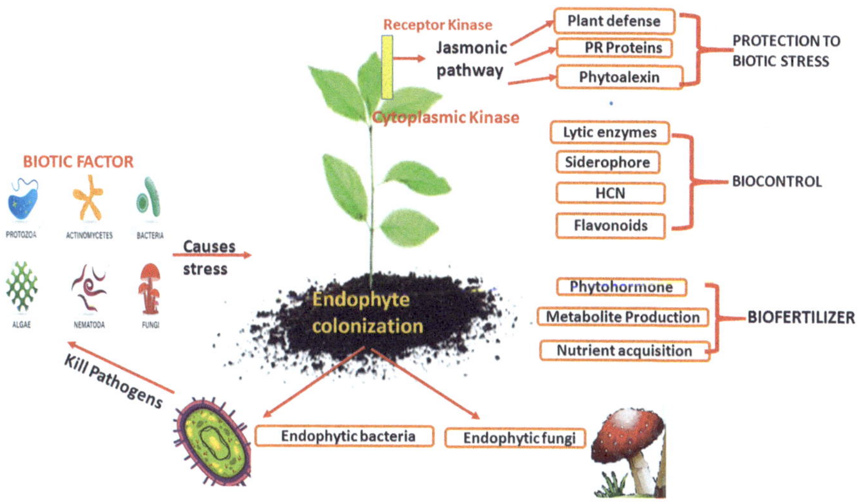

Fig. 1.2 Mechanism of endophytic microbes in plant growth and development

can be broadly categorized into direct and indirect methods (Vuolo et al. 2022; Kumar et al. 2024). Direct mechanisms involve bacteria providing essential nutrients and plant growth resources or modulating plant hormonal concentrations (Meena et al. 2020; Kumar et al. 2022). Alternatively, oblique mechanisms involve microorganisms countering growth inhibition due to plant pathogens, executed via the manufacturing of antibiotics, lytic enzymes, and siderophores, alteration in ethylene stages, induction of systemic resistance, and direct opposition with phytopathogens (Yadav and Meena 2021).

Bacterial endophytes, capable of entering and colonizing plant tissue without causing apparent harm or infection signs, play a crucial role in this symbiotic relationship. Endophytes often interact more effectively with their hosts compared to rhizospheric counterparts. They are generally neither organ- nor host-specific and have been extracted from various tissue types in several plant genotypes, sometimes even coexisting within the same plant (Hulin et al. 2020). It is important to differentiate endophytic bacteria from those causing dormant infections or colonizing senescent plant parts and transient bacteria that do not persist in plant tissues for extended periods.

While much research has focused on rhizospheric PGPB, studies on endophytic PGPB have been comparatively limited. The genetic and biochemical factors contributing to successful endophytic colonization and maintenance remain poorly understood. Nonetheless, numerous studies have explored the use of endophytic PGPB in phytoremediation schemes (Shahid et al. 2023). Notably, endophytic PGPB employs similar growth promotion mechanisms as rhizospheric PGPB, with the added advantage of colonizing the inside of plant tissues, enabling them to promptly and efficiently serve their host (Peng et al. 2023). Evidence suggests that endophytic PGPBs promote plant growth more than non-endophytic bacterial strains across diverse environmental conditions.

1.4.1 Direct Plant Development Systems Utilized by Endophytes

In direct plant development, microorganisms supply or adjust fundamental phytohormone levels, including auxin, cytokinin, ethylene, and gibberellin, or give plants supplements like phosphate, nitrogen, and iron.

1.4.1.1 Gibberellin Biosynthesis

Gibberellins (GAs) play pivotal roles in various metabolic functions crucial for plant development and growth, such as seed germination, stem elongation, formation of fruits, flowering, and senescence. Around 136 GAs have been recognized to date, each named by its solicitation for disclosure. Though the full bacterial gibberellin biosynthesis pathway has made sense, data about GA creation by bacterial endophytes remains limited, with few investigations (Williams et al. 2024). Among bacterial endophytes, *Acetobacter diazotrophicus* and *Herbaspirillum seropedicae* have been accounted for to produce gibberellins (GA1 and GA3) and indole-3-acetic acid (IAA), establishing endophytic relationships with Graminaceae species and promoting growth and yield. Similarly, *Azospirillum lipoferum* strain op 33, identified as an endophyte of grasses, has been recognized as a GA manufacturer in vitro. Notably, gibberellins produced by these bacterial endophytes were detected in various chemically defined media (Gupta et al. 2023a, b, c). Fungi represent another significant assemblage of endophytic microbes capable of producing GAs. Several fungi have been identified as GA producers, facilitating plant growth under various conditions (Paul et al. 2024). For instance, fungal endophytes *Aspergillus fumigatus* and *Scolecobasidium tshawytschae* isolated from salt and drought stress soybean cultivars were found to synthesize physiologically active gibberellins. These fungi promoted plant growth significantly compared to non-endophytic gibberellin-producing fungi, enhancing plant length and biomass. Similarly, fungal endophyte *Aspergillus fumigatus* sp. LH02 facilitated soybean growth under salt stress, increasing shoot length, biomass, and chlorophyll content while altering hormonal levels and metabolome composition to mitigate stress effects (Sethi et al. 2023). Additionally, various fungal endophytes isolated from different plant species have demonstrated the ability to synthesize GAs and promote plant growth under stress conditions, indicating their potential as bioresources for sustainable agriculture.

1.4.1.2 Auxin Biosynthesis

Auxins, pivotal plant hormones, orchestrate various plant growth and development stages. The most prevalent auxin is indole-3-acetic acid (IAA), synthesized in plants, fungi, and bacteria via distinct tryptophan-dependent pathways. The pathway employed by a bacterium to produce IAA, whether genetically or environmentally influenced, can influence the behavior of plant-microbe interactions (Etesami and Glick 2024). Notably, while many plant-beneficial microbes produce IAA through the indole-3-pyruvate pathway, pathogenic bacteria often synthesize IAA through the indole-3-acetamide (IAM) pathway. IAA's roles include elongation, cell division, leaf, root, flower initiation, fruit development, and senescence. Several plant growth-promoting bacteria (PGPB), including gram-positive and gram-negative strains, have been identified as IAA producers (Alhaddad et al. 2024). Among these, genera like *Alcaligenes*, *Rhizobium*, *Pseudomonas*, *Azospirillum*, *Azotobacter*, *Enterobacter*, *Klebsiella*, *Acetobacter*, *Pantoea*, *Herbaspirillum*, *Bacillus*, *Burkholderia*, *Streptomyces*, and *Rhodococcus* are prominent IAA producers. However, it's noteworthy that not all IAA-producing bacteria benefit plants (Bhatt et al. 2022). Phytopathogens often produce IAA at elevated concentrations, potentially affecting plant growth negatively. Several microbial endophytes have been investigated to enhance the growth of plants by biosynthesizing IAA. For instance, studies have shown significant IAA production among bacterial endophytes extracted from the apple tree buds and ginseng stems, with various isolates exhibiting substantial IAA yields. Additionally, investigations into soybean cultivars have revealed that endophytic populations demonstrate higher IAA-producing capabilities than epiphytic bacteria, highlighting the significance of endophytic bacteria in plant growth promotion (Adeleke et al. 2022). Moreover, crop-associated endophytes exhibiting IAA production have been shown to promote the growth of rice seedlings, indicating their potential role in enhancing agricultural productivity.

1.4.1.3 Production of Endophytic ACC Deaminase

The most common microbial enzyme 1-aminocyclopropane-1-carboxylate (ACC) deaminase (E.C. 4.1.99.4) has been extensively investigated in numerous researches (Glick et al. 2007; Glick 2014; Hardoim et al. 2008; Hontzeas et al. 2004a, b; Ose et al. 2003; Rashid et al. 2012; Walsh 1981). Their research has delved into various aspects of biochemistry, enzymology, regulation, mechanism, and prospective applications of this enzyme in plant growth and development.

The ACC deaminase, a multimeric enzyme, relies on pyridoxal 5′-phosphate as a vital cofactor for its enzymatic activity. It facilitates the cleavage of ACC into ammonia and α-ketobutyrate, with ACC serving as the instantaneous precursor of the plant ethylene hormone. This action effectively reduces ethylene levels among higher plants (Gamalero et al. 2023). Ethylene, akin to other phytohormones, holds significant sway over plant growth, development, and stress signaling. Endophytic PGPR expressing ACC deaminase have shown the ability to shield plants from diverse abiotic and biotic stresses while bolstering plant growth under non-stressful conditions (Etesami and Glick 2024). For instance, a cluster of 25 endophytes extracted from tomato, all harboring ACC deaminase, significantly boosted the canola seedling's growth compared to untreated ones. The potential of ACC deaminase-containing endophytes in promoting growth has been underscored in rice, tomato, and ginger plants.

In a recent study by Lowman et al. (2015), *Burkholderia phytofirmans* PsJN—bacterial endophytes—were investigated in a field study of switchgrass across two land sites over 2 years. Results revealed that inoculated switchgrass exhibited increased biomass production, improved tillering and root growth, and greater prior season vigor compared to control plants. Particularly, plants developed on low fertile land displayed superior performance compared with plants treated with bacterial endophytes. The scientists recommended that the observed growth promotion, especially on poor soil, may involve ACC deaminase possessed by the endophytes and the interplay of ethylene and auxins due to the present enzyme action (Rauf et al. 2020).

Moreover, it was found that a mutant of *Burkholderia phytofirmans* PsJN lacking ACC deaminase activity failed to uphold canola seedling growth in root elongation assay (Pal et al. 2022). However, mutant complementing with DNA containing the ACC deaminase gene isolated from wild-type strain restores its growth-promoting ability, emphasizing the significance of the ACC gene for experimental growth support (Rauf et al. 2020). Recent discoveries have also highlighted the role of ACC deaminase-containing endophytes in protecting plants from salt and drought stresses. For instance, the rice endophyte *Pseudomonas stutzeri* A1501 facilitated rice seedling growth in high and moderate salt concentrations and heavy metals (Bosu et al. 2024). Seeds treated with the wild-type strain exhibited significantly better growth than untreated or mutant-treated plants, confirming the pivotal role of ACC deaminase activity under such stresses. Additionally, ACC deaminase-containing endophytes have shown promise in enhancing plant growth under various stress conditions, such as salinity. These endophytes could produce IAA and improve nutrient uptake, benefiting the host under stress.

In conclusion, ACC deaminase is crucial in mediating plant-microbe interactions and promoting plant growth, especially under stressful conditions. Its presence in endophytic bacteria holds immense potential for sustainable agriculture practices, offering solutions to mitigate environmental stresses and enhance crop productivity (Negi et al. 2024).

1.4.1.4 Cytokinin Biosynthesis

Cytokinin constitutes a hormone group pivotal for promoting cell division in plant burgeoning buds, shoots, and roots. These hormones are ubiquitous across diverse plant species, including mosses, fungi, and bacteria. Currently, botanists recognize approximately 200 natural and synthetic cytokinins. Typically, most cytokinins originate from the root meristem and are conveyed to other plant parts via the xylem, the plant's vascular system (Uddin et al. 2024). Cytokinins have been detected in culture filtrate of various bacteria, encompassing *Rhizobium* species, *Azotobacter* sp., *Pantoea agglomerans*, *Pseudomonas fluorescens*, *Rhodospirillum rubrum*, *Bacillus subtilis*, and *Paenibacillus polymyxa*. Despite these bacteria being identified as endophytes in different plants, limited evidence conclusively links bacterial cytokinin production with plant growth promotion (Lastochkina and Bosacchi 2023). Notably, a study documented the separation, recognition, and characterization of bacterial endophytes producing cytokinin-like molecules. In Sambung Nyawa, i.e., *Gynura procumbens*, three bacterial endophytes were isolated and identified as *Acinetobacter calcoaceticus*, *Paenibacillus polymyxa*, and *Pseudomonas resinovorans*. Utilizing the ethylene acetate filtrate from bacterial culture media, cucumber cotyledons were inoculated in a green bioassay, yielding positive outcomes only from *Paenibacillus polymyxa* and *Pseudomonas resinovorans*. This suggests the bacterial endophytes' potential to act as plant growth-promoting agents (Eid et al. 2021). A similar investigation has been reported recently in the cucumber cotyledon greening bioassay to ascertain cytokinin production (Amaresan et al. 2020). After severe drought stress duration, alfalfa plants treated with a modified *S. meliloti* strain exhibited considerably greater size than those treated with parental strains (Diatta et al. 2021). This experiment underscores the potential of rhizobial strains synthesizing elevated cytokinin levels in enhancing plant resistance to severe drought stress despite the relative scarcity of cytokinin-producing PGPB.

1.5 A Model of Endophytic Plant Growth Promotion

A comprehensive model elucidating key aspects of plant growth promotion by plant growth-promoting bacteria (PGPB) has been proposed in previous studies. This model, depicted in Fig. 1.3, outlines the mechanism through which an ACC deaminase- and indole-3-acetic acid (IAA)-producing bacterium enhances plant growth, particularly under stressful surroundings such as drought, salinity, soil toxicity, and flooding with organics and heavy metals, as well as fungal and bacterial pathogens (Dubey et al. 2020). Notably, this mechanism applies to both endophytes and rhizosphere bacteria, although endophytes exhibit faster and more efficient interaction with plants. ACC deaminase acts as an inducible enzyme expressed at very slight levels in the nonexistence of its inducer and substrate, 1-aminocyclopropane-1-carboxylate (ACC). Within the plant tissue, such as grains containing ACC deaminase expressing endophyte gains several advantages: it can (1) use small nutrient molecules formed by the plant, (2) respond rapidly to plant stress, and (3) defend its host plant instantly followed by stress exposure, even before symptoms become evident

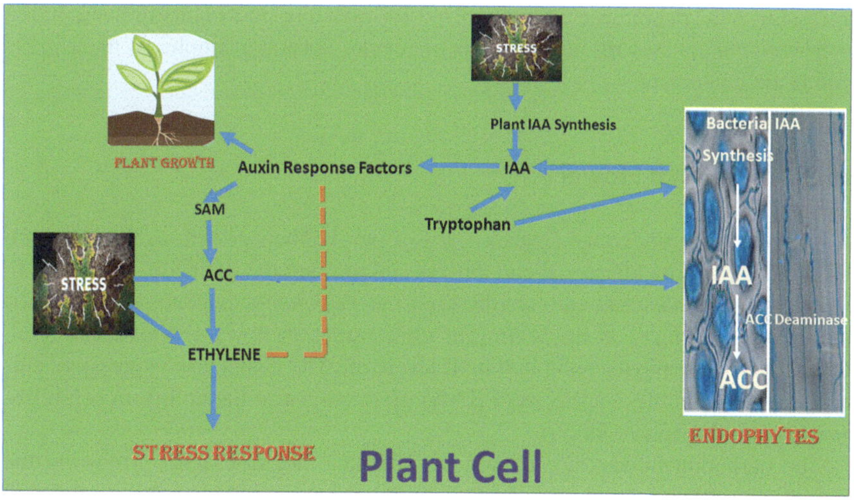

Fig. 1.3 A mechanism of plant growth promotion by interaction between plant tissue and endophytes

(Alzate Zuluaga et al. 2021). Resident endophytes increased and transported the IAA level, promoting plant growth and stimulating ACC synthesis. This is because IAA stimulates the transcription of mRNA that codes ACC synthase in plants, leading to ACC production from S-adenosyl methionine (SAM).

Elevated ACC levels trigger the ACC deaminase synthesis in endophytes, which subsequently breaks down a significant portion of ACC to ammonia and α-ketobutyrate. As a result, ACC levels within plant tissues are reduced, mitigating the production of harmful ethylene. Since ACC oxidases, accountable for ethylene biosynthesis, have a lower Km value for ACC oxidase than ACC deaminase, ACC deaminase must function before considerable ACC oxidase induction occurs, preventing excessive ethylene production. The interplay between IAA construction, ACC deaminase, and stress-associated factors was investigated in a microarray analysis involving the rhizospheric bacterium that produces ACC deaminase, e.g., *Pseudomonas putida* UW4 treatment on canola seeds (Phour and Sindhu 2023). This study revealed that wild-type *P. putida* UW4 upregulates plant genes concerned with auxin signal transduction and downregulates genes encoding proteins that guard plants against environmental stresses, as these proteins become unnecessary under reduced ethylene levels. Briefly, this model showed that bacteria producing both IAA and ACC deaminase promote plant growth by reducing ethylene levels, thereby enhancing IAA signal transduction, ultimately leading to growth promotion as an outcome of IAA functionality.

1.6 Hormonal-Based Endophytic Environmental Stress Control

Given that elevated ethylene levels can induce stress responses in plants, endophytes expressing ACC deaminase are recognized for assisting host plants in overcoming environmental stresses throughout their life cycle. Several examples of these interactions are given here.

1.6.1 Wilting of Flower

Flowers naturally undertake senescence as a part of their growth cycle. Research has demonstrated that bacterial endophytes can prolong the shelf life of flowers. In a specific study, bacterial endophytes from the *Pseudomonas* genus were investigated for their effects on the wilting of mini carnation flowers (Nakkeeran et al. 2021). These endophytes exhibited high levels of ACC deaminase expression, siderophores, and indole-3-acetic acid (IAA) production and the ability to solubilize the inorganic phosphate in vitro.

Mini carnation flowers are known to undertake fast senescence following the accumulation of high ethylene levels, making them susceptible to ethylene-induced wilting. Bacterial endophytes expressing ACC deaminase offer protection to carnation-cut flowers by reducing ACC concentrations in the tissues, thus preventing stress ethylene production and delaying flower senescence for an additional 2–3 days compared to untreated flowers or those treated with ACC deaminase-deficient microbial mutants.

In commercial practices, various cut flowers, such as carnations, are often treated with chemical ethylene inhibitors such as silver thiosulfate to widen their shelf life (Email and Eldeeb 2021). Consequently, environment-friendly utilization of non-hazardous bacterial endophytes presents a smart alternative to chemical treatments in the multibillion-dollar cut flower industry.

1.6.2 Salinity Stress

Salinity poses a significant challenge to agriculture, affecting crop growth worldwide. Bacterial endophytes have emerged as effective solutions for protecting various plants under salt stress conditions. Plants subjected to salinity stress often exhibit elevated ethylene levels that the enzyme ACC deaminase action can mitigate. Two bacterial endophytes isolated from tomato plants—*Pseudomonas migulae* 8R6 and *Pseudomonas fluorescens* YsS6—have demonstrated the ability to give protection and promote growth in tomatoes even under extreme salt stress conditions (185 mM) (Bhupenchandra et al. 2022). These endophytes exhibit traits such as ACC deaminase production, indole-3-acetic acid (IAA) synthesis, siderophore production, and phosphate solubilization in vitro. To isolate the specific contribution of ACC deaminase, mutant strains deficient in ACC deaminase were generated in both endophytes. On the surface, tomato seeds were treated with either mutant or wild-type endophytes or no bacteria as a control (Guo et al. 2020).

Plants treated with wild-type microbial endophytes displayed considerably superior chlorophyll levels, increased biomass, greater fruit yield, and reduced sodium accumulation compared to treated plants with ACC deaminase-deficient mutants or without bacteria at all under salinity stress (Ayuso-Calles et al. 2021). These findings and other field studies highlight bacterial endophytes' capacity to have the enzyme ACC deaminase to enhance plant growth and alleviate stress induced by ethylene accumulation.

1.7 Synergistic Effects of Omics Approaches and Endophyte-Related Hormones

The metabolic attributes of seed-borne associated fungal and bacterial endophytes are contingent upon local environments and harnessed to make possible the advancement of host plants. This enhances the host plant's fitness relative to others, potentially influencing the entire ecosystem.

Seed-borne endophytes represent a fascinating vicinity in plant-microbe interactions, influencing plant fitness, boom, and environmental adaptability. Understanding their modes of vertical transmission is essential for efficiently harnessing their capability in agriculture. Diverse mechanisms are studied through which seed-borne endophytes are transmitted vertically from one era of plants to the next (Vaucheret and Voinnet 2024).

One of the modes of vertical transmission includes endophytes living on the surface or in the seed coat. Seeds can harbor endophytic microorganisms externally, eventually colonizing rising seedlings as they germinate. This transmission mode is facilitated by physical touch among the seed and the encompassing environment, where endophytes may have hooked up a gap throughout seed maturation. Examples include fungal endophytes such as *Claviceps* species in grasses that are known to colonize seeds and persist through generations. Another massive mode includes endophytic colonization of the seed embryo itself. Endophytes can reside inside the developing seed's tissues, including the embryo and endosperm. During seed maturation, these endophytes grow to be included in the plant's reproductive systems, ensuring their transmission to the next era upon seed germination. This inner colonization is discovered in an extensive range of plant species and includes bacteria, fungi, or even viruses tailored to stay in the plant's reproductive organs. Some endophytes are transmitted vertically via inflamed floral components, which include pollen grains or floral tissues (Bard et al. 2024). These endophytes, anthers, stigma, and ovules can colonize the plant's reproductive organs. When seeds increase from fertilized ovules, they may deliver endophytes positioned inside the reproductive systems. This transmission mode is less commonplace but highlights the adaptability of endophytes to distinct plant tissues and reproductive strategies. Endophytes can establish systemic infections in the plant, spreading from roots to shoots and subsequently accomplishing the developing seeds. This transmission mode includes the endophyte's ability to emigrate via vascular tissues and input growing seeds as they shape. Systemic colonization guarantees that endophytes are distributed at some point in the plant, including its reproductive organs, thereby ensuring their presence in seeds.

In some cases, endophytes can be transmitted maternally through the seed coat permeation. This process entails endophytes migrating from the maternal tissues surrounding the seed and colonizing the developing embryo or endosperm. It is based on the seed coat's permeability to allow endophytes to enter the seed's interior, wherein they can establish symbiotic relationships with the developing plant tissues.

Vertical transmission of microbial endophytes has been observed across numerous plant species, such as the isolation of *Mycobacterium* spp. and *Bacillus* sp. harvested from wheat grass seeds grown in 1 year that indicated the presence of the same microbial endophytes in multiple generations. This type of study signifies that seeds serve as a crucial source for mature wheat grass endophyte transmission from generation to generation (Vergine et al. 2024). Furthermore, while fungal endophytes were initially believed to be transmitted horizontally, vertical transmission has been documented in several plant species (Frank et al. 2017). Some studies even report vertical transmission rates exceeding 90% for many fungal endophytes.

1.8 Summary

Multi-omics methods and endophyte-associated hormones represent innovative agricultural strategies to enhance plant growth and productivity. The synergistic integration of these fields holds promise for sustainable agriculture by decreasing reliance on chemical inputs and fostering environmentally friendly practices. This conclusion and perspective will explore their capability, modern-day demanding situations, and destiny guidelines in promoting host plant increase. The utility of omics in plant-endophyte interactions has unveiled complicated mechanisms by which endophytes affect plant physiology. The integration of omics strategies with research on endophyte-associated hormones represents a paradigm shift in agricultural research and exercise. By combining high-throughput molecular analyses with a knowledge of microbial-plant interactions, researchers can uncover novel strategies for optimizing plant growth and fitness. This interdisciplinary approach no longer most effectively expands our essential information on plant biology but also offers sensible solutions for sustainably addressing global food safety challenges. Moreover, the flexibility of omics technology allows researchers to discover the range of endophyte groups throughout exceptional plant species and environments. This know-how is crucial for developing tailor-made techniques to enhance plant resilience and productiveness below varying climatic situations and agricultural practices.

Numerous challenges and opportunities emerge in omics and endophyte-related hormone research. Standardizing experimental protocols and bioinformatic tools for analyzing omics records might be vital to ensure reproducibility and comparison across studies. Furthermore, elucidating the best mechanisms underlying endophyte-mediated hormone signaling pathways will enable centered manipulation of these interactions to maximize agricultural benefits. Regarding practical programs, translating study findings into agricultural practices would require collaboration among researchers, agronomists, and farmers. Field trials and on-farm demonstrations will be pivotal in validating the efficacy of endophyte-based interactions under real-world conditions and ensuring their financial viability.

In conclusion, omics procedures and endophyte-related hormones offer interesting prospects for revolutionizing agricultural practices closer to sustainability and resilience. By harnessing the power of microbial partnerships and molecular technology, we can release new avenues for enhancing crop yields, reducing environmental impacts, and ensuring food security for future generations. Continued efforts in studies and improvement on this promising area could be crucial for knowing these potentials and addressing the demanding situations of world agriculture within the twenty-first century.

Author's Contribution Conceptualization, B.S.S., J.S.D., and B.B.; validation, J.S.D., D.P.J., M.P., and A.M.; formal analysis, B.S.S. and J.S.D.; data curation, B.B., K.B., and S.J.; writing-original draft preparation, B.B., K.B., and R.K.; writing-review and editing, J.S.D., B.B., and P.K.S.; figure, S.J.; supervision, B.S.S., J.S.D., and B.B. All authors have read and agreed to the published version of the chapter.

References

Abu-Elsaoud AM, Abdulmajeed AM, Alhaithloul HAS, Soliman MH (2022) Role of electromagnetic radiation in abiotic stress tolerance. In: Plant abiotic stress physiology. Apple Academic Press, pp 355–391

Adeleke BS, Fadiji AE, Ayilara MS, Igiehon ON, Nwachukwu BC, Babalola OO (2022) Strategies to enhance the use of endophytes as bioinoculants in agriculture. Horticulturae 8:498. https://doi.org/10.3390/horticulturae8060498

Akhter Z, Bi Z, Ali K, Sun C, Fiaz S, Haider FU, Bai J (2021) In response to abiotic stress, DNA methylation confers epigenetic changes in plants. Plants 10:1096. https://doi.org/10.3390/plants10061096, PubMed: 34070712

Alhaddad FA, Bitaar ZM, Abu-Dieyeh MH (2024) Diversity, characterization, and potential applications of bacterial endophytes isolated from the halophyte *Limonium axillare*. J Plant Growth Regul 43:2179–2196. https://doi.org/10.1007/s00344-024-11254-6

Alori ET, Dare MO, Babalola OO (2017) Microbial inoculants for soil quality and plant health. In: Sustainable agriculture reviews, pp 281–307. https://doi.org/10.1007/978-3-319-48006-0_9

Alzate Zuluaga MYA, Milani KML, Miras-Moreno MB, Lucini L, Valentinuzzi F, Mimmo T, Pii Y, Cesco S, Rodrigues EP, de Oliveira ALM (2021) The adaptive metabolomic profile and functional activity of tomato rhizosphere are revealed upon PGPB inoculation under saline stress. Environ Exp Bot 189:104552. https://doi.org/10.1016/j.envexpbot.2021.104552

Amaresan N, Kumar MS, Annapurna K, Kumar K, Sankaranarayanan A (eds) (2020) Beneficial microbes in agro-ecology: bacteria and fungi. Academic Press, Cambridge, MA

Ambrosino L, Colantuono C, Diretto G, Fiore A, Chiusano ML (2020) Bioinformatics resources for plant abiotic stress responses: state of the art and opportunities in the fast evolving-omics era. Plan Theory 9:591. https://doi.org/10.3390/plants9050591. PubMed: 32384671

Amiteye S (2021) Basic concepts and methodologies of DNA marker systems in plant molecular breeding. Heliyon 7:e08093. https://doi.org/10.1016/j.heliyon.2021.e08093. PubMed: 34765757

Ansari WA, Chandanshive SU, Bhatt V, Nadaf AB, Vats S, Katara JL, Sonah H, Deshmukh R (2020) Genome editing in cereals: approaches, applications and challenges. Int J Mol Sci 21:4040. https://doi.org/10.3390/ijms21114040. PubMed: 32516948

Aryal JP, Manchanda N, Sonobe T (2022) Expectations for household food security in the coming decades: a global scenario. In: Future foods. Academic Press, Cambridge, MA, pp 107–131

Ayuso-Calles M, Flores-Félix JD, Rivas R (2021) Overview of the role of rhizobacteria in plant salt stress tolerance. Agronomy 11:1759. https://doi.org/10.3390/agronomy11091759

Balestrini R, Brunetti C, Cammareri M, Caretto S, Cavallaro V, Cominelli E, De Palma M, Docimo T, Giovinazzo G, Grandillo S, Locatelli F, Lumini E, Paolo D, Patanè C, Sparvoli F, Tucci M, Zampieri E (2021) Strategies to modulate specialized metabolism in Mediterranean crops: from molecular aspects to field. Int J Mol Sci 22:2887. https://doi.org/10.3390/ijms22062887. PubMed: 33809189

Bard NW, Cronk QCB, Davies TJ (2024) Fungal endophytes can modulate plant invasion. Biol Rev Camb Philos Soc 99:1652. https://doi.org/10.1111/brv.13085, . PubMed: 38629189

Betsy CJ, Siva C (2023) Molecular markers. In: Fisheries biotechnology and bioinformatics. Springer Nature Singapore, Singapore, pp 141–151

Bhatt K, Suyal DC, Kumar S, Singh K, Goswami P (2022) New insights into engineered plant-microbe interactions for pesticide removal. Chemosphere 309:136635. https://doi.org/10.1016/j.chemosphere.2022.136635. PubMed: 36183882

Bhupenchandra I, Chongtham SK, Devi EL, Ramesh R, Choudhary AK, Salam MD, Sahoo MR, Bhutia TL, Devi SH, Thounaojam AS, Behera C, Harish MN, Kumar A, Dasgupta M, Devi YP, Singh D, Bhagowati S, Devi CP, Singh HR, Khaba CI (2022) Role of biostimulants in mitigating the effects of climate change on crop performance. Front Plant Sci 13:967665. https://doi.org/10.3389/fpls.2022.967665. PubMed: 36340395

Bhupenchandra I, Chongtham SK, Devi AG, Dutta P, Sahoo MR, Mohanty S, Kumar S, Choudhary AK, Devi EL, Sinyorita S, Devi SH, Mahanta M, Kumari A, Devi HL, Josmee RK, Pusparani A, Pathaw N, Gupta S, Meena M, Janmeda P, Seth CS, Arora A, Sahoo A, Swapnil P (2024) Unlocking the potential of arbuscular mycorrhizal fungi: exploring role in plant growth promotion, nutrient uptake mechanisms, biotic stress alleviation, and sustaining agricultural production systems. J Plant Growth Regul 43(9):1–39. https://doi.org/10.1007/s00344-024-11467-9

Bohra A, Kilian B, Sivasankar S, Caccamo M, Mba C, McCouch SR, Varshney RK (2022) Reap the crop wild relatives for breeding future crops. Trends Biotechnol 40:412–431. https://doi.org/10.1016/j.tibtech.2021.08.009. PubMed: 34629170

Bosu S, Rajamohan N, Al Salti S, Rajasimman M, Das P (2024) Biodegradation of chlorpyrifos pollution from contaminated environment-a review on operating variables and mechanism. Environ Res 248:118212. https://doi.org/10.1016/j.envres.2024.118212. PubMed: 38272293

Callister AN, Costa-Neto G, Bradshaw BP, Elms S, Crossa J, Brawner JT (2024) Enviromic prediction enables the characterization and mapping of *Eucalyptus globulus* Labill breeding zones. Tree Genet Genomes 20:3. https://doi.org/10.1007/s11295-023-01636-4

Campos ACAL, van Dijk WFA, Ramakrishna P, Giles T, Korte P, Douglas A, Smith P, Salt DE (2021) 1,135 ionomes reveal the global pattern of leaf and seed mineral nutrient and trace element diversity in Arabidopsis thaliana. Plant J Cell Mol Biol 106:536–554. https://doi.org/10.1111/tpj.15177. PubMed: 33506585

Chandana BS, Mahto RK, Singh RK, Ford R, Vaghefi N, Gupta SK, Yadav HK, Manohar M, Kumar R (2022) Epigenomics as potential tools for enhancing the magnitude of breeding approaches for developing climate resilient chickpea. Front Genet 13:900253. https://doi.org/10.3389/fgene.2022.900253. PubMed: 35937986

Chandran H, Meena M, Sharma K (2020) Microbial biodiversity and bioremediation assessment through omics approaches. Front Environ Chem 1:570326. https://doi.org/10.3389/fenvc.2020.570326

Chandran H, Meena M, Swapnil P (2021) Plant growth-promoting rhizobacteria as a green alternative for sustainable agriculture. Sustainability 13(19):10986. https://doi.org/10.3390/su131910986

Chapman MA, He Y, Zhou M (2022) Beyond a reference genome: pangenomes and population genomics of underutilized and orphan crops for future food and nutrition security. New Phytol 234:1583–1597. https://doi.org/10.1111/nph.18021. PubMed: 35318683

Chen XL, Sun MC, Chong SL, Si JP, Wu LS (2021) Transcriptomic and metabolomic approaches deepen our knowledge of plant–endophyte interactions. Front Plant Sci 12:700200. https://doi.org/10.3389/fpls.2021.700200. PubMed: 35154169

Clark NM, Van den Broeck L, Guichard M, Stager A, Tanner HG, Blilou I, Grossmann G, Iyer-Pascuzzi AS, Maizel A, Sparks EE, Sozzani R (2020) Novel imaging modalities shedding light on plant biology: start small and grow big. Annu Rev Plant Biol 71:789–816. https://doi.org/10.1146/annurev-arplant-050718-100038. PubMed: 32119794

Dai X, Shen L (2022) Advances and trends in omics technology development. Front Med 9:911861. https://doi.org/10.3389/fmed.2022.911861. PubMed: 35860739

Diatta AA, Min D, Jagadish SVK (2021) Drought stress responses in non-transgenic and transgenic alfalfa—current status and future research directions. Adv Agron 170:35–100. https://doi.org/10.1016/bs.agron.2021.06.002

Dikobe T, Masenya K, Manganyi MC (2023) Molecular technologies ending with 'omics': the driving force toward sustainable plant production and protection. F1000Research 12:480. https://doi.org/10.12688/f1000research.131413.1

Dubey RK, Tripathi V, Prabha R, Chaurasia R, Singh DP, Rao CS et al (2020) Unravelling the soil microbiome: perspectives for environmental sustainability. Springer International Publishing, Cham

Eid AM, Fouda A, Abdel-Rahman MA, Salem SS, Elsaied A, Oelmüller R, Hijri M, Bhowmik A, Elkelish A, Hassan SED (2021) Harnessing bacterial endophytes for promotion of plant growth and biotechnological applications: an overview. Plan Theory 10:935. https://doi.org/10.3390/plants10050935. PubMed: 34067154

Email AIA, Eldeeb MB (2021) Effect of some biocides on the vase life of carnation (Dianthus caryophyllus L.) cut flowers. Sci J Flowers Ornam Plants 8:209–221

Etesami H, Glick BR (2024) Bacterial indole-3-acetic acid: a key regulator for plant growth, plant-microbe interactions, and agricultural adaptive resilience. Microbiol Res 281:127602. https://doi.org/10.1016/j.micres.2024.127602. PubMed: 38228017

Fadeel B, Pietroiusti A, Shvedova AA (eds) (2017) Adverse effects of engineered nanomaterials: exposure, toxicology, and impact on human health. Academic Press, Cambridge, MA

Feng NX, Yu J, Zhao HM, Cheng YT, Mo CH, Cai QY, Li YW, Li H, Wong MH (2017) Efficient phytoremediation of organic contaminants in soils using plant–endophyte partnerships. Sci Total Environ 583:352–368. https://doi.org/10.1016/j.scitotenv.2017.01.075. PubMed: 28117167

Frank AC, Saldierna Guzmán JP, Shay JE (2017) Transmission of bacterial endophytes. Microorganisms 5:70. https://doi.org/10.3390/microorganisms5040070. PubMed: 29125552

Gamalero E, Lingua G, Glick BR (2023) Ethylene, ACC, and the plant growth-promoting enzyme ACC deaminase. Biology 12:1043. https://doi.org/10.3390/biology12081043. PubMed: 37626930

Ganeva D, Roumenina E, Dimitrov P, Gikov A, Jelev G, Dragov R, Bozhanova V, Taneva K (2022) Phenotypic traits estimation and preliminary yield assessment in different phenophases of wheat breeding experiment based on UAV multispectral images. Remote Sens 14:1019. https://doi.org/10.3390/rs14041019

George J, Kahlke T, Abbriano RM, Kuzhiumparambil U, Ralph PJ, Fabris M (2020) Metabolic engineering strategies in diatoms reveal unique phenotypes and genetic configurations with implications for algal genetics and synthetic biology. Front Bioeng Biotechnol 8:513. https://doi.org/10.3389/fbioe.2020.00513. PubMed: 32582656

Glick BR (2014) Bacteria with ACC deaminase can promote plant growth and help to feed the world. Microbiol Res 169:30–39. https://doi.org/10.1016/j.micres.2013.09.009. PubMed: 24095256

Glick P, Gangl C, Gibb S, Klumpner S, Weinberg E (2007) Defensive reactions to masculinity threat: more negative affect toward effeminate (not masculine) gay men. Sex Roles 57:55–59. https://doi.org/10.1007/s11199-007-9195-3

Granata I, Manzo M, Kusumastuti A, Guarracino MR (2021) Learning from metabolic networks: current trends and future directions for precision medicine. Curr Med Chem 28:6619–6653. https://doi.org/10.2174/0929867328666201217103148. PubMed: 33334277

Gudi S, Pavan M, Alagappan P, Raigar OP, Halladakeri P, Gowda RSR, Kumar P, Singh G, Tamta M, Susmitha P, Amandeep, Saini DK (2024) Fashion meets science: how advanced breeding approaches could revolutionize the textile industry. Crit Rev Biotechnol:1–27. https://doi.org/10.1080/07388551.2024.2314309

Guo DJ, Singh RK, Singh P, Li DP, Sharma A, Xing YX, Song XP, Yang LT, Li YR (2020) Complete genome sequence of *Enterobacter roggenkampii* ED5, a nitrogen-fixing plant growth promoting endophytic bacterium with biocontrol and stress tolerance properties, isolated from sugarcane root. Front Microbiol 11:580081. https://doi.org/10.3389/fmicb.2020.580081. PubMed: 33072048

Gupta C, Prakash D, Bakar MFA, Kitdamrongtham W (2023a) Omic route to utilize endophytes and their functional potentials in plant growth advancement. In: Plant microbiome for plant productivity and sustainable agriculture. Springer Nature Singapore, Singapore, pp 289–311. https://doi.org/10.1007/978-981-19-5029-2_12

Gupta NK, Shavrukov Y, Singhal RK, Borisjuk N (eds) (2023b) Multiple abiotic stress tolerances in higher plants: addressing the growing challenges. CRC Press, Boca Raton

Gupta S, Verma R, Ravi RK (2023c) Multiomics approach for crop improvement under climate change. In: Sustainable agriculture in the era of the OMICs revolution. Springer International Publishing, Cham, pp 17–36

Hardoim PR, van Overbeek LS, van Elsas JD (2008) Properties of bacterial endophytes and their proposed role in plant growth. Trends Microbiol 16:463–471. https://doi.org/10.1016/j.tim.2008.07.008. PubMed: 18789693

Hontzeas N, Saleh SS, Glick BR (2004a) Changes in gene expression in canola roots induced by ACC-deaminase-containing plant-growth-promoting bacteria. Mol Plant–Microbe Interact 17:865–871. https://doi.org/10.1094/MPMI.2004.17.8.865. PubMed: 15305607

Hontzeas N, Zoidakis J, Glick BR, Abu-Omar MM (2004b) Expression and characterization of 1-aminocyclopropane-1-carboxylate deaminase from the rhizobacterium *Pseudomonas putida* UW4: A key enzyme in bacterial plant growth promotion. Biochim Biophys Acta 1703:11–19. https://doi.org/10.1016/j.bbapap.2004.09.015. PubMed: 15588698

Hulin MT, Jackson RW, Harrison RJ, Mansfield JW (2020) Cherry picking by pseudomonads: after a century of research on canker, genomics provides insights into the evolution of pathogenicity towards stone fruits. Plant Pathol 69:962–978. https://doi.org/10.1111/ppa.13189. PubMed: 32742023

Ijaz S, Iqbal J, Abbasi BA, Yaseen T, Rehman S, Kazi M et al (2024) Role of OMICS-based technologies in plant sciences. In: Omics-based techniques for global food security. Wiley, pp 45–66

Jha UC, Nayyar H, Jha R, Khurshid M, Zhou M, Mantri N, Siddique KHM (2020) Long non-coding RNAs: emerging players regulating plant abiotic stress response and adaptation. BMC Plant Biol 20:466. https://doi.org/10.1186/s12870-020-02595-x. PubMed: 33046001

Jin Y, Chen Z, White JF, Malik K, Li C (2024) Interactions between Epichloë endophyte and the plant microbiome impact nitrogen responses in host Achnatherum inebrians plants. Microbiol Spectrum 12:e0257423. https://doi.org/10.1128/spectrum.02574-23. PubMed: 38488391

Kaki B (2023) Methods for transcriptome analysis in the context of plant-microorganism interactions. Plant-Microbe Interact:43

Karlo M, Boschiero C, Landerslev KG, Blanco GS, Wen J, Mysore KS, Dai X, Zhao PX, de Bang TC (2020) The CLE53–SUNN genetic pathway negatively regulates arbuscular mycorrhiza root colonization in *Medicago truncatula*. J Exp Bot 71:4972–4984. https://doi.org/10.1093/jxb/eraa193. PubMed: 32309861

Katam R, Lin C, Grant K, Katam CS, Chen S (2022) Advances in plant metabolomics and its applications in stress and single-cell biology. Int J Mol Sci 23:6985. https://doi.org/10.3390/ijms23136985. PubMed: 35805979

Kaur B, Sandhu KS, Kamal R, Kaur K, Singh J, Röder MS, Muqaddasi QH (2021) Omics in major cereals: applications, challenges, and prospects. https://doi.org/10.20944/preprints202104.0531.v1

Khan AW, Garg V, Roorkiwal M, Golicz AA, Edwards D, Varshney RK (2020) Super-pangenome by integrating the wild side of a species for accelerated crop improvement. Trends Plant Sci 25:148–158. https://doi.org/10.1016/j.tplants.2019.10.012. PubMed: 31787539

Khatabi B, Gharechahi J, Ghaffari MR, Liu D, Haynes PA, McKay MJ, Mirzaei M, Salekdeh GH (2019) Plant–microbe symbiosis: what has proteomics taught us? Proteomics 19:e1800105. https://doi.org/10.1002/pmic.201800105. PubMed: 31218790

Kitashova A, Brodsky V, Chaturvedi P, Pierides I, Ghatak A, Weckwerth W, Nägele T (2023) Quantifying the impact of dynamic plant-environment interactions on metabolic regulation. J Plant Physiol 290:154116. https://doi.org/10.1016/j.jplph.2023.154116. PubMed: 37839392

Kuang T, Wang X, Qin X, Dai S, Yang P, Li L (eds) (2020) Plant protein and proteome altlas—integrated omics analyses of plants under abiotic stresses. MDPI

Kumar N, Soren KR, Bharadwaj C, Sneha Priya PR, Shrivastava AK, Pal M, Roorkiwal M, Kumar K, Patil BS, Soni A, Nimmy MS, Siddique KHM, Varshney RK (2021) Genome-wide transcriptome analysis and physiological variation modulates gene regulatory networks acclimating salinity tolerance in chickpea. Environ Exp Bot 187:104478. https://doi.org/10.1016/j.envexpbot.2021.104478

Kumar R, Swapnil P, Meena M, Selpair S, Yadav BG (2022) Plant growth-promoting rhizobacteria (PGPR): approaches to alleviate abiotic stresses for enhancement of growth and development of medicinal plants. Sustainability 14(23):15514. https://doi.org/10.3390/su142315514

Kumar GA, Kumar S, Bhardwaj R, Meena M, Swapnil P, Seth CS, Yadav A (2024) Recent advancements in multifaceted roles of flavonoids in plant–rhizomicrobiome interactions. Front Plant Sci 14:1297706. https://doi.org/10.3389/fpls.2023.1297706

Lastochkina O, Bosacchi M (2023) Potential roles of plant growth-promoting microbes in wheat adaptation and tolerance to herbicide and drought stress combination. Turk J Agric For 47:688–712. https://doi.org/10.55730/1300-011X.3121

Liang Q, Lonardi S (2021) Reference-agnostic representation and visualization of pan-genomes. BMC Bioinf 22:502. https://doi.org/10.1186/s12859-021-04424-w. PubMed: 34656081

Lowman JS, Lava-Chavez A, Kim-Dura S, Flinn B, Nowak J, Mei C (2015) Switchgrass field performance on two soils as affected by bacterization of seedlings with *Burkholderia phytofirmans* strain PsJN. Bioenergy Res 8:440–449. https://doi.org/10.1007/s12155-014-9536-3

Meena M, Swapnil P, Divyanshu K, Kumar S, Harish T, Tripathi YN, Zehra A, Marwal A, Upadhyay RS (2020) PGPR-mediated induction of systemic resistance and physiochemical alterations in plants against the pathogens: current perspectives. J Basic Microbiol 60:828–861. https://doi.org/10.1002/jobm.202000370. PubMed: 32815221

Mehari TG, Xu Y, Umer MJ, Shiraku ML, Hou Y, Wang Y, Yu S, Zhang X, Wang K, Cai X, Zhou Z, Liu F (2021) Multi-omics-based identification and functional characterization of Gh_A06G1257 proves its potential role in drought stress tolerance in *Gossypium hirsutum*. Front Plant Sci 12:746771. https://doi.org/10.3389/fpls.2021.746771. PubMed: 34745180

Mohd Saad NS, Severn-Ellis AA, Pradhan A, Edwards D, Batley J (2021) Genomics armed with diversity leads the way in *Brassica* improvement in a changing global environment. Front Genet 12:600789. https://doi.org/10.3389/fgene.2021.600789. PubMed: 33679880

Mondal R, Kumar A, Gnanesh BN (2023) Crop germplasm: current challenges, physiological-molecular perspective, and advance strategies towards development of climate-resilient crops. Heliyon 9:e12973. https://doi.org/10.1016/j.heliyon.2023.e12973. PubMed: 36711267

Nagarajan K, Ibrahim B, Ahmad Bawadikji A, Lim JW, Tong WY, Leong CR, Khaw KY, Tan WN (2021) Recent developments in metabolomics studies of endophytic fungi. J Fungi 8:28. https://doi.org/10.3390/jof8010028. PubMed: 35049968

Naik B, Kumar V, Rizwanuddin S, Chauhan M, Choudhary M, Gupta AK, Kumar P, Kumar V, Saris PEJ, Rather MA, Bhuyan S, Neog PR, Mishra S, Rustagi S (2023) Genomics, proteomics, and metabolomics approaches to improve abiotic stress tolerance in tomato plant. Int J Mol Sci 24:3025. https://doi.org/10.3390/ijms24033025. PubMed: 36769343

Nakkeeran S, Rajamanickam S, Saravanan R, Vanthana M, Soorianathasundaram K (2021) Bacterial endophytome-mediated resistance in banana for the management of Fusarium wilt. 3. Biotech 11:267. https://doi.org/10.1007/s13205-021-02833-5. PubMed: 34017673

Narkhede GW, Kiranmayee KU (2023) Maize improvement using recent omics approaches. In: Maize improvement: current advances in yield, quality, and stress tolerance under changing climatic scenarios. Springer International Publishing, Cham, pp 289–302

Negi R, Sharma B, Kumar S, Chaubey KK, Kaur T, Devi R, Yadav A, Kour D, Yadav AN (2024) Plant endophytes: unveiling hidden applications toward agro-environment sustainability. Folia Microbiol 69:181–206. https://doi.org/10.1007/s12223-023-01092-6. PubMed: 37747637

Niron H, Barlas N, Salih B, Türet M (2020) Comparative transcriptome, metabolome, and ionome analysis of two contrasting common bean genotypes in saline conditions. Front Plant Sci 11:599501. https://doi.org/10.3389/fpls.2020.599501. PubMed: 33362832

Ose T, Watanabe K, Mie T, Honma M, Watanabe H, Yao M, Oikawa H, Tanaka I (2003) Insight into a natural Diels–Alder reaction from the structure of macrophomate synthase. Nature 422:185–189. https://doi.org/10.1038/nature01454. PubMed: 12634789

Pagano A, Kunz L, Dittmann A, Araújo SS, Macovei A, Shridhar Gaonkar S, Sincinelli F, Wazeer H, Balestrazzi A (2023) Changes in *Medicago truncatula* seed proteome along the rehydration–dehydration cycle highlight new players in the genotoxic stress response. Front Plant Sci 14:1188546. https://doi.org/10.3389/fpls.2023.1188546. PubMed: 37409306

Pal G, Saxena S, Kumar K, Verma A, Sahu PK, Pandey A, White JF, Verma SK (2022) Endophytic *Burkholderia*: multifunctional roles in plant growth promotion and stress tolerance. Microbiol Res 265:127201. https://doi.org/10.1016/j.micres.2022.127201. PubMed: 36167006

Pandey P, Tripathi A, Dwivedi S, Lal K, Jhang T (2023) Deciphering the mechanisms, hormonal signaling, and potential applications of endophytic microbes to mediate stress tolerance in medicinal plants. Front Plant Sci 14:1250020. https://doi.org/10.3389/fpls.2023.1250020. PubMed: 38034581

Parveen R, Jain T, Parveen R, Singh S (2023) Genome-wide transcriptome profiling of crop plants. In: Genomics, transcriptomics, proteomics and metabolomics of crop plants. Academic Press, Cambridge, MA, pp 145–171

Paul S, Ingti B, Joshi SR (2024) Utility of endophytes for the enhancement of biofuel production. In: Emerging sustainable technologies for biofuel production. Springer Nature Switzerland, Cham, pp 323–340. https://doi.org/10.1007/978-3-031-52167-6_13

Peng M, Jiang Z, Zhou F, Wang Z (2023) From salty to thriving: plant growth promoting bacteria as nature's allies in overcoming salinity stress in plants. Front Microbiol 14:1169809. https://doi.org/10.3389/fmicb.2023.1169809. PubMed: 37426022

Penna S, Shirani Bidabadi S, Jain SM (2023) Mutation breeding to promote sustainable agriculture and food security in the era of climate change. In: Mutation breeding for sustainable food production and climate resilience. Springer Nature Singapore, Singapore, pp 1–23

Perrone A, Martinelli F (2020) Plant stress biology in epigenomic era. Plant Sci 294:110376. https://doi.org/10.1016/j.plantsci.2019.110376. PubMed: 32234231

Phour M, Sindhu SS (2023) Soil salinity and climate change: microbiome-based strategies for mitigation of salt stress to sustainable agriculture. In: Climate change and microbiome dynamics: carbon cycle feedbacks. Springer International Publishing, Cham, pp 191–243. https://doi.org/10.1007/978-3-031-21079-2_13

Rajczewski AT, Jagtap PD, Griffin TJ (2022) An overview of technologies for MS-based proteomics-centric multi-omics. Expert Rev Proteomics 19:165–181. https://doi.org/10.1080/14789450.2022.2070476. PubMed: 35466851

Rakkammal K, Priya A, Pandian S, Maharajan T, Rathinapriya P, Satish L, Ceasar SA, Sohn SI, Ramesh M (2022) Conventional and omics approaches for understanding the abiotic stress response in cereal crops-an updated overview. Plan Theory 11:2852. https://doi.org/10.3390/plants11212852. PubMed: 36365305

Rashid B, Destrade M, Gilchrist MD (2012) Determination of friction coefficient in unconfined compression of brain tissue. J Mech Behav Biomed Mater 14:163–171. https://doi.org/10.1016/j.jmbbm.2012.05.001. PubMed: 23026694

Rauf M, Awais M, Ud-Din A, Ali K, Gul H, Rahman MM, Hamayun M, Arif M (2020) Molecular mechanisms of the 1-aminocyclopropane-1-carboxylic acid (ACC) deaminase producing *Trichoderma asperellum* MAP1 in enhancing wheat tolerance to waterlogging stress. Front Plant Sci 11:614971. https://doi.org/10.3389/fpls.2020.614971. PubMed: 33537050

Raza A, Tabassum J, Zahid Z, Charagh S, Bashir S, Barmukh R, Khan RSA, Barbosa F, Zhang C, Chen H, Zhuang W, Varshney RK (2021) Advances in 'omics' approaches for improving toxic metals/metalloids tolerance in plants. Front Plant Sci 12:794373. https://doi.org/10.3389/fpls.2021.794373. PubMed: 35058954

Raza A, Tabassum J, Fakhar AZ, Sharif R, Chen H, Zhang C, Ju L, Fotopoulos V, Siddique KHM, Singh RK, Zhuang W, Varshney RK (2023) Smart reprogramming of plants against salinity stress using modern biotechnological tools. Crit Rev Biotechnol 43:1035–1062. https://doi.org/10.1080/07388551.2022.2093695. PubMed: 35968922

Razzaq A, Kaur P, Akhter N, Wani SH, Saleem F (2021) Next-generation breeding strategies for climate-ready crops. Front Plant Sci 12:620420. https://doi.org/10.3389/fpls.2021.620420. PubMed: 34367194

Reshma RS, Das DN (2021) Molecular markers and its application in animal breeding. In: Advances in animal genomics. Academic Press, Cambridge, MA, pp 123–140

Sagare DB, Sandhu N, Yadav S, Singh UM, Alam S, Dixit S et al (2021) Genomic designing for biotic stress resistant rice. In: Genomic designing for biotic stress resistant cereal crops. Springer International Publishing, pp 1–58

Samuels DS, Lybecker MC, Yang XF, Ouyang Z, Bourret TJ, Boyle WK, Stevenson B, Drecktrah D, Caimano MJ (2021) Gene regulation and transcriptomics. Curr Issues Mol Biol 42:223–266. https://doi.org/10.21775/cimb.042.223. PubMed: 33300497

Satam H, Joshi K, Mangrolia U, Waghoo S, Zaidi G, Rawool S, Thakare RP, Banday S, Mishra AK, Das G, Malonia SK (2023) Next-generation sequencing technology: current trends and advancements. Biology 12:997. https://doi.org/10.3390/biology12070997. PubMed: 37508427

Sethi M, Kaur C, Hagroo RP, Singh MP (2023) Endophyte mediated plant health via phytohormones and biomolecules. In: Microbial endophytes and plant growth. Academic Press, Cambridge, MA, pp 151–166

Shafi A, Zahoor I, Habib H (2021) Omics technologies to unravel plant-microbe interactions. In: Plant-microbe dynamics. CRC Press, Boca Raton, pp 201–220

Shahid M, Khan MS, Singh UB (2023) Pesticide-tolerant microbial consortia: potential candidates for remediation/clean-up of pesticide-contaminated agricultural soil. Environ Res 236:116724. https://doi.org/10.1016/j.envres.2023.116724. PubMed: 37500042

Sharma M, Sudheer S, Usmani Z, Rani R, Gupta P (2020) Deciphering the omics of plant-microbe interaction: perspectives and new insights. Curr Genomics 21:343–362. https://doi.org/10.2174/1389202921999200515140420. PubMed: 33093798

Sharma P, Yadav D, Gaur RK (eds) (2022) Bioinformatics in agriculture: next generation sequencing era. Academic Press, Cambridge, MA

Singh A, Rajput V, Singh AK, Sengar RS, Singh RK, Minkina T (2021) Transformation techniques and their role in crop improvements: A global scenario of GM crops. In: Policy issues in genetically modified crops. Academic Press, Cambridge, MA, pp 515–542

Sukumaran S, Reynolds MP, Sansaloni C (2018) Genome-wide association analyses identify QTL hotspots for yield and component traits in durum wheat grown under yield potential, drought, and heat stress environments. Front Plant Sci 9:81. https://doi.org/10.3389/fpls.2018.00081. PubMed: 29467776

Susmitha P, Kumar P, Yadav P, Sahoo S, Kaur G, Pandey MK, Singh V, Tseng TM, Gangurde SS (2023) Genome-wide association study as a powerful tool for dissecting competitive traits in legumes. Front Plant Sci 14:1123631. https://doi.org/10.3389/fpls.2023.1123631. PubMed: 37645459

Suwinski P, Ong C, Ling MHT, Poh YM, Khan AM, Ong HS (2019) Advancing personalized medicine through the application of whole exome sequencing and big data analytics. Front Genet 10:49. https://doi.org/10.3389/fgene.2019.00049. PubMed: 30809243

Thikekar AK, Rathod VS, Panchal VP, Raut SA, Raut RS, Jain KS (2023) A review on-analytical tools in proteomics. J Proteins Proteomics 14:201–221. https://doi.org/10.1007/s42485-023-00108-6

Toppo P, Subba R, Roy K, Mukherjee S, Mathur P (2023) Elucidating the strategies for isolation of endophytic fungi and their functional attributes for the regulation of plant growth and resilience to stress. J Plant Growth Regul 42:1342–1363. https://doi.org/10.1007/s00344-022-10638-w

Tsugawa H, Rai A, Saito K, Nakabayashi R (2021) Metabolomics and complementary techniques to investigate the plant phytochemical cosmos. Nat Prod Rep 38:1729–1759. https://doi.org/10.1039/d1np00014d. PubMed: 34668509

Uddin S, Munir MZ, Larriba E, Pérez-Pérez JM, Gull S, Pervaiz T, Mahmood U, Mahmood Z, Sun Y, Li Y (2024) Temporal profiling of physiological, histological, and transcriptomic dissection during auxin-induced adventitious root formation in tetraploid Robinia pseudoacacia microcuttings. Planta 259:66. https://doi.org/10.1007/s00425-024-04341-1. PubMed: 38332379

Vaucheret H, Voinnet O (2024) The plant siRNA landscape. Plant Cell 36:246–275. https://doi.org/10.1093/plcell/koad253. PubMed: 37772967

Vejan P, Abdullah R, Khadiran T, Ismail S, Nasrulhaq Boyce A (2016) Role of plant growth promoting rhizobacteria in agricultural sustainability-a review. Molecules 21:573. https://doi.org/10.3390/molecules21050573. PubMed: 27136521

Vergine M, Vita F, Casati P, Passera A, Ricciardi L, Pavan S, Aprile A, Sabella E, De Bellis L, Luvisi A (2024) Characterization of the olive endophytic community in genotypes displaying a contrasting response to *Xylella fastidiosa*. BMC Plant Biol 24:337. https://doi.org/10.1186/s12870-024-04980-2. PubMed: 38664617

Vuolo F, Novello G, Bona E, Gorrasi S, Gamalero E (2022) Impact of plant-beneficial bacterial inocula on the resident bacteriome: current knowledge and future perspectives. Microorganisms 10:2462. https://doi.org/10.3390/microorganisms10122462. PubMed: 36557714

Walsh R (1981) Bond dissociation energy values in silicon-containing compounds and some of their implications. Acc Chem Res 14:246–252. https://doi.org/10.1021/ar00068a004

Whitt L, Mahood EH, Ziegler G, Luebbert C, Gillman JD, Norton GJ et al (2023) A comparative approach for selecting orthologous candidate genes underlying signal in genome-wide association studies across multiple species. bioRxiv:2023–2010

Williams A, Sinanaj B, Hoysted GA (2024) Plant–microbe interactions through a lens: tales from the mycorrhizosphere. Ann Bot 133:399–412. https://doi.org/10.1093/aob/mcad191. PubMed: 38085925

Xiao Q, Pan Q, Li J, Zhang J, Yang J (2024) DhuFAP: a platform for gene functional analysis in *Dendrobium huoshanense*. BMC Genomics 25:342. https://doi.org/10.1186/s12864-024-10220-6. PubMed: 38575876

Yadav G, Meena M (2021) Bioprospecting of endophytes in medicinal plants of Thar Desert: an attractive resource for biopharmaceuticals. Biotechnol Rep 30:e00629. https://doi.org/10.1016/j.btre.2021.e00629

Yadav G, Meena M (2022) Biological control of plant diseases by endophytes. In: Shah MP, Deka D (eds) Endophytic association: what, why and how (volume: developments in applied microbiology and biotechnology). Academic Press, Elsevier Inc, pp 119–135. https://doi.org/10.1016/B978-0-323-91245-7.00007-9

Yang S, Gill RA, Zaman QU, Ulhassan Z, Zhou W (2020) Insights on SNP types, detection methods and their utilization in Brassica species: recent progress and future perspectives. J Biotechnol 324:11–20. https://doi.org/10.1016/j.jbiotec.2020.09.018. PubMed: 32979432

Yang Y, Saand MA, Huang L, Abdelaal WB, Zhang J, Wu Y, Li J, Sirohi MH, Wang F (2021) Applications of multi-omics technologies for crop improvement. Front Plant Sci 12:563953. https://doi.org/10.3389/fpls.2021.563953. PubMed: 34539683

Zanini SF, Bayer PE, Wells R, Snowdon RJ, Batley J, Varshney RK, Nguyen HT, Edwards D, Golicz AA (2022) Pangenomics in crop improvement—from coding structural variations to finding regulatory variants with pangenome graphs. Plant Genome 15:e20177. https://doi.org/10.1002/tpg2.20177. PubMed: 34904403

Zenda T, Liu S, Dong A, Li J, Wang Y, Liu X, Wang N, Duan H (2021) Omics-facilitated crop improvement for climate resilience and superior nutritive value. Front Plant Sci 12:774994. https://doi.org/10.3389/fpls.2021.774994. PubMed: 34925418

Proteomics of Fungal Endophytes: Trends and Insight for Biotechnology

2

Juhi Goyal, Priyank Upadhayay, Ankur Das Talukdar, Priti Yadav, Piyashree Banerjee, Anshul Shakya, Namita Ashish Singh, Anoop Kumar, and Nitish Rai

Abstract

Fungal endophytes are endosymbiotic microorganisms that colonize plants and are an abundant source of the secondary metabolites. Although the importance of the fungal endophytes in plant is not fully understood, they actively affect the plant health, plant growth promotion, stress tolerance, community diversity, and surrounding ecosystem. Endophytes hold immense potential for applications in agriculture, environmental remediation, and medicine. Various techniques like genomics, transcriptomics, proteomics, and metagenomics aid in understanding endophyte-plant relationships. Proteomic study has become exceptionally important for the endophyte research as it provides the information about the actual

J. Goyal · P. Upadhayay · A. D. Talukdar · P. Yadav
Department of Biotechnology, Mohanlal Sukhadia University, Udaipur, Rajasthan, India

P. Banerjee · A. Kumar
Department of Biotechnology, University of North Bengal, Darjeeling, West Bengal, India

A. Shakya
Department of Pharmaceutical Sciences, Faculty of Science and Engineering, Dibrugarh University,
Dibrugarh, Assam, India

Department of Pharmacology, School of Clinical Medicine, Faculty of Health Sciences, University of Free State (UFS), Bloemfontein, South Africa

N. A. Singh
Department of Microbiology, Mohanlal Sukhadia University, Udaipur, Rajasthan, India

N. Rai (✉)
Department of Biotechnology, Mohanlal Sukhadia University, Udaipur, Rajasthan, India

Department of Zoology, Lucknow University, Lucknow, Uttar Pradesh, India
e-mail: rai_nitish@lkouniv.ac.in; nitish.rai@mlsu.ac.in

M. Meena et al. (eds.), *Multi-omics Approach to Investigate Endophyte Diversity*, Microorganisms for Sustainability 53, https://doi.org/10.1007/978-981-96-4100-0_2

proteins being produced by the endophyte and their dynamics and functional role in stress tolerance, secondary metabolite production, and plant-fungus relation. Traditional methods of isolation of the endophytes using the culture-based methods are now being replaced by the high-throughput omics methods such as mass spectrometry, chromatography, 2D gel electrophoresis, and microarray. Overall, proteomics is a transformative tool for unraveling the complex world of fungal endophytes. This chapter discusses the role and methods for proteomic study and the potential applications in enhancing the crop production and pest control and improving fertility, stress tolerance, and secondary metabolites' production. By providing a deeper understanding of protein function, it paves the way for developing innovative strategies to harness the power of these fascinating organisms for a healthier and more sustainable future.

Keywords

Endophytic · Microbiome · Endophytes · Genomics · Metagenomics · Metatranscriptomics

2.1 Introduction

The Greek terms "endon" (inside) and "phyte" (plant) are the origin of the word "endophyte." An organism that survives inside a host plant without causing harm is known as an endophyte. J.H.F. Link—a German botanist—named them "Entophytae" in 1809, referring to a group of rather parasitic organisms that resided inside plants (Subudhi et al. 2019). A vast array of nonpathogenic microorganisms, including bacteria, fungi, viruses, and actinomycetes, inhabit plants. These microorganisms are collectively referred to as endophytes. In many ways, these endosymbiotic individuals show endophytism by helping the host plant through shielding it from both abiotic stress (nutrients, water, salinity, heat, and herbivory) and biotic stress (pathogenic infections). These endosymbionts obtain energy from the host tissue in exchange, in the form of carbon. Additionally, there is emerging but restricted evidence suggesting that endophytes could play a role in shaping population dynamics, influencing plant community diversity, and contributing to ecosystem functionality. Recent research has emphasized that nonpathogenic fungi, particularly endophytes found within plant microbiomes, play an important role in influencing the expression of diseases in their host plants (Busby et al. 2016). Plants employ induced genetic defenses to resist pathogens, as elucidated by Jones and Dangl (2006).

The term "endophyte" was coined by De Bary and originally referred to "any organisms occurring within plant tissues" or "all organisms inhabiting plant organs that can colonise internal plant tissues at some stage in their lives without harming the host" (Redlin and Carris 1996). A class of fungus known as endophytic fungi inhabits the tissues of all healthy plants, either throughout the duration of the plant's life cycle or just a phase of it, without producing any symptoms of illness or morphological changes (Sunitha et al. 2013). They live in biological interaction with

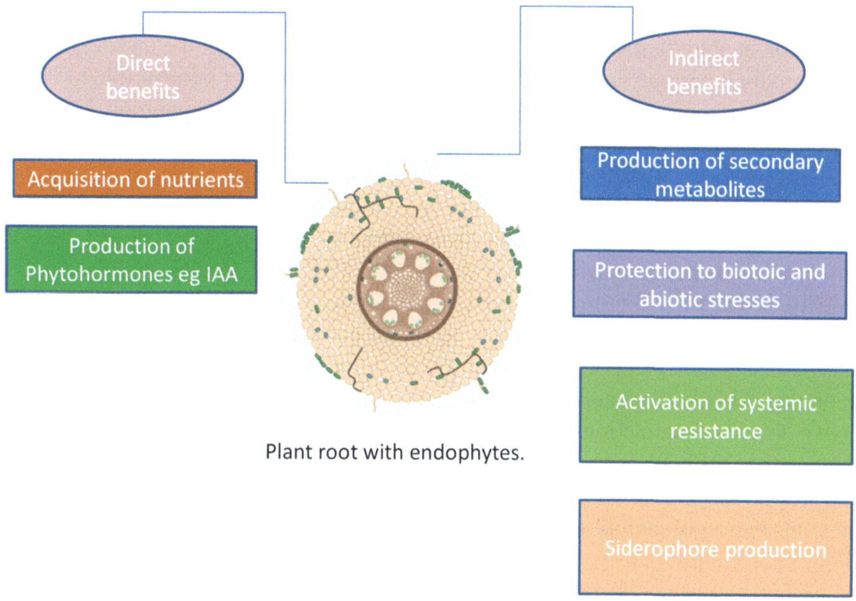

Fig. 2.1 Advantages conferred to plants through endophytic fungal colonization. (Image generated through BioRender)

living plants in the mycelia condition. They have been found to be connected to a diverse range of plants, such as algae, grasses, trees, and herbaceous plants. They are among the crucial elements of biodiversity. They are very varied, with the ascomycetes family making up the majority. Telemorphic states are absent in endophytic fungus (Arnold 2005; Kumaresan and Suryanarayanan 2002). Endophytes have limited competition with other microorganisms, a sheltered habitat, and the nutrients necessary for fungal development and survival inside the plant. In return for this protective environment, fungi enhance plant fitness through various mechanisms (Khan et al. 2015; Lugtenberg et al. 2016; Chitnis et al. 2020). The fungi are able to do so using a variety of mechanisms either directly or indirectly as shown in Fig. 2.1 (Berg 2009). The most significant direct methods for promoting growth are nutrient absorption and the synthesis of phytohormones, and resistance to biotic and abiotic stressors, such as the fight against infections, is regarded as an indirect factor in growth promotion (Hardoim et al. 2015, Souza and dos Santos 2017; Chandran et al. 2021; Kumar et al. 2022a). Interaction with endophytic fungi directly benefits plants by increasing their uptake of nutrients and phytohormone levels, which are directly associated with increased biomass production, enhanced root system growth, greater plant height and weight, improved reproduction, and higher yield. These advantages allow them to be called biofertilizers (Bamisile et al. 2018); Kumar et al. 2022b). The activation of the immune responses and the production of various primary and secondary metabolites are methods by which indirectly the endophytic fungi help the plants.

Certain diseases cause plants to lack an immune response, yet endophytic fungi that reside in plant tissues strengthen the plants' defenses against the invaders. Plant fungal endophytes can create chemicals that are identical to or comparable to those generated by their host plants (Yadav and Meena 2021). They commonly increase plant biomass under stressful conditions (Melotto et al. 2008; Hoffman and Arnold 2008; Firáková et al. 2007). Endophytic fungi play a crucial role in initiating the biological breakdown of dead or decaying host plants, which is essential for biogeochemical cycles (Hawksworth and Rossman 1997; Yadav and Meena 2021). However, it has become evident that endophytes also contribute to modulating the severity of disease symptoms as discussed by Stone et al. (2000). Endophytes have notably been found to either suppress (through pathogen competition) or amplify (through pathogen support) the severity of plant diseases in functional assays involving vulnerable plants, aggressive pathogens, and an abiotic environment conducive to these interactions (Freeman and Rodriguez 1993; Arnold et al. 2003; Meena et al. 2017; Bhupenchandra et al. 2024). While being more diverse and abundant than pathogens in the plant microbiome, as evidenced by Ganley et al. (2004), the contribution of endophytes to plant defense is often overlooked. The goal is to establish a comprehensive foundation that can guide future research endeavors focused on exploring the ecological and evolutionary significance of disease modulation by endophytes in natural environments. Additionally, insights from this synthesis may be instrumental in exploring the potential application of endophytes as biocontrol agents in agricultural plant systems, aligning with the perspectives presented by Dangl et al. (2013) and Ledford (2015).

A widely accepted working definition of endophytism is encapsulated by the notion that "infections are subtle, with the host tissues remaining at least temporarily symptomless, and the microbial colonization is confirmed to occur internally," as articulated by Stone et al. (2000). In general, this description is suitable and consistent with the standard practice of separating plant tissues' endophytes that have been surface-sterilized and are asymptomatic. However, it is crucial to acknowledge certain nuances associated with this methodology: (1) epiphytes that are resistant to surface sterilization may be mistakenly identified as endophytes; (2) some fungi present on the surface of the plant also inhabit internal plant tissues, making rigid distinctions between epiphytes and endophytes less categorical, as shown in Fig. 2.2; and (3) endophytes coexist alongside pathogens within infected plant tissues, suggesting that studies on endophytes that exclude diseased tissues might unintentionally overlook endophytes that impact disease progression.

The "endophytic continuum" model proposes that the nature of the plant-fungus interaction, which may range from mutualism to parasitism, depends on factors like the specific fungal species involved, the host's genetic background, and the environment (Rajamanikyam et al. 2017). Plants represent vast and varied ecological niches for endophytic organisms, a phenomenon extensively documented by numerous researchers. Traditionally, researchers have classified endophytes into two groups—clavicipitaceous (C) and nonclavicipitaceous (NC)—based on phylogenetic considerations and biological life cycle characteristics. However, research shows that NC endophytes can be further classified into three distinct functional groups. These

Fig. 2.2 Oomycota sporulation inside plant leaf. (Image generated through BioRender)

differences are determined by factors like how the host is colonized and transmitted, the diversity of organisms within the plant, and the benefits to the host plant's fitness. Using this refined framework, we compare the life histories, interactions with hosts, and potential roles in the ecophysiology of plants between C endophytes and NC endophytes (Rodriguez et al. 2009). Fungal endophytes comprise a wide range of species, each with different symbiotic and ecological roles. To understand the effect of endophytes in plants, studies have employed criteria such as colonization and transmission patterns, host range, tissue specificity, and fitness benefits offered that distinguish endophytes into four separate functional classes. Although the exact effect of each class on plants can differ greatly, the majority of endophytes remain inadequately characterized. As researchers progressively use Koch's postulates to examine host range, colonization and transmission patterns, and symbiotic/ecological roles, our understanding of endophytes is poised to evolve. In addition, a range of molecular techniques (such as denaturing gradient gel electrophoresis (DGGE) and terminal-restriction fragment length polymorphism (T-RFLP)) and biochemical methods (like stable isotope profiling (SIP) and metabolic incorporation of nucleotide analogs such as bromodeoxyuridine (BrdU)) are being used to improve the characterization of endophytes and their functions in plant ecophysiology, as discussed in reviews by Johansson et al. (2004) and Vandenkoornhuyse et al. (2007). For instance, molecular techniques enable the assessment of endophyte community structure across landscapes, as shown in studies like those conducted by Maciá-Vicente et al. (2008). On the other hand, biochemical methods can identify the metabolically active members of the community. These insights contribute greatly to research on the systematics and taxonomy of these relatively understudied fungi.

As our understanding of the role of endophytes in expression of plant genes advances, it becomes feasible to analyze its expression patterns in order to assess the symbiotic status of plant communities. Ultimately, it might be possible to evaluate the metabolic activity and community structure of every fungal symbiont associated with plants in different landscapes. This knowledge could pave the way for the creation of new tools to evaluate changes in ecosystems caused by natural variations, climate change, and other human-induced factors impacting environmental changes. Endophytes have proven to be valuable source for novel metabolites, enzymes of industrial importance, and stress-relief agents for host plants. However, numerous features of endophytic biology remain enigmatic. Their complex and continuous interactions with the host plant and other components of the host microbiota are what give endophytes their unique role. Viewing plant microbiomes as integrated systems enables the analysis and synthesis of these intricate interactions. Contemporary genomic investigations utilizing metaomics and conducting comparative studies offer promise in elucidating the ambiguities surrounding endophytism. A better understanding of host invasion mechanisms and the functions of endophytes could be applied to improve agricultural practices, especially in promoting biocontrol, bioremediation, and plant growth. Techniques such as genome sequencing, comparative genomics, microarray analysis, next-generation sequencing, metagenomics, and metatranscriptomics are either being utilized or hold potential for unraveling the dynamics of the plant-endophyte relationship. It is imperative to explore modern techniques and approaches to comprehensively investigate endophytes and their presumed roles in host plant ecology. This chapter underscores the utility of "omics" tools in exploring the function of endophytes in the microbiome of plants (Kaul et al. 2016).

Analysis of endophytic genomes has revealed genes essential for their lifestyle, including those involved in nitrogen fixation, phytohormone synthesis (e.g., IAA, GA), mineral uptake (e.g., Fe, P), stress resilience, adhesion, and traits related to colonization (Fouts et al. 2008; Firrincieli et al. 2015; Martínez-García et al. 2015). These genetic traits elucidate the microbes' role in nutrient cycling and their capacity to inhabit the plant endosphere. A comprehensive examination of endophytic microorganisms' entire genomes has unveiled genetic characteristics that either directly or indirectly impact various biological activities and colonization preferences. This analysis helps identify specific genes involved in endophytic secretory systems, surface attachment and insertion elements, transport mechanisms, plant growth promotion, antibiotic resistance mechanisms, and other associated metabolic processes. Such investigations have enhanced our knowledge of the ecological roles and evolutionary development of endophytes. Noteworthy genetic elements like genes coding for N-acyl homoserine lactone synthases and hydrolases, hyperadherence factors, and fusaric acid resistance proteins emphasize the biotechnological capabilities of the endophytic bacterium *Pantoea ananatis* (Megías et al. 2016). Endophytic fungi belonging to the order *Sebacinales* have garnered significant attention for their potential in enhancing plant growth and improving stress tolerance (Weiss et al. 2011). *Piriformospora indica*, a member of the *Sebacinales* order, serves as a model for studying symbiotic relations. Genome sequencing of *P. indica* has revealed its potential as a plant probiotic (Qiang XiaoYu et al. 2012). Moreover, studying proteomic analysis complements these genomic insights by providing a comprehensive understanding of the molecular events underlying microbial physiology and functionality.

Proteomic analysis is a potent technique that can resolve thousands of proteins, allowing for a systematic understanding of the molecular events that take place within an organism's various physiological states, the level of protein expression of particular genes, and the provision of spatially specific information on protein expression through cellular fractionation (Pillai 2017). While metaproteomics primarily focuses on the investigation of functional expression of community genes and the interpretation of activities during sampling, proteomics analyzes the total amount of proteins found in an organism. Metaproteomics involves the direct identification and analysis of the current functions of the microbial population within an environmental sample. It provides an evaluation of the microbial functional profile in real time. Furthermore, advances in bioinformatics and computing offer a more reliable source for protein identification. The four essential steps in metaproteomic analysis are, (1) protein isolation and purification, concentration; (2) denaturation and reduction; (3) protein separation, digestion, MS analysis; and (4) protein identification based on spectroscopic data (Schneider and Riedel 2010).

It is essential for the sample protein in metaproteomics to have both quality and quantity characteristics. Metaproteomic analysis of endophytes has been performed through two different methods: a comparative analysis of their fingerprinting through two-dimensional gel electrophoresis is conducted to understand the influence of any parameter on secondary metabolite production. Additionally, the direct lysis method, which extracts the total protein from the endosphere (the microhabitat where the plant-endophyte interaction takes place) under various environmental conditions, is also depicted in Fig. 2.3 (Maron et al. 2007). With the AMD biofilm system, the first metaproteomic analysis was carried out (Ram et al. 2005). During symbiosis, proteins released by fungal hyphae are crucial to the plant cell. These molecules are employed in the establishment and maintenance of a fruitful symbiotic interaction between them. Proteomics seems to be a potent technique for expanding plant-microbe interactions and gaining a global perspective. Plant proteomes infected and noninfected can be compared using two-dimensional electrophoresis (2-DE), which also allows for the detection of fungal and plant proteins that are differentially regulated in this symbiosis (Zhou et al. 2006; Li et al. 2012).

To enhance genome annotation, characterize the activity of individual proteins and non-coding RNAs, identify phosphorylated proteomes, and evaluate cellular reactivity to various environmental conditions, proteomic studies are crucial for protein-level research. Translational data about protein modification and functionally translated expression can be found in the proteome. This could be connected to other high-throughput methods such as transcriptomics and metabolomics, which offer data on protein interaction networks, posttranslational alterations of protein structure, and functionally translated proteins. Proteomics has a multifaceted function in shaping bioinoculants to increase crop production. The plant-fungus symbiosis has received less attention at the protein level in the past. Differential protein expression was found in the relationship between *Brassica napus* and the endophyte *Piriformospora indica.* These processes include metabolism, symbiotic signaling,

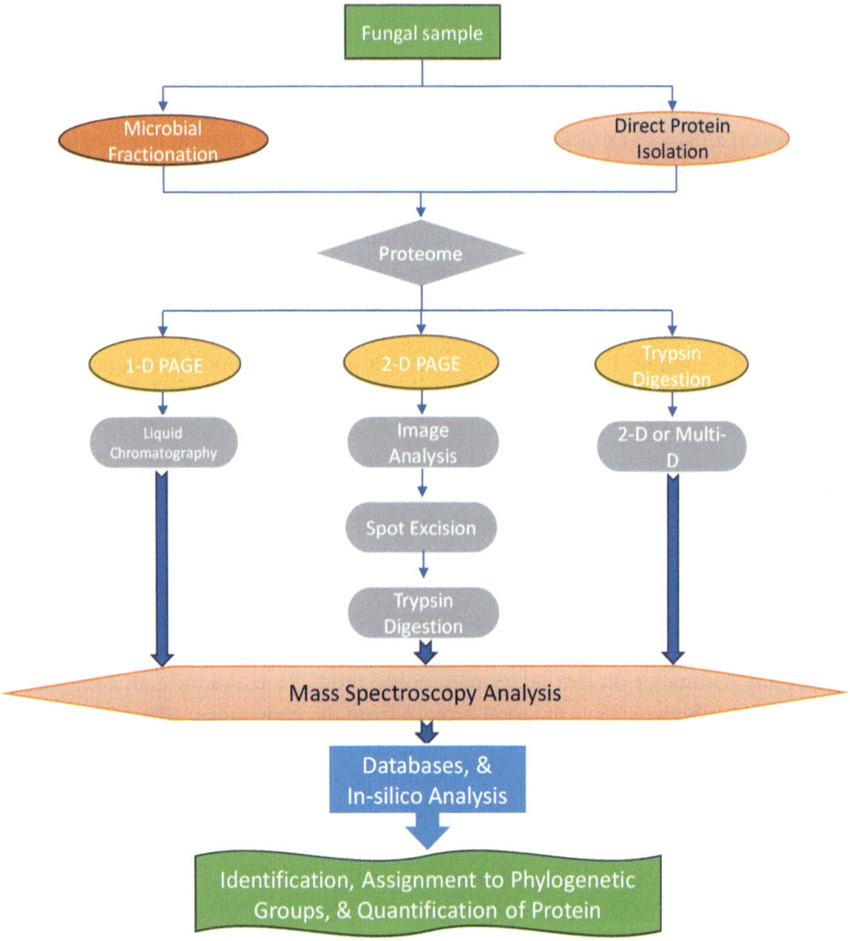

Fig. 2.3 Flowchart showing the metaproteomic analysis of endophyte

defense responses, nutrient uptake, energy production, and biosynthesis of essential metabolites. During the plant's symbiotic growth phase, they are in charge of cellular homeostasis, root architecture alteration, and cell remodeling. This was revealed during the integrated proteomic analysis (Shrivastava et al. 2018).

In a different investigation, co-inoculation of consortia increased plant growth and nutrient uptake more than individual inoculation of endophytic *Piriformospora indica* and *Azotobacter chroococcum* on *Oryza sativa*. Additionally, proteome research showed that proteins involved in the metabolism of phosphorus and nitrogen were upregulated, which increased plants' absorption of these nutrients. The relationship and control of gene and expression of proteins in primary and secondary metabolic processes have also been studied by comparative transcriptomic and proteomic analysis (Bandyopadhyay et al. 2022). To study the fungal endophyte

Gilmaniella sp. AL12 inoculation on *Atractylodes lancea*, Yuan et al. used transcriptomic and proteomic analysis. The results of the study demonstrated the endophyte's advantageous contribution, which included increased biomass, sesquiterpenoid content, and primary metabolism as a result of higher photosynthesis, an expanding sink, and improved host metabolic flux. Reduced plant immune response as a result of the relationship could improve plant-endophyte interaction. These studies show how omics technology can be applied to enhance comprehension of gene-protein networks and metabolic pathways. Additionally, the combination of omics technology with other analytical techniques, like mathematical modeling and other in silico approaches, can aid in the understanding of biological processes related to plant-microbe interactions (Yuan et al. 2019).

Mass spectrometry (MS) is now the dominant technology in proteomics, acting as the central platform for most proteomic analyses. Metaproteomics leverages the advanced capabilities of high-performance MS to comprehensively analyze the entire set of proteins expressed by a microbial community within an environmental sample. Proteins can be extracted from the microenvironment using either direct or indirect lysis techniques (Maron et al. 2007).

Gaining a deeper understanding of metagenomes from diverse environments is vital for leveraging the full potential of metaproteomics in characterizing endophytic microbial communities. Collaborative research efforts, technological advancements, and data sharing initiatives will contribute to addressing this knowledge gap and advancing our understanding of endophyte-plant interactions. Metaproteogenomics integrates both the protein and genetic information of environmental samples, enabling the identification of a broader range of proteins and their functions compared to proteomics alone. This approach involves a combined analysis of the metagenome (genetic material) and metaproteome (protein content) from the same sample. Knief et al. (2012) utilized metaproteogenomics to investigate microbial populations in the phyllosphere and rhizosphere of rice plants. Their findings revealed that although both microenvironments contained nifH genes, which are associated with nitrogen fixation, the expression of these genes was observed only in the rhizosphere.

Applying such an approach to study the endosphere—the interior tissues of plants—could yield valuable insights into the functionality of endophytes. By simultaneously examining the genetic potential (metagenome) and actual protein expression (metaproteome) of endophytic microbial communities, researchers can obtain more comprehensive data on their roles and activities within the plant host. This integrated approach has the potential to uncover novel functions, metabolic pathways, and interactions that contribute to plant health, disease resistance, and growth promotion mediated by endophytes. Indeed, employing microarray technology facilitates the comprehensive exploration and enhanced understanding of the intricate network of relationships among endophytes, their host plants, and other related microorganisms. Utilizing research on plant-pathogen interactions as a basis, researchers can gain valuable insights into the dynamics of plant-endophyte relationships.

Advanced techniques, with their high accuracy, can be effectively applied to both bacterial and fungal endophytes, unveiling their genetic and metabolic potential, ecological roles, and evolutionary trajectories. This comprehensive understanding extends to elucidating the diverse microbial communities comprising the plant microbiome and their significance within natural ecosystems.

By harnessing the biotechnological potential of endophytes more efficiently and sustainably, researchers can unlock novel applications in agriculture, environmental remediation, and pharmaceuticals. Furthermore, elucidating the roles of endophytes in plant health and ecosystem functioning contributes to the development of strategies for enhancing agricultural productivity, fostering ecological sustainability, and addressing global challenges such as food security and climate change. Metaproteogenomics addresses this challenge by combining information from both the metagenome and metaproteome of the same sample. This allows for a more comprehensive analysis, as it not only identifies proteins based on available reference genomes but also enables the prediction and discovery of novel proteins and functions encoded within the metagenome. Consequently, researchers can gain deeper understanding into the metabolic pathways, regulatory mechanisms, and interactions involved in the endophytic lifestyle.

Overall, the integration of metagenomic and metaproteomic data through metaproteogenomics offers a powerful approach to unraveling the intricacies of endophytic microbial communities. By overcoming the limitations of traditional proteomic approaches, metaproteogenomics provides a more complete understanding of microbial adaptation and function within plant hosts, paving the way for innovative strategies in agriculture, biotechnology, and environmental science. Modern genome-based research has been transformed by the microarray technology, which offers tools for analyzing gene expression particular to the genome, profiling endophyte genes, and exploring interactions between host plants and symbionts through transcriptomic analysis. Felitti et al. (2006) and Barnett et al. (2004) utilized a dual-genome Symbiosis Chip-based tool to investigate symbiotic interactions. This method allows for the concurrent examination of gene expression in both members of the association, making it especially valuable for investigating endophyte-host interactions. Barnett et al. (2004) focused on examining the coordinated differentiation and reactions initiated by the exchange of signals between two symbiotic partners—the α-proteobacterium *Sinorhizobium meliloti* and its legume partner *Medicago truncatula*—during nodule formation. To support this investigation, a custom Affymetrix GeneChip was developed, encompassing the full genome of *S. meliloti* along with around 10,000 probe sets for *M. truncatula*.

This innovative approach allowed researchers to unravel the intricate molecular mechanisms underlying symbiotic relationships, shedding light on the gene expression dynamics and signaling pathways involved in endophyte-host interactions. The use of microarray technology in such studies serves as an essential tool for deciphering the complexities of symbiotic associations and holds great potential for advancing our understanding of plant-microbe interplay.

2.2 Conventional Approaches to Studying Fungal Endophytes and Their Challenges

Generally, the techniques utilized in studying endophytes are fairly straightforward, although they entail complex protocols that must be carefully adhered to in order to avoid misinterpretation. After collecting plant material, the initial step involves superficially disinfecting the sampled tissue to eliminate the epiphytic microbiota. This is then followed by utilizing culture-dependent approaches and/or independent methods to isolate endophytic fungi. The primary method for isolating the endophytic fungi is the culture-dependent method, which involves the fragmentation and distribution of the sample on the culture medium. The fungi samples are allowed to grow under a controlled environment, and after proper growth, pure colonies are cultured and molecular and morphological identification is carried out.

The limitations of culture-based methods in capturing the true diversity of fungi from a particular habitat are widely acknowledged (Wu et al. 2019; Lücking et al. 2020; Fan et al. 2020). This is due to the inability to cultivate certain microorganisms under laboratory conditions (Wu et al. 2019). Relying solely on morphological characteristics for identification can lead to misclassification, as similar structural features are shared among fungi of different species (Lücking et al. 2020), for instance, hyphae composed of single or multiple cells encased in a tubular cell wall. Addressing these limitations often involves alterations to culture media, including the addition of specific substrates to enhance fungal species diversity (Mourad et al. 2018), alongside the incorporation of molecular techniques for accurate detection (Noriler et al. 2018).

Regardless of improvements in endophytic fungal research techniques, significant hurdles persist throughout commonly employed protocols, necessitating well-defined strategies to mitigate the risk of misinterpretation. Some common errors that occur in the study of plant endophytic fungi are the misinterpretation of the plant species, low sampling, problem with processing, surface distillation plant sample preparation, culturing of the fungi, and phenotypic and molecular identification (Dos Reis et al. 2022).

2.3 Proteomics in Deciphering Plant-Endophyte Interaction

Proteomics is pivotal in decoding the intricate interplay between fungal endophytes and their host plants, furnishing invaluable knowledge into the underlying biochemical mechanisms and prospective implementations of these associations (Vincent et al. 2012; Verma et al. 2024). It delves into the comprehensive examination of all proteins synthesized by a cell, tissue, or organism within a specific timeframe. It is an effective instrument for determining, characterizing, and quantifying proteins expressed by fungal endophytes, providing insights into their interactions and roles (Vincent et al. 2020; Jain et al. 2021). Fungal endophytes impart a bundle of benefits to their host plant. These include alleviating stresses in plants caused by abiotic and biotic factors, promoting plant growth, increasing nutrient acquisition, enhancing crop yield, improving disease resistance, elevating phytohormones, and inducing secondary metabolites (Ancheeva et al. 2020; Saad and Badry 2020; Fadiji and Babalola 2020; Baron and Rigobelo 2022; Poveda et al. 2022; Saini et al. 2024). To explore such effects, currently, proteomics is applied to few areas of endophytic fungal research, which is discussed in the following section.

2.3.1 Abiotic Stress

Pinus taeda, a significant forest tree species prized for its rapid growth and high oleoresin yield, is commonly cultivated in plantations across warm and wet regions of southern China. However, the occurrence of drought, at times severe and prolonged, poses a challenge in these areas. Recently, Wu et al. (2024) evaluated the drought tolerance of *P. taeda* and the potential benefits of utilizing root endophytic fungus *Serendipita indica* by conducting proteomic and metabolomic analysis of needles. Pots were used to grow seedlings of *P. taeda*, which were inoculated with *S. indica*, and subjected to drought stress for 53 days. The findings revealed significant alterations in the levels of several metabolites in the *S. indica*-inoculated *P. taeda* seedling needles under drought conditions, particularly flavonoids and organic acids. Notably, various metabolites, including eriocitrin, vitamin C, trans-aconitic acid, uric acid, vitamin A, alpha-ketoglutaric acid, coumalic acid, stachydrine, itaconic acid, citric acid, etc., exhibited upregulation of more than threefold in inoculated seedlings in comparison to un-inoculated ones under drought stress. Furthermore, proteomic analysis identified specific differential proteins, including H9VDW5 and H9X056, exclusively present in the needles of seedlings inoculated with *S. indica*. Moreover, the protein H9VNE7 exhibited an over 11-fold upregulation in fungus-inoculated seedlings compared to non-inoculated ones under drought stress. Interestingly, *S. indica* inoculation also enhanced the enrichment of proteins induced under water-deficient conditions, including dehydrins, LP3-1, LP3-2, and LP3-3. Furthermore, under drought conditions, inoculation with *S. indica* induced significant changes in biosynthesis and metabolism pathways. Positive relationships were observed between the buildup of certain metabolites and the enhancement of proteins in *P. taeda* in drought situations. *Piriformospora indica*, a root fungal endophyte, plays a pivotal part in enhancing the growth and survival of plants across various species, particularly under abiotic stresses like drought, which significantly impact crop productivity. Ghaffari et al. (2019) explored the effect of *P. indica* on moisture-stressed barley plant leaves to study the drought stress response using proteomic and metabolomic approaches. Upon moderate drought imposition, 145 proteins were altered in plants treated with *P. indica* than 104 in untreated plants. Many of these responsive proteins played crucial roles in the primary metabolism of barley, particularly in eliminating oxidative stress-induced damage. Elevated activity of the photosystem and electron transfer chain and a buildup of proteins imparting protection against photorespiration, energy modulation, and autophagy were observed upon colonization of plant roots by *P. indica*. Such findings demonstrate substantial variations in the metabolome and proteome of fungus-treated seedlings following drought circumstances, providing insights for further investigation into the roles of proteins and metabolites, particularly those associated with drought stress tolerance.

Soil salinity poses a significant challenge in agricultural settings, impacting crop productivity. Microbial inoculants have emerged as a promising strategy to mitigate plant salinity stress. Zeng et al. (2023) explored the efficacy of the biocontrol strain *Pseudomonas alcaliphila* Ej2 in alleviating salt stress in rice plants. Through a combination of greenhouse and field experiments, rice proteomic analysis, and microbial community profiling, the study methodically examined the action mechanism of Ej2. The results revealed that Ej2 significantly improves root and shoot length, as well as dry and fresh weight, in rice plants under salinity stress compared to untreated controls. Additionally, Ej2 exhibits the capacity to manage rice blast disease and enhance overall rice growth. Furthermore, proteomic analysis revealed significant alterations in the rice proteomic profile following Ej2 inoculation, particularly in pathways related to the biosynthesis of unsaturated fatty acids, plant-pathogen interactions, and metabolism. These findings collectively provide a strong indication that Ej2 inoculation induces changes in the rice endophytic microbiome and proteomic profiles, ultimately promoting plant development under salt stress situation. Sepehri et al. (2021) studied the proteome pattern of barley (*Hordeum vulgare* L.) leaf blades in plants colonized by *Serendipita indica*, aiming to elucidate the molecular mechanisms underlying *S. indica*-mediated salt stress alleviation. The experiment involved two treatment factors: salinity levels (0 and 300 mmol/L NaCl) and fungus (non-inoculated and *S. indica*-inoculated barley). In non-inoculated control plants, shoot dry matter production decreased by 84% following exposure to 300 mmol/L NaCl, a reduction twofold greater than observed in *S. indica*-inoculated plants. However, *S. indica*-inoculated plants exhibited significantly lower sodium accumulation in the shoots. Proteomic analysis revealed a notable rise in the abundance of photosynthesis-related proteins. A significant upregulation of 26 identified proteins out of 42 was observed. The study emphasized that considerable salt stress resistance and enhanced plant growth induced by *S. indica* are correlated with the upregulation of various proteins associated with carbohydrate metabolism and photosynthesis.

2.3.2 Biotic Stress

The molecular interplay between fungal endophytes *Beauveria bassiana*, *L.* cf. *psalliotae*, and *Lecanicillium dimorphum* and the date palm *Phoenix dactylifera* L. was explored by Gómez-Vidal et al. (2009) using proteomic methodologies. Two independent bioassays were conducted, analyzing field date palms inoculated with these fungi 15 and 30 days post-inoculation. Additionally, in vitro date palms were treated with either *B. bassiana* or *L.* cf. *psalliotae* revealing qualitative and quantitative differences in protein accumulation that were equated to un-inoculated and fungi-inoculated palms. Notably, in-field date palms, proteins associated with plant defense mechanisms, and stress responses were upregulated in *P. dactylifera* leaves following endophytic colonization by fungi. Moreover, proteins linked to photosynthesis and energy metabolism exhibited alterations due to fungi colonization. Furthermore, it was suggested that endophytic colonization by these fungi modifies energy metabolism and plant defense responses, potentially influencing the expression of proteins linked to cell division in in vitro palms at the proteomic level.

Aerobic metabolism naturally generates reactive oxygen species (ROS) as byproducts within cellular compartments. These ROS are produced within cellular compartments during aerobic metabolism and play a vital role in various biological processes such as activating pathways that suppress pathogen proliferation and acting as signaling molecules. However, they are also considered byproducts due to their potential to cause oxidative damage. Excessive ROS can be detrimental to plants, causing damage to cellular structures and interfering with physiological processes, ultimately leading to reduced yield. ROS can also degrade essential biomolecules such as proteins, enzymes, and nucleic acids. Plants possess their defense mechanisms to counteract excessive ROS and maintain cellular homeostasis. Notably, microbes, particularly endophytes, have been identified to play an essential role in maintaining ROS balance under both biotic stresses through various mechanisms. Endophytes can produce antioxidant compounds themselves and also stimulate the host plant biochemical network to enhance ROS elimination (Sahu et al. 2022).

2.3.3 Enhance Growth and Yield

Rice is a staple food for numerous communities worldwide, particularly in regions such as Asia, parts of Africa, and Latin America. Presently, farmers resort to chemical interventions to combat rice diseases and enhance yield, albeit with probable adverse impacts on human safety and ecological balance. To overcome such issues, the role of fungal endophytes in enhancing plant growth is emerging (Grabka et al. 2022; Baron and Rigobelo 2022). Leetanasaksakul et al. (2024) determined the plant-promoting attributes of the endophytic fungal strain DMKU-R3G3, isolated from organic rice roots, through proteomic analysis. Based on morphological features and phylogenetic examination, the DMKU-R3G3 was identified as *Aspergillus cejpii.* Treatment with *A. cejpii* DMKU-R3G3 significantly improved rice growth, evident from a prominent rise in shoot height, fresh weight, and root length. Furthermore, the chlorophyll concentration of *A. cejpii*-inoculated rice plants escalated by 1.78 times compared to the non-inoculated group. Additionally, proteomic analysis unveiled rice's response to endophytic fungal colonization, with the production of auxin-responsive proteins that regulate indole acetic acid levels and the increased expression of proteins associated with chlorophyll biosynthesis, leading to enhanced chlorophyll production. The findings suggested that *A. cejpii* strain DMKU-R3G3 holds the ability to increase plant growth through phytohormone production under greenhouse conditions. Thus, the endophytic *A. cejpii* DMKU-R3G3 emerges as a prospective eco-friendly growth enhancer for sustainable rice farming.

Paphiopedilum barbigerum, the rarest and most endangered orchid species in China, holds significant ornamental importance. Germination of mature seeds of *P. barbigerum* is challenging due to the lack of endosperm, relying heavily on mycorrhizal fungi for germination and subsequent growth. Tian et al. (2024) explored the symbiotic germination of *P. barbigerum* with *Epulorhiza* sp. FQXY019, a mycorrhizal fungus, at 90 days post germination and the underlying mechanism by the application of transcriptomics and proteomics. Transcriptomic sequencing identified 10,961 genes with differential expression, of which 2599 were found to be upregulated and 8402 downregulated between the control and FQXY019-treated groups. Furthermore, tandem mass tag-based quantitative proteomics identified 537 differentially expressed proteins, of which 220 were upregulated and 317 downregulated, enriched in processes such as photosynthesis and fatty acid biosynthesis.

The fungal endophyte *Gilmaniella* sp. AL12 demonstrates the capacity to create a mutually valuable relationship with the medicinally significant herb *Atractylodes lancea*, leading to enhanced plant development and accumulation of sesquiterpenoids, a phenomenon termed "double promotion." Yuan et al. (2019) applied transcriptomic and proteomic analyses to examine the effects of *Gilmaniella* sp. AL12 inoculation on plants, comparing those inoculated with *Gilmaniella* sp. AL12 to those free of the endophyte. The study revealed that genes and proteins related to primary metabolism, including carbon fixation, energy metabolism, and carbohydrate metabolism, were significantly upregulated following *Gilmaniella* sp. *AL12* inoculation. Additionally, *Gilmaniella* sp. AL12 upregulates genes associated in terpene skeleton biosynthesis, including those annotated as β-caryophyllene synthase and β-farnesene synthase. This study suggests that endophyte-plant associations could elevate production of biomass and sesquiterpenoid content by augmenting photosynthesis, expanding metabolism, and boosting sesquiterpenoid biosynthesis pathway in *A. lancea*.

Shrivastava et al. (2018) assessed the interface between the endophyte *Piriformospora indica* and *Brassica napus* plants, by employing integrated proteomic analyses to shed light on its potential impact on symbiosis and total plant growth and development. Utilizing an LC-MS/MS-based label-free approach (quantitative), differential proteomics in plants treated with *P. indica* was reported as compared to control plants. A total of 8123 proteins were detected, from which 46 exhibited significant changes, i.e., 34 downregulated while 12 upregulated. Through in silico investigations using GO and KEGG pathway analysis, the identified differentially expressed proteins were mapped, revealing significant enrichment in gene sets associated with symbiotic signaling, energy production, stress/defense responses, nutrient acquisition, metabolic processes, and biosynthesis of essential metabolites.

2.4 Technological Advances in Fungal Endophyte Proteomics

Recent advancements in technology have transformed the study of fungal endophyte proteomics, greatly enhancing our understanding of their interactions with host plants. The advent of high-throughput sequencing technologies has allowed researchers to comprehensively analyze the entire proteome of fungal endophytes, uncovering novel proteins and metabolic pathways involved in their symbiotic relationship with plants. Moreover, cutting-edge mass spectrometry techniques have enabled the identification and characterization of fungal endophyte proteins with high precision and sensitivity (Verma et al. 2024). These advancements have enabled the identification of bioactive compounds synthesized by endophytes, opening new paths for the creation of innovative pharmaceuticals and agricultural solutions. The field of proteomics has seen significant advancements in both gel-based and gel-free methodologies in the study of fungal endophytes. Gel-based techniques, such as 2D gel electrophoresis, have been used to separate and visualize complex mixtures of proteins, while gel-free techniques, like shotgun proteomics, have allowed for comprehensive proteomic analysis without the constraints of gel-based methods (Lee et al. 2020). These technological advancements have greatly expanded our understanding of fungal endophyte proteomics, laying the foundation for uncovering novel natural products and advancing the development of cutting-edge biotechnological innovations.

2.4.1 Mass Spectrometry (MS)-Based Techniques for Proteomic Analysis

MS has revolutionized the field of proteomics by offering high sensitivity, specificity, and throughput in protein identification and quantification. The principles of MS involve the ionization of proteins or peptides, followed by their separation based on mass-to-charge ratio and detection (Shuken 2023). Liquid chromatography coupled with tandem mass spectrometry (LC-MS/MS) has become a fundamental technique in proteomics, allowing for comprehensive analysis of complex protein mixtures with unparalleled depth and accuracy (Li and Zhu 2020).

Concerning fungal endophyte research, MS-based proteomics has provided unprecedented insights into the proteomes of these symbiotic organisms. By employing sophisticated sample preparation techniques and advanced MS instrumentation, researchers can profile the complete repertoire of proteins expressed by fungal endophytes during different growth stages or under various environmental conditions (Gupta et al. 2023). This detailed characterization enables the detection of essential proteins involved in nutrient acquisition, stress response, and secondary metabolite biosynthesis, shedding light on the molecular mechanisms underlying fungal endophyte-host interactions.

Protein identification using mass spectrometry typically occurs through whole-protein analysis, known as "top-down" proteomics, or through the analysis of peptides generated enzymatically or chemically, referred to as "bottom-up" proteomics (Fig. 2.4) (Han et al. 2008). Bottom-up approaches, where peptide detection is employed to deduce protein presence, are considered the standard for large-scale or high-throughput analysis of complex samples, notably LC-MS/MS-based approaches, crucial for analyzing the proteomic composition of complex samples such as fungal endophytes (Dupree et al. 2020). This methodology encompasses two primary workflows: "sort-then-break" and "break-then-sort." In the former, offline protein fractionation is performed before protein digestion, followed by peptide analysis using techniques such as peptide mass fingerprinting (PMF) or LC coupled with tandem mass spectrometry (LC-MS/MS). The latter approach, known as shotgun proteomics, has garnered popularity due to its capacity to collect tandem mass spectra for numerous peptides, subsequently compared against databases to identify proteins (Neagu et al. 2022). Multidimensional protein identification technology (MudPIT) serves as a widely employed shotgun proteomic technique, enabling comprehensive analysis of diverse samples, including whole organism, tissue samples, cell lysates, subcellular fractions, and sub-proteomes. Recent advancements in MudPIT include enhanced peptide separation techniques and the integration of novel mass spectrometry technologies, thereby augmenting the precision and comprehensiveness of proteomic analysis (Delahunty and Yates 2007).

These techniques are particularly beneficial for studying fungal endophytes. By employing shotgun proteomics, researchers can comprehensively analyze the proteomes of fungal endophytes, elucidating the intricate relationships between these organisms and their host plants. Through MudPIT and other multidimensional separation techniques, a wide range of proteins involved in nutrient acquisition, stress response, and secondary metabolite biosynthesis can be identified, offering important understanding of the molecular processes driving fungal endophyte-host interactions.

Top-down proteomics represents a promising advancement in the study of fungal endophytes, as it involves the gas-phase ionization of intact proteins, followed by high-resolution mass measurement and direct fragmentation inside the mass spectrometer (Cupp-Sutton and Wu 2020). This approach offers a comprehensive characterization of protein isoforms and posttranslational modifications, which are critical for understanding the molecular intricacies of fungal endophyte-host interactions. While top-down proteomics initially faced challenges in analyzing large proteins, recent developments such as pre-folding dissociation and middle-down strategies have significantly expanded its applicability (Po and Eyers 2023). These innovations allow researchers to tackle complex biological problems associated with fungal endophytes, including the deciphering of intricate signaling networks and the elucidation of key molecular players involved in host colonization and symbiotic interactions.

Quantitative proteomics, on the other hand, serves as a valuable tool for studying dynamic changes in protein abundances within fungal endophytes. Facilitated by stable isotope labeling and label-free methods, quantitative proteomics enables researchers to measure alterations in protein expression levels under various growth conditions or in response to environmental stimuli (Pan and Aebersold 2007).

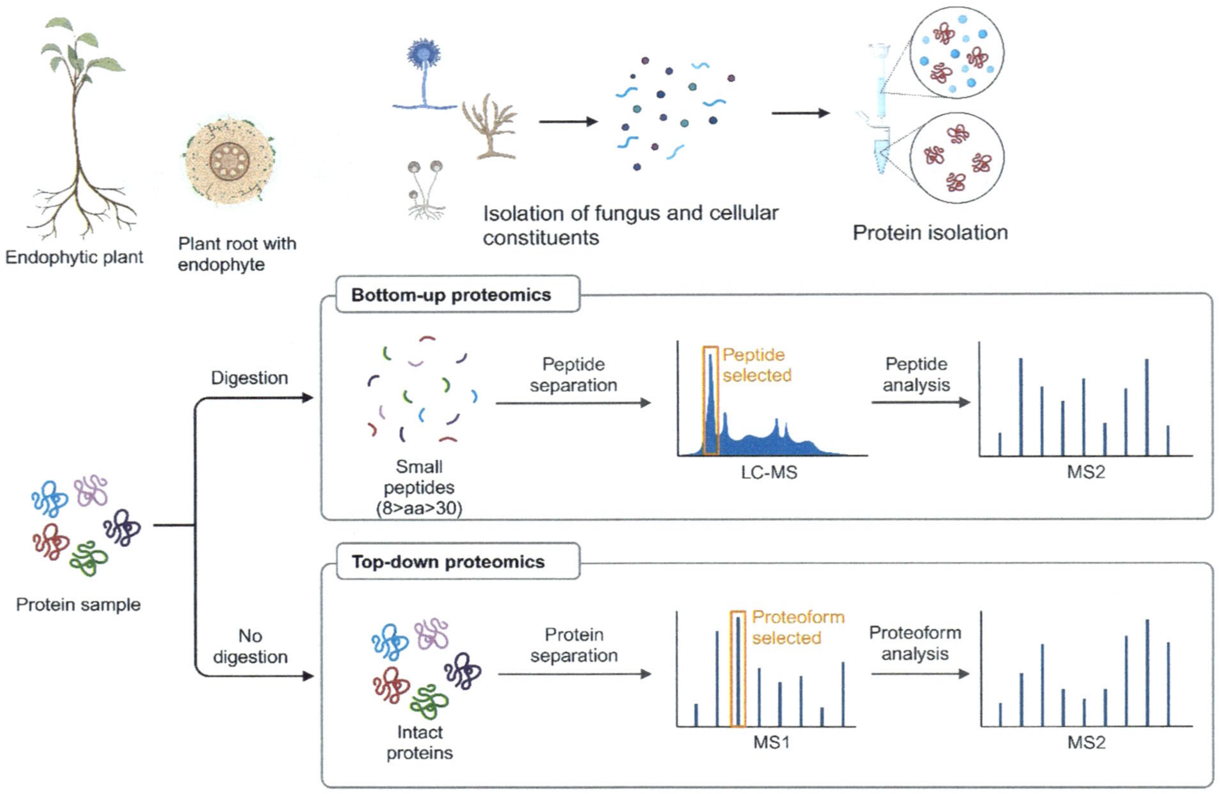

Fig. 2.4 Schematic representation of mass spectrometry-based techniques for proteomic analysis of fungal endophytes

Metabolic labeling, post-biosynthetic labeling, and absolute quantification using synthetic peptides are among the strategies employed for protein quantification in fungal endophytes. By offering both relative and absolute quantifications of proteins, these techniques provide crucial insights into the regulation of cellular processes, signaling cascades, and disease mechanisms within fungal endophytes. This quantitative approach enhances our understanding of the molecular adaptations of fungal endophytes to their host environment and aids in the identification of potential targets for biotechnological interventions aimed at improving plant health and agricultural productivity (Bantscheff et al. 2007).

2.4.2 Advancements in Gel-Based and Gel-Free Proteomics

Advancements in proteomic techniques have greatly advanced our comprehension of fungal endophytes, symbiotic microorganisms residing within plant tissues, and their intricate interactions with host plants. Before these advancements, proteomic analyses in fungal endophyte research were predominantly reliant on conventional gel-based techniques, notably two-dimensional gel electrophoresis (2DE) (Magdeldin et al. 2014). Despite its widespread use, 2DE posed several limitations, including a narrow dynamic range, poor resolution of low-abundance proteins, and challenges in detecting membrane-associated proteins. Consequently, these constraints hindered the comprehensive characterization of the fungal endophyte proteome, restricting our understanding of the molecular mechanisms that control fungal endophyte-host interactions.

In response to the limitations of gel-based techniques, gel-free proteomic strategies, particularly liquid chromatography coupled with tandem mass spectrometry (LC-MS/MS), have emerged as powerful alternatives in fungal endophyte research. Unlike gel-based methods, LC-MS/MS bypasses gel-based separation, allowing for more in-depth and high-throughput analyses of complex proteomes such as those of fungal endophytes (Tazik et al. 2020). This approach facilitates the identification and quantification of a broader spectrum of proteins, including low-abundance and membrane-associated species, thereby offering a more comprehensive understanding of fungal endophyte biology and their interactions with host plants. Recent years have witnessed significant progress in both gel-based and gel-free proteomic methodologies, enabling novel insights into fungal endophyte proteomes (Abdallah et al. 2012). In the realm of gel-based proteomics, advancements have been made in gel electrophoresis techniques, including the adoption of immobilized pH gradient strips for enhanced resolution and reproducibility. Additionally, improvements in staining and detection methods, such as the utilization of fluorescent dyes and mass spectrometry-compatible stains, have facilitated enhanced visualization and identification of proteins in gel-based proteomic analyses.

Conversely, gel-free proteomics has seen notable advancements in preparation of samples, chromatographic separation, and mass spectrometry instrumentation. Techniques such as filter-aided sample preparation (FASP) and in-solution digestion have streamlined workflows and increased protein recovery, overcoming challenges encountered in gel-based approaches (Wiśniewski et al. 2011). Furthermore, the arrival of ultra-high-performance liquid chromatography (UHPLC) systems has enabled faster and more efficient peptide separation, leading to improved throughput and sensitivity in gel-free proteomic analyses (Dong and Zhang 2014). Moreover, enhancements in mass spectrometry instrumentation, including higher resolution, faster scanning speeds, and improved sensitivity, have broadened the analytical capabilities of gel-free proteomics, allowing for more precise and comprehensive characterization of fungal endophyte proteomes (Shuken 2023). These advancements have not only revolutionized our understanding of fungal endophyte proteomes but have also presented new opportunities for agricultural and biotechnological applications. By elucidating the molecular mechanisms underlying fungal endophyte-host interactions, proteomic studies have identified key proteins involved in symbiosis, pathogenesis, and biocontrol. Furthermore, proteomic analyses have facilitated the discovery of potential targets for enhancing plant health, stress tolerance, and productivity through biotechnological interventions.

2.4.3 High-Throughput Proteomics for Large-Scale Studies and Its Impact on Biotechnological Research

High-throughput proteomics, characterized by the ability to analyze a large number of proteins rapidly and efficiently, has revolutionized biotechnological research by enabling comprehensive exploration of complex biological systems (Cui et al. 2022). This approach, which leverages advanced instrumentation and analytical techniques, has had a profound impact on understanding cellular processes, elucidating disease mechanisms, and identifying potential targets for therapeutic intervention and biotechnological applications. One of the key aspects of high-throughput proteomics is its capacity for large-scale studies, allowing researchers to analyze thousands of proteins simultaneously. This capability is particularly advantageous for examining intricate biological systems like signaling networks, metabolic pathways, and protein-protein interactions. High-throughput proteomic techniques, such as liquid chromatography coupled with tandem mass spectrometry (LC-MS/MS), offer unparalleled sensitivity, throughput, and coverage of the proteome (Gobena et al. 2024). By generating large volume of data in a relatively brief period, these methods facilitate comprehensive profiling of protein expression, posttranslational modifications, and protein-protein interactions on an extensive level.

The influence of high-throughput proteomics on biotechnological research is multifaceted and far-reaching. For instance, in drug discovery and development, high-throughput proteomic screening enables the identification of potential drug targets and biomarkers for various diseases. By analyzing protein expression patterns in healthy versus diseased tissues, researchers can pinpoint proteins associated with specific pathological conditions, paving the way for the development of targeted therapies and diagnostic tools (Dueñas et al. 2022). Additionally, high-throughput proteomics allows for the systematic analysis of drug-protein interactions, elucidating the mechanisms of action and potential side effects of pharmaceutical compounds. Furthermore, high-throughput proteomics is crucial in understanding microbial physiology and exploiting microorganisms for biotechnological applications. For example, in the field of industrial microbiology, high-throughput proteomic techniques enable the optimization of microbial strains for the production of biofuels, pharmaceuticals, and bioplastics (Montaño López et al. 2022). By identifying key proteins involved in metabolic pathways, stress responses, and protein secretion, researchers can engineer microbial strains with enhanced productivity, robustness, and efficiency (Adegboye et al. 2021). Moreover, high-throughput proteomics simplifies the discovery of novel enzymes and metabolic pathways with biotechnological potential, expanding the repertoire of tools available for bioprocessing and biocatalysis.

In agricultural biotechnology, high-throughput proteomics plays a key role in developing crop varieties with enhanced traits, such as increased yield, stress tolerance, and nutritional quality. By analyzing the proteomes of plants under different environmental conditions, researchers can identify proteins associated with desirable traits and breed or engineer crop varieties with enhanced agronomic performance (Eldakak et al. 2013). Additionally, high-throughput proteomics aids in the studying of plant-microbe interactions, facilitating the development of sustainable agricultural practices such as biofertilization, biocontrol, and phytoremediation (Diwan et al. 2022). Overall, high-throughput proteomics has emerged as a powerful tool for large-scale studies in biotechnological research, driving advancements in drug discovery, industrial biotechnology, agricultural biotechnology, and personalized medicine. By providing comprehensive insights into the proteome and its dynamics, high-throughput proteomic techniques offer unprecedented opportunities for understanding biological complexity, accelerating innovation, and translating scientific discoveries into real-world applications.

2.5 Exploiting Secondary Metabolites for Pharmaceuticals

Fungal endophytes produce various metabolites that are part of their intricate interaction with host plant. These metabolites exhibit diverse chemical structures and biological activities, making them valuable target for drug discovery and development. Endophytic fungi have evolved complex metabolic pathways to adapt to their unique ecological niches within host plants. Consequently, they produce a range of secondary metabolites, including alkaloids, terpenoids, polyketides, and peptides, many of which possess potent pharmacological properties (Tiwari and Bae 2022). These bioactive compounds have demonstrated antimicrobial, anticancer, antiviral, antioxidant, and immunomodulatory activities, among others, making them attractive candidates for novel drug discovery.

One of the most renowned examples is Taxol, a widely used anticancer drug originally isolated from the endophytic fungus *Taxomyces andreanae* residing within the Pacific yew tree (*Taxus brevifolia*) (Stierle et al. 1993). This discovery underscores the therapeutic potential of fungal endophytes and highlights the importance of exploring their secondary metabolites for therapeutic purposes. In

Table 2.1 Bioactive compound of endophytic fungi and their host plant

Sl. no.	Host plant	Fungal endophyte	Bioactive compound	Application	References
1	*T. brevifolia*	*Taxomyces andreanae*	Taxol	Anticancer	Stierle et al. (1993)
2	*Terminalia morobensis*	*Pestalotiopsis microspora*	Isopestacin	Antifungal, antioxidant	Strobel and Daisy (2003)
3	*Coleus forskohlii*	*Rhizoctonia bataticola*	Forskolin	Anti-HIV, antitumor	Mir et al. (2015)
4	*Capsicum annuum*	*Alternaria alternata*	Capsaicin	Cardioprotective	Devari et al. (2014)
5	*Camptotheca acuminata*	*Fusarium solani*	Camptothecin	Anticancer	Ran et al. (2017)
6	*Terminalia arjuna*	*Alternaria tenuissima*	Paclitaxel	Anticancer	Ismaiel et al. (2017)
7	*Taxodium distichum*	*Alternaria fumigatus*	Paclitaxel	Anticancer	Ismaiel et al. (2017)
8	*Salix alba*	*Aspergillus niger*	Salicylic	Analgesic, anti-inflammatory	Lubbers et al. (2021)
9	*Triticum aestivum*	*Aspergillus terreus*	Lovastatin	Lowering of cholesterol	Boruta and Bizukojc (2017)
10	*Huperzia serrata*	*Acremonium* sp.	Huperzine A	Improves cognitive function	Ohba et al. (2015)
11	*Hypericum perforatum*	*Chaetomium globosum*	Hypericin	Antidepressant	Kusari et al. (2008)

addition to Taxol, numerous other secondary metabolites from fungal endophytes have demonstrated promising pharmaceutical applications. For instance, camptothecin, isolated from endophytic fungi like *Fusarium solani*, exhibits potent anticancer activity by inhibiting topoisomerase I, a key enzyme involved in DNA replication (Ran et al. 2017). Furthermore, paclitaxel, another anticancer agent that inhibits cell proliferation by blocking mitosis, is produced by various endophytic fungi, including *Alternaria tenuissima* and *Alternaria fumigatus* (Ismaiel et al. 2017). There are various other bioactive compounds produced by fungal endophytes that have potential for pharmaceutical application as shown in Table 2.1.

The pharmaceutical industry is increasingly turning to biotechnological methods to unlock the vast chemical diversity harbored by fungal endophytes. Utilizing techniques such as genome mining, metagenomics, metabolomics, and synthetic biology, researchers can now identify, isolate, and characterize novel bioactive metabolites with significant therapeutic potential (Mishra et al. 2021). Moreover, current developments in fermentation and metabolic engineering have revolutionized the generation of bioactive compounds by fungal endophytes. These technologies enable the scalable synthesis of valuable pharmaceuticals in controlled laboratory environments, overcoming the challenges associated with extracting natural products from wild sources. This sustainable approach ensures consistent access to bioactive compounds for pharmaceutical development.

The exploration of secondary metabolites derived from fungal endophytes holds great promise for pharmaceutical advancements. By delving into nature's vast chemical repertoire, researchers can uncover new drugs with applications in addressing a broad range of human diseases, including cancer, infectious diseases, and neurological disorders. Fungal endophytes serve as a rich source of bioactive secondary metabolites, offering exciting prospects for drug discovery and development (Singh et al. 2021). Through interdisciplinary collaboration and innovative biotechnological strategies, the pharmaceutical industry can fully harness the potential of these remarkable microorganisms to address unmet medical needs and enhance global health outcomes.

2.6 Plant Growth Promotion Activities of Endophytes

2.6.1 Bioinoculants

To increase crop productivity, bioinoculants are bioformulations made of specific microbial strains on an appropriate carrier. When utilized as bioinoculants, fungal endophytes offer the host a number of advantages, including immune response-based pathogen defense, mineralization of vital nutrients, and enhanced plant development. As bioinoculants for sustainable agriculture, fungal endophytes are a polyphyletic group of fungi that are primarily *ascomycetous* with *Fusarium*, *Aspergillus*, *Alternaria*, *Trichoderma*, *Penicillium*, *Cladosporium*, *Colletotrichum*, and *Talaromyces* being the predominant species (Khalil et al. 2021; Grabka et al. 2022). Their capacity to yield a wide range of bioactive secondary metabolites, including terpenoids, tannins, saponins, steroids, quinones, phenolics, and alkaloids, has drawn the majority of research attention to them. In addition to having many additional qualities of biotechnological relevance, these metabolites are repositories of powerful and new sources of antibacterial, anti-insect, anticancer, antioxidative, and antifungal chemicals (Suryanarayanan 2013; Kamat et al. 2022). By strengthening the soil's structure and assisting the host in overcoming a variety of biotic and abiotic stressors, they are meant to promote plant health, resulting in either up- or downregulation of proteins belonging to nine distinct functional categories. It may have greater protein import and transport to the chloroplast for the light and dark responses, as well as more effective nutrient distribution, leading to improved photosynthesis (Chi et al. 2010).

2.6.2 Biocontrol Agents

The first filamentous fungus for which mass spectrometry (liquid chromatography-tandem mass spectrometry (LC-MS-MS) and matrix-assisted laser desorption/ionization time-of-flight mass spectrometry (MALDI-TOF)) has been used to improve the method of protein identification after classical two-dimensional electrophoresis carried out on a whole cell extract is a biocontrol strain of *T. harzianum* (dela Cruz et al. 2023). A significant risk to the development and yield of rice and other graminaceous crops is *Magnaporthe oryzae*. Strong antagonistic effects against *M. oryzae* have been observed for the rice endophyte *Streptomyces hygroscopicus* OsiSh-2. Physiological, biochemical, and label-free proteomic analyses were employed in a study to examine OsiSh-2's global response to *M. oryzae* toxins or Mo toxins. OsiSh-2 controlled its growth rate in response to Mo toxins, enabling a quick shift to the stationary growth phase, which is when higher antibiotic activity was attained. Furthermore, in response to Mo toxin stress, OsiSh-2 induced the production of reactive oxygen species (ROS), hydrolase activity, and biofilm formation (Liu et al. 2021).

2.6.3 Importance of Proteomics of Endophytic Fungi in Sustainable Agriculture

Acquisition of nutrients: The uptake of certain macronutrients, including K, N, Na, and Mg, as well as micronutrients, including Zn, Fe, and Cu, is enhanced by endophytic fungi. The host plant receives more of these nutrients as a result (Rana et al. 2020). Behie et al. (2012) first documented the endophytic colonization of bean plants (*Phaseolus vulgaris*) and switchgrass (*Panicum virgatum*) by the entomopathogenic fungus *Metarhizium robertsii*. Additionally, they discussed the fungal invasion and the subsequent transfer of nitrogen (N) from larvae of *Galleria mellonella*. According to reports, one study focused on the function of *P. indica*'s phosphorus transporter protein in maintaining phosphorus homeostasis in maize plants (Yadav et al. 2010).

It has been demonstrated that the well-known endophyte *Serendipita (=Piriformospora) indica* is capable of delivering nutrients, such as phosphates, to host plants (Card et al. 2016). For vital growth processes, plants require both nitrogen and phosphorus, which are frequently supplied by mycorrhizal fungus. The cellular proteomes of closely related *Serendipita vermifera* subsp. *bescii* NFPB0129 and *Serendipita vermifera* MAFF305830 were examined in an in vitro investigation under circumstances of (N) ammonium and (P) phosphate starvation. The findings suggest that *S. bescii* and *S. vermifera* have the metabolic capacity to assimilate nitrogen from organic nitrogen sources (Ray et al. 2019).

Numerous nonleguminous species are also infected and colonized by rhizobial endophytes, in addition to leguminous plants. It has been demonstrated that rhizobia infect many rice species and move from plant roots to aerial tissues including leaf sheaths and leaves using the green fluorescent protein tagging approach. It was discovered that their interaction encouraged the growth of rice. The combined effect of improved photosynthesis and stress tolerance is the encouragement of growth. To understand the molecular mechanism, the protein profiles of three distinct tissues (leaf sheath, root, and leaf) of rice inoculated with *Sinorhizobium meliloti* 1021 were analyzed using gel-based comparative proteomic techniques.

Production of phytohormones: Numerous phytohormones, including auxins, gibberellins (GAs), and cytokinins, are produced by endophytic fungus. Fungi produce auxins mostly in the form of indole-3-acetic acid (IAA). Waqas et al. (2012) reported on the synthesis of IAA and GA by the endophytic fungus *Penicillium* sp. and *Phoma glomerata*. It has also been reported that the endophytic fungus *Paecilomyces formosus* (=*P. maximus*) strain LHL10, which was isolated from cucumber plants, produces GAs and IAAs (Khan et al. 2012).

Induction of systemic resistance: Endophytic fungi trigger the activation of induced systemic resistance (ISR) pathways, enhancing the plant's ability to reinforce its defense mechanisms. This may also occasionally coincide with acquired systemic resistance (ASR), as both of these routes enhance plant growth and shield plants from different pathogens (Berg 2009, Busby et al. 2016; Chadha et al. 2015). As endophytes, *Beauveria bassiana*, *Lecanicillium dimorphum*, and *L.* cf. *psalliotae* are entomopathogenic fungi that can live and colonize living palm tissue. Proteomic methods were used to study the molecular interactions between these biocontrol agent fungi and the *Phoenix dactylifera L.* date palm. When entomopathogenic fungi colonized field date palms endophytically, *P. dactylifera* leaves produced proteins involved in plant defense or stress response. Colonization of entomopathogenic fungi also impacted photosynthesis and energy metabolism-related proteins. In vitro palms infected with these fungi developed a protein that resembled a heavy chain of myosin. This implies that these entomopathogenic fungi's endophytic colonization affects the energy metabolism and defense mechanisms of field date palms as well as potentially the proteome level expression of proteins linked to cell division in in vitro palms (Gómez-Vidal et al. 2009).

A standardized procedure for effectively extracting cellular proteins from endophytic fungi in order to produce two-dimensional gel electrophoresis was presented in a study. Using this approach, they were able to acquire quantitative proteome information for *P. indica*. To better understand the role of endophytes under biotic stress, the team also compared protein fingerprints extracted from their study. Research on transporter proteins involved in interactions between endophytic fungi and plants is rarely conducted at a high-throughput level (Yadava et al. 2015). *Methylobacterium* sp. was identified to be the primary source of the overexpression and induction of proteins related to carbohydrates, including four chitinases, in another study. Class 1b chitinase was not present in E+/As and was significantly upregulated in E+ shoots, indicating that these enzymes are involved in the defense reaction against bacterial infection. Major pathogenesis-related proteins called chitinases are involved in symbiosis, growth and development regulation, and stress tolerance (Hamid et al. 2013). *Methylobacterium* sp. upregulates chitinases, suggesting that they are also important in mutualistic symbiosis (Alcántara-Martínez et al. 2018).

Antibiotic and secondary metabolite production: A prominent example of a secondary metabolite produced by an endophytic fungus is the alkaloids synthesized by *Epichloë* species in various grass species. Not only do endophytic fungi help their hosts by supplying an abundance of beneficial substances but they also stimulate the production of defense molecules in plants. Among other substances with activity against illnesses and herbivores, they are great producers of volatile organic compounds (VOCs), alkaloids, steroids, terpenoids, peptides, polyketones, flavonoids, quinols, phenols, and chlorinated compounds (Card et al. 2016; Lugtenberg et al. 2016; Latz et al. 2018; Kaddes et al. 2019). It has been reported that *Nodulisporium* sp. produces nodulisporic acid. Endophytes produce various bioactive metabolites, including polyketides, terpenoids, flavonoids, cytochalasins, alkaloids, and steroids (Guo et al. 2008). Cell division, glucose transport, HIV-1 protease, and cytoskeleton are examples of cellular targets, and bacteria, fungus, and cancer cells are examples of activity targets (Gimenez et al. 2007; Hawksworth and Rossman 1997; Stierle et al. 1993). The first anticancer medication to be worth a billion dollars, paclitaxel, is among the most well-known and intriguing instances. Overharvesting of the yew *Taxus brevifolia*, the parent plant, threatened to endanger its life, due to its widespread use in the 1990s (Strobel and Daisy 2003).

Production of siderophores: Iron is a micronutrient that is essential for the healthy functioning of all living cells (Rana et al. 2020, Turbat et al. 2020). Siderophores are small high-affinity iron-chelating compounds that are secreted by various microorganisms such as bacteria and fungi. The endophytic fungi produce these compounds for the uptake of ferric ions from the rhizosphere to their host plants (Chowdappa et al. 2020, Ibiang et al. 2020). Endophytic strains of *Macrophomina phaseolina* SS1L10, *M. phaseolina* SS1R10, and *Trichoderma koningii ST-KKU1* were reported to produce siderophores by Suebrasri et al. (2020). Furthermore, for recombinant endophytic strains of *Trichoderma harzianum* that colonize beans (*P. vulgaris*), the manufacture of siderophores is explained (Eslahi et al. 2020).

2.6.4 Importance of Proteomic Endophytic Fungi in Environmental Management

Heavy metal augmentation: Applying the metal-resisting endophytic *Penicillium funiculosum* LHL06 to soybeans increased their growth under conditions of combined metal toxicity and improved their roots' ability to withstand metal stress. By lowering metal accumulation and dispersion in soybean tissues, LHL06's capacity to mitigate metal toxicity and release IAA and GAs under metal-induced stress influenced host growth. Additionally, LHL06 inoculation affected the host roots' hormonal regulation and antioxidant system to lessen the harmful impacts of metal poisoning. Stress mitigation-related genes, including G6PDH, GmGST3, GmGST8, and GmSOD1[Cu-Zn], as well as proteins, including glutathione S-transferase, isoflavone reductase, and chalcone isomerase, were induced by inoculation of LHL06. These results represent the first report showing *Penicillium funiculosum*'s phytohormone production under metal toxicity, and they point that LHL06 presents a promising solution for enhancing plant growth and development in soils contaminated with heavy metals and producing safer food for human consumption (Bilal et al. 2019).

2.6.5 Biotic and Abiotic Stress Tolerance

Salt stress tolerance: Both under normal and salt stress circumstances, *P. indica* had a considerable effect on the biomass of the shoots and barley growth. As a dependable indication of salt stress resistance, *P. indica* increased the foliar potassium (K^+)/sodium (Na^+) ratio in colonized plants to control ion accumulation. *P. indica* increased the amount of calcium (Ca^{2+}) and probably had an impact on the transduction of the stress signal. Subsequently, a proteomic analysis using two-dimensional electrophoresis of the barley leaf sheath revealed 968 protein spots. Of these sites that were found, 72 protein spots showed a substantial change in abundance in response to *P. indica* root colonization and salt treatment. Fifty-one proteins were identified by mass spectrometry analysis of responsive proteins. These proteins were involved in photosynthesis, signal transduction, energy production, protein translation and degradation, cell antioxidant defense, and cell wall organization, among other functional areas. The results showed that *P. indica* altered the physiological and proteomic responses of the plant host, thereby triggering a systemic reaction to salt stress as shown in Fig. 2.5 (Alikhani et al. 2013).

Drought stress tolerance: A mutualistic root endophytic fungus called *Piriformospora indica* helps its hosts by promoting plant development and raising yield in both favorable and unfavorable environments. It has been demonstrated that root colonization by *P. indica* enhances tolerance to water stress through a general and nonspecific mechanism that is not limited to specific plant species. Two proteome-based studies have been carried out on this organism. It is well known that

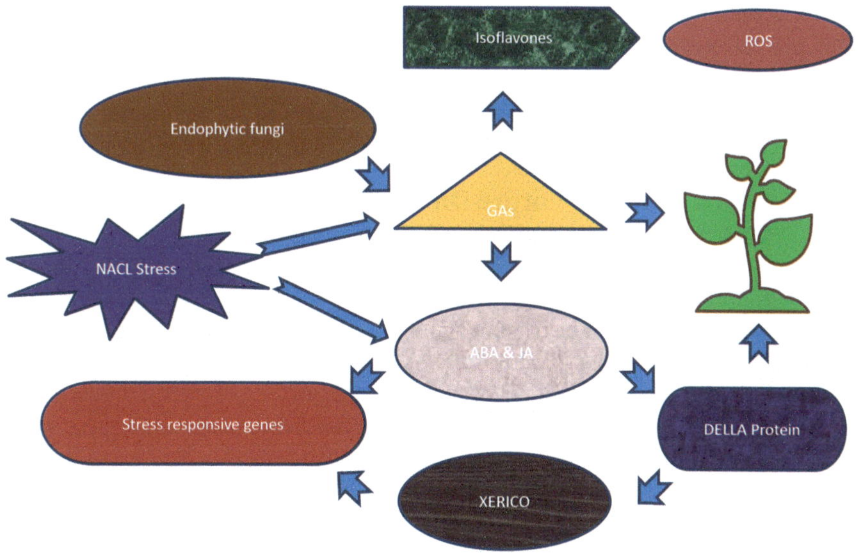

Fig. 2.5 The mitigation reduction of salt stress by the use of phytohormones using endophytic fungi

this species can stimulate plant growth in both normal and stressful environments. The effects of this endophyte on barley plants at different phases of drought stress have been demonstrated (Ghabooli et al. 2013). Using well-treated and water-deficient barley leaves colonized by *P. indica*, proteomic analysis identified 726 consistently detected protein locations. Forty-five differently accumulating proteins involved in metabolisms, signal transduction, photosynthesis, reactive oxygen scavenging, and plant defense responses were identified by mass spectrometry analysis. Interestingly, *P. indica* increased the levels of proteins critical for energy transport, antioxidant defense, and photosynthesis (Ghabooli et al. 2013).

Rehabilitation of soil in mining sites: *Dioclea apurensis* Kunth is indigenous to the eastern Amazon's ferruginous rocky outcrops or canga. Native cangas feature one of the world's greatest iron ore resources and are regarded as hotspots of ecological variety. *D. apurensis* can flourish there in post-mining sites where its establishment in repairing minelands (RM) is anticipated to be facilitated by molecular processes and rhizospheric interactions with soil microorganisms. In order to characterize the primary processes that permit the development and establishment in post-mining environments, a study was carried out to analyze the root proteome profile and the bacterial and fungal communities associated with the rhizosphere of *D. apurensis* growing in canga. The findings demonstrated that greater amounts of proteins in canga were associated with responses to oxidative stress, drought, excess iron, and phosphate deficiencies, and this helped to explain the plant's high establishment rates in RM. The rhizospheric selectivity of microorganisms was more pronounced in canga.

The microbial community structure was predominantly different between the two habitats, suggesting that *D. apurensis* is capable of forming associations with beneficial soil microorganisms without limitations, despite its specific preferences. Thus, its ability to deal with advantageous soil-borne microbes can also be improved upon or credited to its superior performance in RM. Utilizing native plants with these traits will improve the recovery process (do Nascimento et al. 2022).

2.7 Challenges and Future Directions in Fungal Endophyte Proteomics

Various avenues of agriculture innovation, plant disease management, secondary metabolite production, and environmental management have opened with the integration of recently developed omics technologies in plant endophyte research. The traditional approaches were fairly straightforward and involved field studies or experiments conducted in controlled environments such as greenhouses and small-/large-scale in vitro studies. Culturing methods and various bioactivity and toxicological studies are also part of traditional approaches of endophyte study. The traditional approaches have led to the majority of the finding in this domain; high-throughput methodologies, including whole-genome sequencing, RNA-seq, metagenomics, two-dimensional fluorescence difference gel electrophoresis (2D-DIGE), and mass spectrometry, are driving rapid advancements in omics research within the field of plant endophytes. Insights gleaned from omics hold potential advantages across various aspects, including the creation of economically significant multi-resistant plants (Chetia et al. 2019).

Proteomics is essential for studying fungal endophytes as it allows for the identification, characterization, and quantification of their proteins. This knowledge is essential for elucidating protein functions, interactions, and their roles within the host-microbe system.

The integration of 2DE with MS has been a cornerstone of proteomic analysis. However, modern MS-based approaches have emerged as powerful tools for comprehensively mapping the proteomes of fungal endophytes. Still, limitations of analysis and limited information about plant-endophyte relationships pose a challenge in achieving the goals. Through proper analysis of proteomic data, valuable insights into the interactions between host plants and endophytes, including biological pathways and posttranslational modifications, can be gained (Chetia et al. 2019).

To ensure the reliability of environmental proteomic analyses, supplementary approaches for data validation are utilized. These may include transcriptomic analyses or phenotypic testing. The multi-pronged approach is necessary to overcome the inherent challenges of analyzing complex environmental samples, such as (1) high organism/protein complexity, (2) over- or underrepresentation of certain organisms/proteins, and (3) heterogeneity of organic and inorganic pollutants. By incorporating these validation strategies, researchers can enhance the accuracy and robustness of their proteomic findings in the context of fungal endophyte research.

2.8 Integrating Multi-omics Approaches for a Comprehensive Understanding

Omics technologies like genomics, transcriptomics, proteomics, metabolomics, and other molecular techniques have played a major role in developing our understanding of the endophytic fungi. Fungal endophytes represent an exceptional reservoir of versatile bioactive compounds. These metabolites serve various functions, including combating both biotic and abiotic stresses, as well as exhibiting antiviral, anti-carcinogenic, and antimicrobial properties. Omics techniques, such as genomics, transcriptomics, proteomics, and metabolomics, play a crucial role in the investigation of novel bioactive compounds synthesized by fungal endophytes and their potential applications. These techniques are pivotal in deciphering the intricate dynamics between fungal endophytes and their host plants, offering valuable insights into both the molecular mechanisms at play and the potential applications stemming from these associations (Verma et al. 2024).

Genomics has emerged as a crucial tool for gaining a comprehensive understanding of the genomic landscape of endophytic fungal strains. This knowledge aids researchers in identifying novel genes responsible for synthesizing potent metabolic compounds, achieved through the integration of genomics with metabolomics (Sagita et al. 2021). Genomics has emerged as a cornerstone for understanding endophytic fungi. High-throughput sequencing technologies, such as those offered by Illumina, PacBio, and Oxford Nanopore, have accelerated whole-genome sequencing efforts for various endophytic fungi, including species of *Alternaria*, *Bipolaris*, *Purpureocillium*, and others (Bálint et al. 2014; Sagita et al. 2021). The genomes of various endophytic fungi such as *Alternaria alternata* JS-1623, *Alternaria* sp. MG1, *Amphirosellinia*, *Colletotrichum gloeosporioides*, *Simplicillium aogashimaense*, *Bipolaris* sp. *aff.* *sorokiniana*, *Bipolaris* sp. *aff.* *sorokiniana*, *Alternaria alternata*, *Curvularia trifolii*, *Purpureocillium lilacinum* PLFJ-1, *Purpureocillium lilacinum* TERIBC1, and many more (Prasad et al. 2015; Kim et al. 2017; Meena et al. 2018; Uchida et al. 2019; Park et al. 2020; Jauregui et al. 2020) are now sequenced. While these efforts have expanded our knowledge, a significant portion of secondary metabolite biosynthetic pathways remains undiscovered due to limited genomic information and the challenges associated with experimentally validating predicted gene clusters.

Beyond genomics, metagenomics offers a powerful approach for directly analyzing the genomes present in environmental samples, offering an extensive overview of the microbial community. Transcriptomics, employing techniques like RNA-seq and microarrays, investigates gene expression patterns at the RNA level, revealing the dynamic shifts in gene activity during host interactions. For instance, studies have explored pyrimidine metabolism in *Epulorhiza* sp. and antifungal gene expression in *Epichloë festucae*, shedding light on the molecular mechanisms underlying these interactions.

Proteomics complements these approaches by providing insights into the proteins produced by endophytic fungi. Finally, integrated omics approaches, combining genomics, transcriptomics, and metabolomics, offer a holistic view of endophyte biology, as exemplified by studies investigating cellulose biodegradation and biofuel production in *Ascocoryne sarcoides*.

2.9 Emerging Trends and Future Prospects in the Field

Endophytic fungi show great promise as producers of biologically significant metabolites, which play an important role in both treating and improving human health. However, some of these bioactive metabolites are often found in low quantities, presenting a challenge in their isolation during the initial stages of research. Consequently, metabolomics has emerged as a robust approach for analyzing the comprehensive profile of metabolites from minimal sample quantities. The advancement of analytical tools has notably propelled progress in metabolomics, especially in the context of endophytic fungi.

There is a clear need for further exploration into endophytic fungi through comprehensive metabolomic analyses integrating both mass spectrometry (MS) and nuclear magnetic resonance (NMR). This approach can facilitate high-throughput metabolomic studies with enhanced detection and identification of metabolites. Continuous research in metabolomics of endophytic fungi is essential to continually expand the chemical database of fungal metabolites, aiding in biomarker discovery.

Although metabolomics of endophytic fungi is still evolving, the next decade holds great promise for significant advancements, particularly by integrating liquid chromatography-high-resolution mass spectrometry (LC-HRMS) and NMR in study designs for direct comparison and correlation of metabolite data. Furthermore, the utilization of statistical tools, software, machine learning, and neural networks can greatly assist in the analysis and visualization of metabolomic experiments.

Future research should aim to explore the mechanisms underlying host plant-endophyte interactions, genetic manipulations of endophytic fungi, and modification of biosynthetic pathways to improve the production of natural bioactive metabolites from fungal endophytes. This concerted effort will contribute to the broader understanding and exploitation of the bioactive potential of endophytic fungi for various applications.

2.10 Conclusion

This chapter offers an in-depth exploration of the multifaceted roles of endophytic fungi in promoting host plant health, stress tolerance, and soil rehabilitation. Endophytic fungi, belonging to diverse taxonomic groups such as ascomycetous species, exhibit remarkable capabilities in producing bioactive secondary metabolites with pharmaceutical applications, contributing to plant growth promotion and biocontrol against pathogens. Proteomic techniques have been instrumental in clarifying the molecular mechanisms behind the interactions between endophytic fungi and their host plants, providing valuable understandings into protein expression regulation and the synthesis of proteins that enhance plant growth. Furthermore, the integration of omics technologies, including genomics, transcriptomics, and metabolomics, has facilitated a deeper understanding of the complex dynamics within endophytic fungi-plant systems. The future prospects of this research field lie in leveraging the potential of endophytic fungi for sustainable agriculture, biotechnology, and pharmaceutical applications, with a focus on developing multi-resistant plants and exploring the bioactive properties of endophytic fungal metabolites. Overall, this chapter underscores the significant contributions of endophytic fungi to ecosystem health and highlights the pivotal role of proteomics and omics technologies in advancing our understanding and utilization of these fascinating microorganisms.

References

Abdallah C, Dumas-Gaudot E, Renaut J, Sergeant K (2012) Gel-based and gel-free quantitative proteomics approaches at a glance. Int J Plant Genom 2012:494572. https://doi.org/10.1155/2012/494572

Adegboye MF, Ojuederie OB, Talia PM, Babalola OO (2021) Bioprospecting of microbial strains for biofuel production: metabolic engineering, applications, and challenges. Biotechnol Biofuels 14:5. https://doi.org/10.1186/s13068-020-01853-2

Alcántara-Martínez N, Figueroa-Martínez F, Rivera-Cabrera F et al (2018) An endophytic strain of *Methylobacterium* sp. increases arsenate tolerance in *Acacia farnesiana* (L.) Willd: a proteomic approach. Sci Total Environ 625:762–774. https://doi.org/10.1016/j.scitotenv.2017.12.314

Alikhani M, Khatabi B, Sepehri M et al (2013) A proteomics approach to study the molecular basis of enhanced salt tolerance in barley (*Hordeum vulgare* L.) conferred by the root mutualistic fungus *Piriformospora indica*. Mol Bio Syst 9:1498. https://doi.org/10.1039/c3mb70069k

Ancheeva E, Daletos G, Proksch P (2020) Bioactive secondary metabolites from endophytic fungi. Curr Med Chem 27:1836–1854. https://doi.org/10.2174/0929867326666190916144709

Arnold AE (2005) Diversity and ecology of fungal endophytes in tropical forests. In: Biodiversity of fungi: their role in human life, pp 49–68

Arnold AE, Mejía LC, Kyllo D, Rojas EI, Maynard Z, Robbins N, Herre EA (2003) Fungal endophytes limit pathogen damage in a tropical tree. Proc Natl Acad Sci 100(26):15649–15654

Bálint M, Schmidt P-A, Sharma R, Thines M, Schmitt I (2014) An Illumina metabarcoding pipeline for fungi. Ecol Evol 4:2642–2653. https://doi.org/10.1002/ece3.1107

Bandyopadhyay P, Yadav BG, Kumar SG et al (2022) Piriformospora indica and *Azotobacter chroococcum* consortium facilitates higher Acquisition of N, P with improved carbon allocation and enhanced plant growth in *Oryza sativa*. J Fungi (Basel) 8:453. https://doi.org/10.3390/jof8050453

Bantscheff M, Schirle M, Sweetman G et al (2007) Quantitative mass spectrometry in proteomics: a critical review. Anal Bioanal Chem 389:1017–1031. https://doi.org/10.1007/s00216-007-1486-6

Barnett MJ, Toman CJ, Fisher RF, Long SR (2004) A dual-genome symbiosis chip for coordinate study of signal exchange and development in a prokaryote–host interaction. Proc Natl Acad Sci 101:16636–16641

Baron NC, Rigobelo EC (2022) Endophytic fungi: a tool for plant growth promotion and sustainable agriculture. Mycology 13:39–55. https://doi.org/10.1080/21501203.2021.1945699

Bamisile BS, Dash CK, Akutse KS, Keppanan R, Wang L (2018) Fungal endophytes: Beyond herbivore management. Front Microbiol 9. https://doi.org/10.3389/fmicb.2018.00544

Behie SW, Zelisko PM, Bidochka MJ (2012) Endophytic insect-parasitic fungi translocate nitrogen directly from insects to plants. Science 336:1576–1577. https://doi.org/10.1126/science.1222289

Berg G (2009) Plant–microbe interactions promoting plant growth and health: perspectives for controlled use of microorganisms in agriculture. Appl Microbiol Biotechnol 84:11–18. https://doi.org/10.1007/s00253-009-2092-7

Bhupenchandra I, Chongtham SK, Devi AG et al (2024) Unlocking the potential of arbuscular mycorrhizal fungi: exploring role in plant growth promotion, nutrient uptake mechanisms, biotic stress alleviation, and sustaining agricultural production systems. J Plant Growth Regul 43(9):1–39. https://doi.org/10.1007/s00344-024-11467-9

Bilal S, Shahzad R, Khan AL et al (2019) Phytohormones enabled endophytic *Penicillium funiculosum* LHL06 protects *Glycine max* L. from synergistic toxicity of heavy metals by hormonal and stress-responsive proteins modulation. J Hazard Mater 379:120824. https://doi.org/10.1016/j.jhazmat.2019.120824

Boruta T, Bizukojc M (2017) Production of lovastatin and itaconic acid by *Aspergillus terreus*: a comparative perspective. World J Microbiol Biotechnol 33:34. https://doi.org/10.1007/s11274-017-2206-9

Busby PE, Ridout M, Newcombe G (2016) Fungal endophytes: modifiers of plant disease. Plant Mol Biol 90:645–655

Card S, Johnson L, Teasdale S, Caradus J (2016) Deciphering endophyte behaviour: the link between endophyte biology and efficacious biological control agents. FEMS Microbiol Ecol 92:fiw114. https://doi.org/10.1093/femsec/fiw114

Chadha N, Mishra M, Rajpal K, Bajaj R, Choudhary DK, Varma A (2015) An ecological role of fungal endophytes to ameliorate plants under biotic stress. Arch Microbiol 197:869–881. https://doi.org/10.1007/s00203-015-1130-3

Chandran H, Meena M, Swapnil P (2021) Plant growth-promoting rhizobacteria as a green alternative for sustainable agriculture. Sustain For 13(19):10986. https://doi.org/10.3390/su131910986

Chetia H, Kabiraj D, Bharali B, Ojha S, Barkataki MP, Saikia D, Singh T, Mosahari PV, Sharma P, Bora U (2019) Exploring the benefits of endophytic fungi via omics. In: Singh BP (ed) Advances in endophytic fungal research: present status and future challenges. Springer, Cham, pp 51–81

Chi F, Yang P, Han F et al (2010) Proteomic analysis of rice seedlings infected by *Sinorhizobium meliloti* 1021. Proteomics 10:1861–1874. https://doi.org/10.1002/pmic.200900694

Chitnis VR, Suryanarayanan TS, Nataraja KN, Prasad SR, Oelmüller R, Shaanker RU (2020) Fungal Endophyte-Mediated Crop Improvement: The Way Ahead. Front Plant Sci 11. https://doi.org/10.3389/fpls.2020.561007

Chowdappa S, Jagannath S, Konappa N, Udayashankar AC, Jogaiah S (2020) Detection and characterization of antibacterial siderophores secreted by endophytic fungi from *Cymbidium aloifolium*. Biomolecules 10:1412. https://doi.org/10.3390/biom10101412

Cui M, Cheng C, Zhang L (2022) High-throughput proteomics: a methodological mini-review. Lab Investig 102:1170–1181. https://doi.org/10.1038/s41374-022-00830-7

Cupp-Sutton KA, Wu S (2020) High-throughput quantitative top-down proteomics. Mol Omics 16:91–99. https://doi.org/10.1039/c9mo00154a

Dangl JL, Horvath DM, Staskawicz BJ (2013) Pivoting the plant immune system from dissection to deployment. Science 341(6147):746–751

dela Cruz TEE, Behr JH, Geistlinger J, Grosch R, Witzel K (2023) Monitoring of an applied beneficial *Trichoderma* strain in root-associated soil of field-grown maize by MALDI-TOF MS. Microorganisms 11:1655. https://doi.org/10.3390/microorganisms11071655

Delahunty CM, Yates JR (2007) MudPIT: multidimensional protein identification technology. Biotechniques 43(5) 563, 565, 567 passim

Devari S, Jaglan S, Kumar M et al (2014) Capsaicin production by *Alternaria alternata*, an endophytic fungus from *Capsicum annum*, LC-ESI-MS/MS analysis. Phytochemistry 98:183–189. https://doi.org/10.1016/j.phytochem.2013.12.001

Diwan D, Rashid MM, Vaishnav A (2022) Current understanding of plant-microbe interaction through the lenses of multi-omics approaches and their benefits in sustainable agriculture. Microbiol Res 265:127180. https://doi.org/10.1016/j.micres.2022.127180

do Nascimento SV, de Oliveira CPH, Herrera H et al (2022) Proteomic profiling and rhizosphere-associated microbial communities reveal adaptive mechanisms of *Dioclea apurensis* Kunth in eastern Amazon's rehabilitating minelands. Plants (Basel) 11:712. https://doi.org/10.3390/plants11050712

Dong MW, Zhang K (2014) Ultra-high-pressure liquid chromatography (UHPLC) in method development. TrAC Trends Anal Chem 63:21–30. https://doi.org/10.1016/j.trac.2014.06.019

Dos Reis JBA, Lorenzi AS, do HMM V (2022) Methods used for the study of endophytic fungi: a review on methodologies and challenges, and associated tips. Arch Microbiol 204:675. https://doi.org/10.1007/s00203-022-03283-0

Dueñas ME, Peltier-Heap RE, Leveridge M et al (2022) Advances in high-throughput mass spectrometry in drug discovery. EMBO Mol Med 15:e14850. https://doi.org/10.15252/emmm.202114850

Dupree EJ, Jayathirtha M, Yorkey H et al (2020) A critical review of bottom-up proteomics: the good, the bad, and the future of this field. Proteomes 8:14. https://doi.org/10.3390/proteomes8030014

Eldakak M, Milad SIM, Nawar AI, Rohila JS (2013) Proteomics: a biotechnology tool for crop improvement. Front Plant Sci 4. https://doi.org/10.3389/fpls.2013.00035

Eslahi N, Kowsari M, Motallebi M, Zamani MR, Moghadasi Z (2020) Influence of recombinant *Trichoderma* strains on growth of bean (*Phaseolus vulgaris* L) by increased root colonization and induction of root growth related genes. Sci Hort 261:108932. https://doi.org/10.1016/j.scienta.2019.108932

Fadiji AE, Babalola OO (2020) Exploring the potentialities of beneficial endophytes for improved plant growth. Saudi J Biol Sci 27:3622–3633. https://doi.org/10.1016/j.sjbs.2020.08.002

Fan G, Liu X, Sun S, Shi C, Du X, Han K, Yang B, Fu Y, Liu M, Seim I, Zhang H, Xu Q, Wang J, Su X, Shao L, Zhu Y, Shao Y, Zhao Y, Wong AKC, Zhuang D, Chen W, Zhang G, Yang H, Xu X, Tsui SK-W, Liu X, Lee SM-Y (2020) The chromosome level genome and genome-wide association study for the agronomic traits of *Panax notoginseng*. iScience 23:101538. https://doi.org/10.1016/j.isci.2020.101538

Felitti S, Shields K, Ramsperger M, Tian P, Sawbridge T, Webster T, Spangenberg G (2006) Transcriptome analysis of *Neotyphodium* and *Epichloë* grass endophytes. Fungal Genet Biol 43:465–475

Firáková S, Šturdíková M, Múčková M (2007) Bioactive secondary metabolites produced by microorganisms associated with plants. Biologia 62:251–257. https://doi.org/10.2478/s11756-007-0044-1

Firrincieli A, Otillar R, Salamov A, Schmutz J, Khan Z, Fleck ND, Doty SL (2015) Genome sequence of the plant growth promoting endophytic yeast Rhodotorula graminis WP1. Front Microbiol 6:154315

Fouts DE, Tyler HL, DeBoy RT, Daugherty S, Ren Q, Badger JH, Methé BA (2008) Complete genome sequence of the N2-fixing broad host range endophyte *Klebsiella pneumoniae* 342 and virulence predictions verified in mice. PLoS Genet 4:e1000141

Freeman S, Rodriguez RJ (1993) Genetic conversion of a fungal plant pathogen to a nonpathogenic, endophytic mutualist. Science 260(5104):75–78

Ganley RJ, Brunsfeld SJ, Newcombe G (2004) A community of unknown, endophytic fungi in western white pine. Proc Natl Acad Sci 101:10107–10112

Ghabooli M, Khatabi B, Ahmadi FS et al (2013) Proteomics study reveals the molecular mechanisms underlying water stress tolerance induced by *Piriformospora indica* in barley. J Proteome 94:289–301. https://doi.org/10.1016/j.jprot.2013.09.017

Ghaffari MR, Mirzaei M, Ghabooli M et al (2019) Root endophytic fungus *Piriformospora indica* improves drought stress adaptation in barley by metabolic and proteomic reprogramming. Environ Exp Bot 157:197–210. https://doi.org/10.1016/j.envexpbot.2018.10.002

Gimenez C, Cabrera R, Reina M, Gonzalez-Coloma A (2007) Fungal endophytes and their role in plant protection. COC 11:707–720. https://doi.org/10.2174/138527207780598765

Gobena S, Admassu B, Kinde MZ, Gessese AT (2024) Proteomics and its current application in biomedical area: concise review. Sci World J 2024:4454744. https://doi.org/10.1155/2024/4454744

Gómez-Vidal S, Salinas J, Tena M, Lopez-Llorca LV (2009) Proteomic analysis of date palm (*Phoenix dactylifera* L.) responses to endophytic colonization by entomopathogenic fungi. Electrophoresis 30:2996–3005. https://doi.org/10.1002/elps.200900192

Grabka R, d'Entremont TW, Adams SJ et al (2022) Fungal endophytes and their role in agricultural plant protection against pests and pathogens. Plant 11:384. https://doi.org/10.3390/plants11030384

Guo B, Wang Y, Sun X, Tang K (2008) Bioactive natural products from endophytes: a review. Prikl Biokhim Mikrobiol 44:153–158

Gupta A, Meshram V, Gupta M et al (2023) Fungal endophytes: microfactories of novel bioactive compounds with therapeutic interventions, a comprehensive review on the biotechnological developments in the field of fungal endophytic biology over the last decade. Biomol Ther 13:1038. https://doi.org/10.3390/biom13071038

Hamid R, Khan MA, Ahmad M, Ahmad MM, Abdin MZ, Musarrat J, Javed S (2013) Chitinases: An update. J Pharm Bioallied Sci 5:21. https://doi.org/10.4103/0975-7406.106559

Han X, Aslanian A, Yates JR (2008) Mass spectrometry for proteomics. Curr Opin Chem Biol 12:483–490. https://doi.org/10.1016/j.cbpa.2008.07.024

Hardoim PR, van Overbeek LS, Berg G et al (2015) The hidden world within plants: ecological and evolutionary considerations for defining functioning of microbial endophytes. Microbiol Mol Biol Rev 79:293–320. https://doi.org/10.1128/MMBR.00050-14

Hawksworth DL, Rossman AY (1997) Where are all the undescribed fungi? Phytopathology 87:888–891. https://doi.org/10.1094/PHYTO.1997.87.9.888

Hoffman MT, Arnold AE (2008) Geographic locality and host identity shape fungal endophyte communities in cupressaceous trees. Mycol Res 112:331–344. https://doi.org/10.1016/j.mycres.2007.10.014

Ibiang SR, Sakamoto K, Kuwahara N (2020) Performance of tomato and lettuce to arbuscular mycorrhizal fungi and *Penicillium pinophilum* EU0013 inoculation varies with soil, culture media of inoculum, and fungal consortium composition. Rhizosphere 16. https://doi.org/10.1016/j.rhisph.2020.100246

Ismaiel AA, Ahmed AS, Hassan IA et al (2017) Production of paclitaxel with anticancer activity by two local fungal endophytes, *Aspergillus fumigatus* and *Alternaria tenuissima*. Appl Microbiol Biotechnol 101:5831–5846. https://doi.org/10.1007/s00253-017-8354-x

Jain A, Singh HB, Das S (2021) Deciphering plant-microbe crosstalk through proteomics studies. Microbiol Res 242:126590. https://doi.org/10.1016/j.micres.2020.126590

Jauregui R, Johnson LJ, Teasdale SE (2020) Draft genome sequence of *Simplicillium aogashimaense* 72-15.1, a putative endophyte of *Brachiaria brizantha*. Microbiol Resour Announc 9:e00366–e00320. https://doi.org/10.1128/MRA.00366-20

Johansson JF, Paul LR, Finlay RD (2004) Microbial interactions in the mycorrhizosphere and their significance for sustainable agriculture. FEMS Microbiol Ecol 48(1):1–13

Jones JD, Dangl JL (2006) The plant immune system. Nature 444:323–329

Kaddes A, Fauconnier M-L, Sassi K, Nasraoui B, Jijakli M-H (2019) Endophytic fungal volatile compounds as solution for sustainable agriculture. Molecules 24:1065. https://doi.org/10.3390/molecules24061065

Kamat S, Dixit R, Kumari M (2022) Endophytic microbiome in bioactive compound production and plant disease management. In: Kumar A (ed) Microbial biocontrol: food security and post harvest management. Springer, Cham, pp 79–128

Kaul S, Sharma T, Dhar MK (2016) "Omics" tools for better understanding the plant–endophyte interactions. Front Plant sci 7:183603

Khalil AMA, Hassan SE-D, Alsharif SM et al (2021) Isolation and characterization of fungal endophytes isolated from medicinal plant *Ephedra pachyclada* as plant growth-promoting. Biomol Ther 11:140. https://doi.org/10.3390/biom11020140

Khan AL, Hamayun M, Radhakrishnan R, Waqas M, Kang S-M, Kim Y-H, Shin J-H, Choo Y-S, Kim J-G, Lee I-J (2012) Mutualistic association of *Paecilomyces formosus* LHL10 offers thermotolerance to *Cucumis sativus*. Antonie van Leeuwenhoek 101:267–279. https://doi.org/10.1007/s10482-011-9630-x

Khan AL, Hussain J, Al-Harrasi A, Al-Rawahi A, Lee IJ (2015) Endophytic fungi: resource for gibberellins and crop abiotic stress resistance. Crit Rev Biotechnol 35(1):62–74. https://doi.org/10.3109/07388551.2013.800018

Kim JA, Jeon J, Park S-Y, Kim K-T, Choi G, Lee H-J, Kim Y, Yang H-S, Yeo J-H, Lee Y-H, Kim S (2017) Genome sequence of an endophytic fungus, *Fusarium solani* JS-169, which has antifungal activity. Genome Announc 5:e01071–e01017. https://doi.org/10.1128/genomeA.01071-17

Knief C, Delmotte N, Chaffron S, Stark M, Innerebner G, Wassmann R, Vorholt JA (2012) Metaproteogenomic analysis of microbial communities in the phyllosphere and rhizosphere of rice. ISME J 6:1378–1390

Kumar R, Swapnil P, Meena M, Selpair S, Yadav BG (2022a) Plant growth-promoting rhizobacteria (PGPR): approaches to alleviate abiotic stresses for enhancement of growth and development of medicinal plants. Sustain For 14(23):15514. https://doi.org/10.3390/su142315514

Kumar S, Chandra R, Keswani C et al (2022b) Trichoderma viride—mediated modulation of oxidative stress network in potato challenged with *Alternaria solani*. J Plant Growth Regul 42:1919–1936. https://doi.org/10.1007/s00344-022-10669-3

Kumaresan V, Suryanarayanan TS (2002) Endophytes assemblages in young mature and senescent leaves of *Rhizophora apiculata*: evidence for the role of endophytes in mangrove litter degradation. Fungal Divers 9:81–91

Kusari S, Lamshöft M, Zühlke S, Spiteller M (2008) An endophytic fungus from *Hypericum perforatum* that produces hypericin. J Nat Prod 71:159–162. https://doi.org/10.1021/np070669k
Latz MAC, Jensen B, Collinge DB, Jørgensen HJL (2018) Endophytic fungi as biocontrol agents: elucidating mechanisms in disease suppression. Plant Ecol Divers 11:555–567. https://doi.org/10.1080/17550874.2018.1534146
Ledford H (2015) Plant dwellers take the limelight. Nature 523(7559):137–138
Lee PY, Saraygord-Afshari N, Low TY (2020) The evolution of two-dimensional gel electrophoresis – from proteomics to emerging alternative applications. J Chromatogr A 1615:460763. https://doi.org/10.1016/j.chroma.2019.460763
Leetanasaksakul K, Roytrakul S, Kittisenachai S et al (2024) Exploring the impact of endophytic fungus *Aspergillus cejpii* DMKU-R3G3 on Rice: plant growth promotion and molecular insights through proteomic analysis. Agronomy 14:498. https://doi.org/10.3390/agronomy14030498
Li H, Holguin O, Wang J, Schaub T, Wang J, Hao C, Geng G, Creamer R (2012) Proteomic analysis of the endophytic fungus *Undifilum oxytropis*. Afr J Biotechnol 11:10484–10493. https://doi.org/10.5897/AJB11.4081
Li J, Zhu H-J (2020) Liquid chromatography-tandem mass spectrometry (LC-MS/MS)-based proteomics of drug-metabolizing enzymes and transporters. Molecules 25:2718. https://doi.org/10.3390/molecules25112718
Liu Y, Chen N, Gao Y, Bu Z, Niu S, Wang Y, Liu X, Zhu Y (2021) Physiological, biochemical and proteomic insight into the response system of *Streptomyces hygroscopicus* OsiSh-2 to rice blast fungus toxins. Applied Soil Ecology 167:104058. https://doi.org/10.1016/j.apsoil.2021.104058
Lubbers RJM, Dilokpimol A, Visser J et al (2021) Discovery and functional analysis of a salicylic acid hydroxylase from *Aspergillus niger*. Appl Environ Microbiol 87:e02701–e02720. https://doi.org/10.1128/AEM.02701-20
Lücking R, Aime MC, Robbertse B, Miller AN, Ariyawansa HA, Aoki T, Cardinali G, Crous PW, Druzhinina IS, Geiser DM, Hawksworth DL, Hyde KD, Irinyi L, Jeewon R, Johnston PR, Kirk PM, Malosso E, May TW, Meyer W, Öpik M, Robert V, Stadler M, Thines M, Vu D, Yurkov AM, Zhang N, Schoch CL (2020) Unambiguous identification of fungi: where do we stand and how accurate and precise is fungal DNA barcoding? IMA Fungus 11:14. https://doi.org/10.1186/s43008-020-00033-z
Lugtenberg BJJ, Caradus JR, Johnson LJ (2016) Fungal endophytes for sustainable crop production. FEMS Microbiol Ecol 92:fiw194. https://doi.org/10.1093/femsec/fiw194
Maciá-Vicente JG, Jansson HB, Abdullah SK, Descals E, Salinas J, Lopez-Llorca LV (2008) Fungal root endophytes from natural vegetation in Mediterranean environments with special reference to *Fusarium* spp. FEMS Microbiol Ecol 64:90–105
Magdeldin S, Enany S, Yoshida Y et al (2014) Basics and recent advances of two dimensional-polyacrylamide gel electrophoresis. Clin Proteomics 11:16. https://doi.org/10.1186/1559-0275-11-16
Maron PA, Ranjard L, Mougel C, Lemanceau P (2007) Metaproteomics: a new approach for studying functional microbial ecology. Microb Ecol 53:486–493
Martínez-García P M, Ruano-Rosa D, Schilirò E, Prieto P Ramos C, Rodríguez-Palenzuela P & Mercado-Blanco J (2015) Complete genome sequence of *Pseudomonas fluorescens* strain PICF7, an indigenous root endophyte from olive (*Olea europaea* L.) and effective biocontrol agent against *Verticillium dahliae*. Stand Genomic Sci 10, 1–7
Meena M, Swapnil P, Zehra A et al (2017) Antagonistic assessment of *Trichoderma* spp. by producing volatile and non-volatile compounds against different fungal pathogens. Arch Phytopathol Plant Protec 50(13–14):629–648. https://doi.org/10.1080/03235408.2017.1357360
Meena N, Vasundhara M, Reddy MS, Suravajhala P, Raghavender US, Medicherla KM (2018) Draft genome sequence of a fungus (*Fusarium tricinctum*) cultured from a Monoisolate native to the Himalayas. Genome Announc 6:e00365–e00318. https://doi.org/10.1128/genomeA.00365-18
Megías E, Megías M, Ollero FJ, Hungria M (2016) Draft genome sequence of *Pantoea ananatis* strain AMG521, a rice plant growth-promoting bacterial endophyte isolated from the *Guadalquivir marshes* in southern Spain. Genome Announc 4:10–1128

Melotto M, Underwood W, He SY (2008) Role of stomata in plant innate immunity and foliar bacterial diseases. Annu Rev Phytopathol 46:101–122. https://doi.org/10.1146/annurev.phyto.121107.104959

Mir R, Kaushik PS, Chowdery RA, Anuradha M (2015) Elicitation of forskolin in cultures of *Rhizactonia bataticola*-a phytochemical synthesizing endophytic fungi. Int J Pharm Pharm Sci 7:185–189

Mishra RC, Kalra R, Dwivedi N, Goel M (2021) Exploring endophytes using "omics": an approach for sustainable production of bioactive metabolites. In: Prasad R, Nayak SC, Kharwar RN, Dubey NK (eds) Mycoremediation and environmental sustainability. Springer, Cham, pp 349–376

Montaño López J, Duran L, Avalos JL (2022) Physiological limitations and opportunities in microbial metabolic engineering. Nat Rev Microbiol 20:35–48. https://doi.org/10.1038/s41579-021-00600-0

Mourad EF, Sarhan MS, Daanaa H-SA, Abdou M, Morsi AT, Abdelfadeel MR, Elsawey H, Nemr R, El-Tahan M, Hamza MA, Abbas M, Youssef HH, Abdelhadi AA, Amer WM, Fayez M, Ruppel S, Hegazi NA (2018) Plant materials are sustainable substrates supporting new technologies of plant-only-based culture media for *in vitro* culturing of the plant microbiota. Microbes Environ 33:40–49. https://doi.org/10.1264/jsme2.ME17135

Neagu A-N, Jayathirtha M, Baxter E et al (2022) Applications of tandem mass spectrometry (MS/MS) in protein analysis for biomedical research. Molecules 27:2411. https://doi.org/10.3390/molecules27082411

Noriler SA, Savi DC, Aluizio R, Palácio-Cortes AM, Possiede YM, Glienke C (2018) Bioprospecting and structure of fungal endophyte communities found in the Brazilian biomes, Pantanal, and Cerrado. Front Microbiol 9:1526. https://doi.org/10.3389/fmicb.2018.01526

Ohba T, Yoshino Y, Ishisaka M et al (2015) Japanese *Huperzia serrata* extract and the constituent, huperzine A, ameliorate the scopolamine-induced cognitive impairment in mice. Biosci Biotechnol Biochem 79:1838–1844. https://doi.org/10.1080/09168451.2015.1052773

Pan S, Aebersold R (2007) Quantitative proteomics by stable isotope labeling and mass spectrometry. Methods Mol Biol 367:209–218. https://doi.org/10.1385/1-59745-275-0:209

Park S-Y, Jeon J, Kim JA, Jeon MJ, Jeong M-H, Kim Y, Lee Y, Chung H, Lee Y-H, Kim S (2020) Draft genome sequence of *Alternaria alternata* JS-1623, a fungal endophyte of *Abies koreana*. Mycobiology 48:240–244. https://doi.org/10.1080/12298093.2020.1756134

Pillai TG (2017) Pathogen to endophytic transmission in fungi- a proteomics approach. SOJMID 5:1–5. https://doi.org/10.15226/sojmid/5/3/00173

Po A, Eyers CE (2023) Top-down proteomics and the challenges of true Proteoform characterization. J Proteome Res 22:3663–3675. https://doi.org/10.1021/acs.jproteome.3c00416

Poveda J, Baptista P, Sacristán S, Velasco P (2022) Editorial: beneficial effects of fungal endophytes in major agricultural crops. Front Plant Sci 13:1061112. https://doi.org/10.3389/fpls.2022.1061112

Prasad P, Varshney D, Adholeya A (2015) Whole genome annotation and comparative genomic analyses of bio-control fungus *Purpureocillium lilacinum*. BMC Genomics 16:1004. https://doi.org/10.1186/s12864-015-2229-2

Qiang XiaoYu QX, Weiss M, Kogel KH, Schäfer P (2012) *Piriformospora indica*-a mutualistic basidiomycete with an exceptionally large plant host range. Mol Plant Pathol 13(5):508–518

Rajamanikyam M, Vadlapudi V, Amanchy R, Upadhyayula SM (2017) Endophytic fungi as novel resources of natural therapeutics. Braz Arch Biol Technol 60. https://doi.org/10.1590/1678-4324-2017160542

Ram RJ, Verberkmoes NC, Thelen MP et al (2005) Community proteomics of a natural microbial biofilm. Science 308:1915–1920

Ran X, Zhang G, Li S, Wang J (2017) Characterization and antitumor activity of camptothecin from endophytic fungus fusarium solani isolated from Camptotheca acuminate. Afr Health Sci 17:566–574. https://doi.org/10.4314/ahs.v17i2.34

Rana KL, Kour D, Kaur T, Devi R, Yadav AN, Yadav N, Dhaliwal HS, Saxena AK (2020) Endophytic microbes: biodiversity, plant growth-promoting mechanisms and potential applications for agricultural sustainability. Antonie van Leeuwenhoek 113:1075–1107. https://doi.org/10.1007/s10482-020-01429-y

Ray P, Abraham PE, Guo Y et al (2019) Scavenging organic nitrogen and remodelling lipid metabolism are key survival strategies adopted by the endophytic fungi, *Serendipita vermifera* and *Serendipita bescii* to alleviate nitrogen and phosphorous starvation in vitro. Environ Microbiol Rep 11:548–557. https://doi.org/10.1111/1758-2229.12757

Redlin SC, Carris LM (eds) (1996) Endophytic fungi in grasses and woody plants: systematics, ecology, and evolution. APS Press, St. Paul, Minn

Rodriguez RJ, White JF Jr, Arnold AE, Redman ARA (2009) Fungal endophytes: diversity and functional roles. New Phytol 182:314–330

Saad MMG, Badry HH (2020) Phytohormones producing fungal endophytes enhance nutritional status and suppress pathogenic fungal infection in tomato. J Agric Sci Technol 22:1383–1395

Sagita R, Quax WJ, Haslinger K (2021) Current state and future directions of genetics and genomics of endophytic fungi for bioprospecting efforts. Front Bioeng Biotechnol 9:649906. https://doi.org/10.3389/fbioe.2021.649906

Sahu PK, Jayalakshmi K, Tilgam J et al (2022) ROS generated from biotic stress: effects on plants and alleviation by endophytic microbes. Front Plant Sci 13:1042936. https://doi.org/10.3389/fpls.2022.1042936

Saini P, Verma A, Tiwari H et al (2024) Omics-based approaches in studying fungal endophytes and their associated secondary metabolites. In: Singh BP, Abdel-Azeem AM, Gautam V et al (eds) Endophytic fungi: the hidden sustainable jewels for the pharmaceutical and agricultural industries. Springer, Cham, pp 209–227

Schneider T, Riedel K (2010) Environmental proteomics: analysis of structure and function of microbial communities. Proteomics 10:785–798. https://doi.org/10.1002/pmic.200900450

Sepehri M, Ghaffari MR, Khayam Nekoui M et al (2021) Root endophytic fungus *Serendipita indica* modulates barley leaf blade proteome by increasing the abundance of photosynthetic proteins in response to salinity. J Appl Microbiol 131:1870–1889. https://doi.org/10.1111/jam.15063

Shrivastava N, Jiang L, Li P et al (2018) Proteomic approach to understand the molecular physiology of symbiotic interaction between *Piriformospora indica* and *Brassica napus*. Sci Rep 8:5773. https://doi.org/10.1038/s41598-018-23994-z

Shuken SR (2023) An introduction to mass spectrometry-based proteomics. J Proteome Res 22:2151–2171. https://doi.org/10.1021/acs.jproteome.2c00838

Singh A, Singh DK, Kharwar RN et al (2021) Fungal endophytes as efficient sources of plant-derived bioactive compounds and their prospective applications in natural product drug discovery: insights, avenues, and challenges. Microorganisms 9:197. https://doi.org/10.3390/microorganisms9010197

Souza B dos S, Santos TT dos (2017) Endophytic fungi in economically important plants: ecological aspects, diversity and potential biotechnological applications. J Bioenergy Food Sci 4:113–126. https://doi.org/10.18067/jbfs.v4i2.121

Stierle A, Strobel G, Stierle D (1993) Taxol and taxane production by *Taxomyces andreanae*, an endophytic fungus of Pacific yew. Science 260:214–216. https://doi.org/10.1126/science.8097061

Stone JK, Bacon CW, White JF Jr (2000) An overview of endophytic microbes: endophytism defined. In: Microbial endophytes. CRC Press, pp 17–44

Strobel G, Daisy B (2003) Bioprospecting for microbial endophytes and their natural products. Microbiol Mol Biol Rev 67:491–502. https://doi.org/10.1128/MMBR.67.4.491-502.2003

Subudhi E, Sahoo RK, Dey S et al (2019) Unraveling plant-endophyte interactions: an omics insight. In: Jha S (ed) Endophytes and secondary metabolites. Springer, Cham, pp 249–267

Suebrasri T, Harada H, Jogloy S, Ekprasert J, Boonlue S (2020) Auxin-producing fungal endophytes promote growth of sunchoke. Rhizosphere 16:100271. https://doi.org/10.1016/j.rhisph.2020.100271

Sunitha VH, Devi DN, Srinivas C (2013) Extracellular enzymatic activity of endophytic fungal strains isolated from medicinal plants. World J Agric Sci 9(1):01–09

Suryanarayanan TS (2013) Endophyte research: going beyond isolation and metabolite documentation. Fungal Ecol 6:561–568. https://doi.org/10.1016/j.funeco.2013.09.007

Tazik Z, Rahnama K, White JF et al (2020) LC-MS based identification of stylosin and tschimgine from fungal endophytes associated with ferula ovina. Iran J Basic Med Sci 23:1565–1570. https://doi.org/10.22038/ijbms.2020.46334.10703

Tian F, Wang J, Ding F et al (2024) Comparative transcriptomics and proteomics analysis of the symbiotic germination of *Paphiopedilum barbigerum* with *Epulorhiza* sp. FQXY019. Front Microbiol 15:1358137. https://doi.org/10.3389/fmicb.2024.1358137

Tiwari P, Bae H (2022) Endophytic fungi: key insights, emerging prospects, and challenges in natural product drug discovery. Microorganisms 10:360. https://doi.org/10.3390/microorganisms10020360

Turbat A, Rakk D, Vigneshwari A, Kocsubé S, Thu H, Szepesi Á, Bakacsy L, D. Škrbić B, Jigjiddorj E-A, Vágvölgyi C, Szekeres A (2020) Characterization of the plant growth-promoting activities of endophytic fungi isolated from *Sophora flavescens*. Microorganisms 8:683. https://doi.org/10.3390/microorganisms8050683

Uchida R, Kondo A, Yagi A, Nonaka K, Masuma R, Kobayashi K, Tomoda H (2019) Simpotentin, a new potentiator of amphotericin B activity against *Candida albicans*, produced by *Simplicillium minatense* FKI-4981. J Antibiot (Tokyo) 72:134–140. https://doi.org/10.1038/s41429-018-0128-x

Vandenkoornhuyse P, Mahé S, Ineson P, Staddon P, Ostle N, Cliquet JB, Young JPW (2007) Active root-inhabiting microbes identified by rapid incorporation of plant-derived carbon into RNA. Proc Natl Acad Sci 104:16970–16975

Verma V, Srivastava A, Garg SK et al (2024) Incorporating omics-based tools into endophytic fungal research. Biotechnol Notes 5:1–7. https://doi.org/10.1016/j.biotno.2023.12.006

Vincent D, Tan K-C, Cassidy L et al (2012) Proteomic techniques for plant-fungal interactions. Methods Mol Biol 835:75–96. https://doi.org/10.1007/978-1-61779-501-5_5

Vincent D, Rafiqi M, Job D (2020) The multiple facets of plant–fungal interactions revealed through plant and fungal secretomics. Front Plant Sci 10:1626. https://doi.org/10.3389/fpls.2019.01626

Waqas M, Khan AL, Kamran M, Hamayun M, Kang S-M, Kim Y-H, Lee I-J (2012) Endophytic fungi produce gibberellins and indoleacetic acid and promotes host-plant growth during stress. Molecules 17:10754–10773. https://doi.org/10.3390/molecules170910754

Weiss M, Sýkorová Z, Garnica S, Riess K, Martos F, Krause C, Redecker D (2011) Sebacinales everywhere: previously overlooked ubiquitous fungal endophytes. PloS One 6:e16793

Wiśniewski JR, Zielinska DF, Mann M (2011) Comparison of ultrafiltration units for proteomic and N-glycoproteomic analysis by the filter-aided sample preparation method. Anal Biochem 410:307–309. https://doi.org/10.1016/j.ab.2010.12.004

Wu B, Hussain M, Zhang W, Stadler M, Liu X, Xiang M (2019) Current insights into fungal species diversity and perspective on naming the environmental DNA sequences of fungi. Mycology 10:127–140. https://doi.org/10.1080/21501203.2019.1614106

Wu C, Yang Y, Wang Y et al (2024) Colonization of root endophytic fungus *Serendipita indica* improves drought tolerance of *Pinus taeda* seedlings by regulating metabolome and proteome. Front Microbiol 15:1294833. https://doi.org/10.3389/fmicb.2024.1294833

Yadava PK, Bhuyan SK, Bandyopadhyay P (2015) Extraction of proteins for two-dimensional gel electrophoresis and proteomic analysis from an endophytic fungus. Protocol Exchange. https://doi.org/10.1038/protex.2015.084

Yadav G, Meena M (2021) Bioprospecting of endophytes in medicinal plants of Thar Desert: an attractive resource for biopharmaceuticals. Biotechnol Rep 30:e00629. https://doi.org/10.1016/j.btre.2021.e00629

Yadav V, Kumar M, Deep DK, Kumar H, Sharma R, Tripathi T, Tuteja N, Saxena AK, Johri AK (2010) A phosphate transporter from the root endophytic fungus *Piriformospora indica* plays a role in phosphate transport to the host plant. J Biol Chem 285:26532–26544. https://doi.org/10.1074/jbc.M110.111021

Yuan J, Zhang W, Sun K et al (2019) Comparative transcriptomics and proteomics of *Atractylodes lancea* in response to endophytic fungus *Gilmaniella* sp. AL12 reveals regulation in plant metabolism. Front Microbiol 10:1208. https://doi.org/10.3389/fmicb.2019.01208

Zeng Q, Man X, Huang Z et al (2023) Effects of rice blast biocontrol strain *Pseudomonas alcaliphila* Ej2 on the endophytic microbiome and proteome of rice under salt stress. Front Microbiol 14:1129614. https://doi.org/10.3389/fmicb.2023.1129614

Zhou W, Eudes F, Laroche A (2006) Identification of differentially regulated proteins in response to a compatible interaction between the pathogen *Fusarium graminearum* and its host, *Triticum aestivum*. Proteomics 6:4599–4609. https://doi.org/10.1002/pmic.200600052

Functional Metagenomics of *Avicennia marina* Root-Associated Microbiomes

3

Gomathi Selvam, Ambikapathy Varatharasu, Deepan Baskaran, Guhanraj Radhamanalan, Prakash Pandiyan, Babu Selvarajan, and Panneerselvam Annamalai

Abstract

Mangroves are unique, and endangered, coastal ecosystems that play a vital role in the tropical and subtropical environments. A comprehensive description of the microbial communities in these ecosystems is currently lacking, and additional studies are required to have a complete understanding of the functioning and persistence of mangroves worldwide. *Avicennia marina*, also known as a gray mangrove, grows in multiple areas across the globe. It is known for its particularly pronounced tolerance to extreme and harsh environmental conditions, such as salinity, aridity, water temperatures, and frost frequency. Metagenomics has provided new insights into the microbial communities that associate with *A. marina* roots, revealing the functional capabilities of these microbes and their crucial role in plant growth, nutrient acquisition, stress tolerance, and environmental resilience. In this chapter, we discussed about the complete analysis of the functional roles of root-associated microbial communities in *A. marina*, based on recent metagenomic studies. Microbial functions such as nitrogen fixation, phosphorus solubilization, organic matter degradation, and heavy metal detoxification help promote plant growth and improve soil fertility.

Keywords

Mangroves · *Avicennia marina* · Metagenomics · Microbes · Environment

G. Selvam (✉) · A. Varatharasu · D. Baskaran · P. Pandiyan · B. Selvarajan · P. Annamalai
Department of Botany, A.V.V.M. Sri Pushpam College (Autonomous) Poondi (Affiliated to Bharathidasan University, Tiruchirappalli), Thanjavur, Tamil Nadu, India

G. Radhamanalan
Department of Microbiology, Bharathidasan University, Tiruchirappalli, India

M. Meena et al. (eds.), *Multi-omics Approach to Investigate Endophyte Diversity*, Microorganisms for Sustainability 53, https://doi.org/10.1007/978-981-96-4100-0_3

3.1 Introduction

Avicennia marina (gray mangrove) plays a vital role in coastal ecosystems, contributing to shoreline stabilization and nutrient cycling and acting as a buffer against storms and tides. Resilience in high-salinity, low-oxygen environments make to model species for understanding plant-microbe interactions in extreme habitats. Metagenomics has provided new insights into the microbial communities that associate with *A. marina* roots, revealing the functional capabilities of these microbes and their crucial role in plant growth, nutrient acquisition, stress tolerance, and environmental resilience (Alzubaidy et al. 2016).

This chapter presents a complete analysis of the functional roles of root-associated microbial communities in *A. marina*, based on recent metagenomic studies. Microbial functions such as nitrogen fixation, phosphorus solubilization, organic matter degradation, heavy metal detoxification, and plant growth promotion.

3.2 Nutrient Acquisition by Root-Associated Microbes

3.2.1 Nitrogen Fixation

Nitrogen is an essential nutrient for plant growth, yet it is often a limiting factor in saline and coastal soils where *A. marina* thrives. Root-associated microbes provide a solution to this limitation by fixing atmospheric nitrogen into bioavailable forms for the plant. Metagenomic studies have identified key genes associated with nitrogen fixation, particularly the nif gene cluster responsible for encoding nitrogenase enzymes that catalyze this process.

Studies on *A. marina* root microbiomes reveal an abundance of nitrogen-fixing bacteria, particularly from genera such as *Rhizobium* and *Bradyrhizobium*. The detection of the nifH gene, a highly conserved marker for nitrogen fixation, confirms the functional role of these bacteria in nitrogen cycling within the mangrove ecosystem (Basak et al. 2016).

3.2.2 Phosphorus Solubilization

Phosphorus is another critical nutrient that is often poorly available in coastal soils. Microbial communities associated with *A. marina* roots play a key role in solubilizing phosphorus through the production of organic acids and enzymes such as phosphatases. Metagenomic analyses have detected genes involved in phosphate solubilization, indicating the capacity of these microbes to convert insoluble forms of phosphorus into forms accessible to the plant (Andreote et al. 2012).

3.3 Stress Tolerance Mechanisms in *A. marina* Root Microbiomes

3.3.1 Salt Stress Tolerance

Avicennia marina is well-adapted to saline environments, but its success in these habitats is also linked to its association with salt-tolerant microbial communities. Metagenomic studies have identified several key genes involved in salt tolerance, including those responsible for the synthesis of osmoprotectants like trehalose, betaine, and proline. Additionally, genes coding for sodium efflux pumps (e.g., nhaA) and transporters that help in maintaining cellular ionic balance have been detected in the microbiome of *A. marina* (Andreote et al. 2012). These microbial salt tolerance mechanisms complement the plant's own physiological adaptations, such as salt secretion through specialized glands, thereby enhancing the overall resilience of *A. marina* in high-salinity environments.

3.3.2 Heavy Metal Detoxification

Mangroves, including *A. marina*, are often exposed to heavy metal contamination in coastal regions. The root-associated microbial community plays an important role in detoxifying these metals, allowing the plant to survive in contaminated soils. Metagenomic analysis of *A. marina* roots has identified genes such as merA (mercury resistance) and arsC (arsenic resistance), which are involved in the reduction and transformation of heavy metals into less toxic forms. These detoxification processes not only help protect the plant but also contribute to the overall bioremediation potential of mangrove ecosystems, making them essential for cleaning polluted coastal regions.

3.4 Organic Matter Degradation and Nutrient Cycling

3.4.1 Cellulose and Lignin Degradation

Mangrove ecosystems are rich in organic material, including fallen leaves and wood debris, which require efficient decomposition. The microbial communities associated with *A. marina* roots possess a wide array of enzymes capable of breaking down complex organic polymers such as cellulose, hemicellulose, and lignin. Metagenomic studies have identified genes encoding cellulases, hemicellulases, and ligninases, highlighting the role of these microbes in organic matter degradation and nutrient cycling (Yang et al. 2023).

3.4.2 Sulfur Cycling

Sulfur cycling is another critical process facilitated by root-associated microbes in coastal ecosystems. Metagenomic studies of *A. marina* microbiomes have uncovered genes involved in both sulfur oxidation and reduction, particularly the sox gene cluster, which plays a role in the oxidation of sulfur compounds. These processes are vital for the stability and health of mangrove ecosystems, where sulfur is a key element in soil chemistry.

3.5 Plant Growth-Promoting Microbes (PGPR)

3.5.1 Phytohormone Production

Microbial communities associated with *A. marina* roots play a direct role in promoting plant growth through the production of phytohormones such as auxins, gibberellins, and ethylene inhibitors. Metagenomic studies have identified genes involved in auxin biosynthesis (e.g., iaaM and ipdC), which stimulate root growth and enhance nutrient uptake.

3.5.2 Quorum Sensing and Biofilm Formation

Quorum sensing, a cell-to-cell communication mechanism, is prevalent in the microbial communities associated with *A. marina* roots. Metagenomic analysis has detected quorum sensing genes (luxI and luxR), which are involved in the regulation of biofilm formation and microbial behavior. Biofilms help protect the microbes and the plant from environmental stresses, including desiccation, high salinity, and pathogen attack (Alsharif et al. 2024) (Table 3.1).

3.5.2.1 KEGG (Kyoto Encyclopedia of Genes and Genomes) Database Functional Analysis on Endophytes

KEGG (Kyoto Encyclopedia of Genes and Genomes) database, *Avicennia marina*, commonly known as the gray mangrove and the metabolic and genetic pathways that are active in this salt-tolerant species.

3.5.2.2 Salt Tolerance Pathways

Osmoprotectant Production: KEGG may provide insights into pathways that lead to the biosynthesis of osmoprotectants like proline or glycine betaine, which help *A. marina* manage osmotic stress in high-salinity conditions.

Table 3.1 Functional roles of root-associated microbial communities in *A. marina*, based on recent metagenomic studies

Endophyte function	Metagenomic function	References
Root-associated microbes help fix atmospheric nitrogen, making it available for plant uptake	Identification of nitrogen-fixing genes (*nifH*) in *A. marina* root-associated bacterial communities	Basak et al. (2016) and Andreote et al. (2012)
Microbial communities associated with *A. marina* help solubilize phosphorus, improving nutrient availability	Genes for phosphorus metabolism and solubilization enzymes detected in the rhizosphere microbiome	Yang et al. (2023)
A. marina thrives in saline environments, aided by salt-tolerant microbes that help with osmoregulation	Identification of salt tolerance genes (e.g., sodium efflux pumps, osmoprotectants) in microbial genomes	Andreote et al. (2012)
Decomposition of organic material is aided by microbes producing cellulases, hemicellulases, and ligninases	Enzymes for cellulose, lignin, and hemicellulose degradation identified, supporting nutrient cycling	Muñoz-García et al. (2022)
Microbes associated with *A. marina* detoxify heavy metals, allowing the plant to thrive in contaminated environments	Genes for heavy metal detoxification (e.g., *merA*, *arsC*) identified, contributing to phytoremediation	Dai et al. (2024)
Sulfur cycling is crucial in coastal environments, and associated microbes play a role in sulfur reduction and oxidation	Presence of sulfur metabolism genes (e.g., *sox* genes) in *A. marina* root microbiomes	Basak et al. (2016)
Microbes can degrade hydrocarbons, important for detoxifying pollutants in mangrove ecosystems	Hydrocarbon-degrading genes (e.g., *alkB*, *CYP450*) identified, enabling bioremediation in polluted soils	Dai et al. (2024)
Microbes enhance plant growth by producing phytohormones like auxins, gibberellins, and ethylene inhibitors	Auxin and ethylene inhibition pathways identified in root-associated bacterial genomes of *A. marina*	Priya et al. (2018)
Microbial communities help *A. marina* cope with environmental stresses such as salinity and heavy metals	Metagenomic analysis reveals stress-responsive genes (e.g., heat shock proteins, antioxidant enzymes)	Simões et al. (2015)
Microbes communicate and form biofilms to enhance their survival and interactions with the plant	Quorum sensing genes (*luxI*, *luxR*) and biofilm formation pathways detected in root-associated bacteria	Basak et al. (2016)
Redundant microbial functions ensure ecosystem stability under environmental fluctuations	Multiple microbial taxa with overlapping nutrient cycling and stress tolerance genes found in metagenomes	Yang et al. (2023)
Microbes produce antimicrobial compounds that protect the plant from pathogens	Pathways for antimicrobial compound production detected in root-associated microbial genomes	Dai et al. (2024)

Ion Transport and Homeostasis: Sodium (Na^+) and chloride (Cl^-) ion transporters, like HKT and SOS1, which play a critical role in salt exclusion and tolerance, might be present in the organism's KEGG pathways. This could include channels, pumps, and transporters involved in ionic homeostasis (Haldar and Nazareth 2018).

3.5.2.3 Oxidative Stress Response

Reactive Oxygen Species (ROS) Detoxification: The KEGG pathways related to antioxidative enzymes such as superoxide dismutase (SOD), catalase, and glutathione peroxidase might reveal how *A. marina* copes with oxidative stress induced by saline environments.

3.5.2.4 Primary and Secondary Metabolism

Lipid Metabolism: Since *A. marina* may alter its membrane lipid composition to adapt to high salinity, the pathways associated with fatty acid biosynthesis and modification might be of interest.

Phenylpropanoid Pathway: KEGG pathways related to secondary metabolite production (e.g., phenylpropanoids, flavonoids) could be essential for stress resistance, pathogen defense, and UV protection (Liao et al. 2020).

3.5.2.5 Photosynthesis and Carbon Metabolism

C4 or CAM Pathways: Given its environmental adaptation, studying carbon fixation pathways in *A. marina* through KEGG can help understand its efficiency in photosynthesis, especially under saline or low-water conditions (Yang et al. 2023).

3.5.2.6 Hormonal Regulation

ABA (Abscisic Acid) Biosynthesis and Signal Transduction: Abscisic acid is a key hormone in plant stress response. The pathways related to ABA might provide insight into how *A. marina* manages drought and salinity stress (Ma et al. 2022).

3.5.2.7 Microbial Interactions

Endophyte-Plant Interaction Pathways: KEGG may also provide clues about how endophytes interact with *A. marina* microbiome, helping to understand mutualistic relationships or defense mechanisms (Shang et al. 2024) (Table 3.2 and Fig. 3.1).

Although *A. marina* may not have been directly annotated in KEGG pathways, studying related species or the plant metabolic pathways from closely related genomes could reveal much about its unique adaptations. It uses KEGG modules to explore potential orthologs and pathway similarities.

3.6 Conclusion

Metagenomic studies of *Avicennia marina* root-associated microbial communities have uncovered a wealth of functional traits that contribute to the health and resilience of this keystone species. These microbes play critical roles in nutrient cycling,

Table 3.2 Endophyte functional prediction in KEGG pathway and orthology

Pathway	Function	KEGG KO identifier
Glycolysis/gluconeogenesis	Conversion of glucose to pyruvate (energy metabolism)	K00844 (hexokinase)
TCA cycle	Energy production via oxidation of acetyl-CoA	K01647 (citrate synthase)
Fatty acid biosynthesis	Synthesis of fatty acids (membrane adaptation)	K00626 (acetyl-CoA carboxylase)
Proline biosynthesis	Osmoprotectant production (response to salt stress)	K01940 (pyrroline-5-carboxylate reductase)
Glycine betaine synthesis	Osmoprotectant production under salt stress	K00152 (choline dehydrogenase)
ROS detoxification	Detoxification of reactive oxygen species	K04565 (superoxide dismutase)
Abscisic acid biosynthesis	Hormonal regulation of stress responses	K14133 (9-cis-epoxycarotenoid dioxygenase)
Salt stress response (SOS pathway)	Sodium ion exclusion and homeostasis	K13433 (salt overly sensitive 1, SOS1)
Nitrogen metabolism	Assimilation of ammonia and nitrate	K00265 (glutamine synthetase)
Phosphate metabolism	Solubilization and uptake of phosphate	K08357 (phosphate transporter)
Calvin cycle (carbon fixation)	Carbon dioxide fixation during photosynthesis	K01601 (ribulose bisphosphate carboxylase)
Photosynthesis (light reaction)	Capture of light energy for photosynthesis	K02689 (photosystem II)
Auxin biosynthesis	Plant hormone for growth regulation	K04149 (AUX/IAA transcriptional regulator)

stress tolerance, and bioremediation, making *A. marina* a model system for understanding plant-microbe interactions in extreme environments. The potential for harnessing these microbial functions in biotechnological applications presents exciting opportunities for sustainable agriculture, ecosystem restoration, and climate change mitigation. The continued exploration of *A. marina* microbiomes through advanced metagenomic and functional genomic approaches promises to reveal even deeper insights into the intricate relationships between plants and their associated microbial communities.

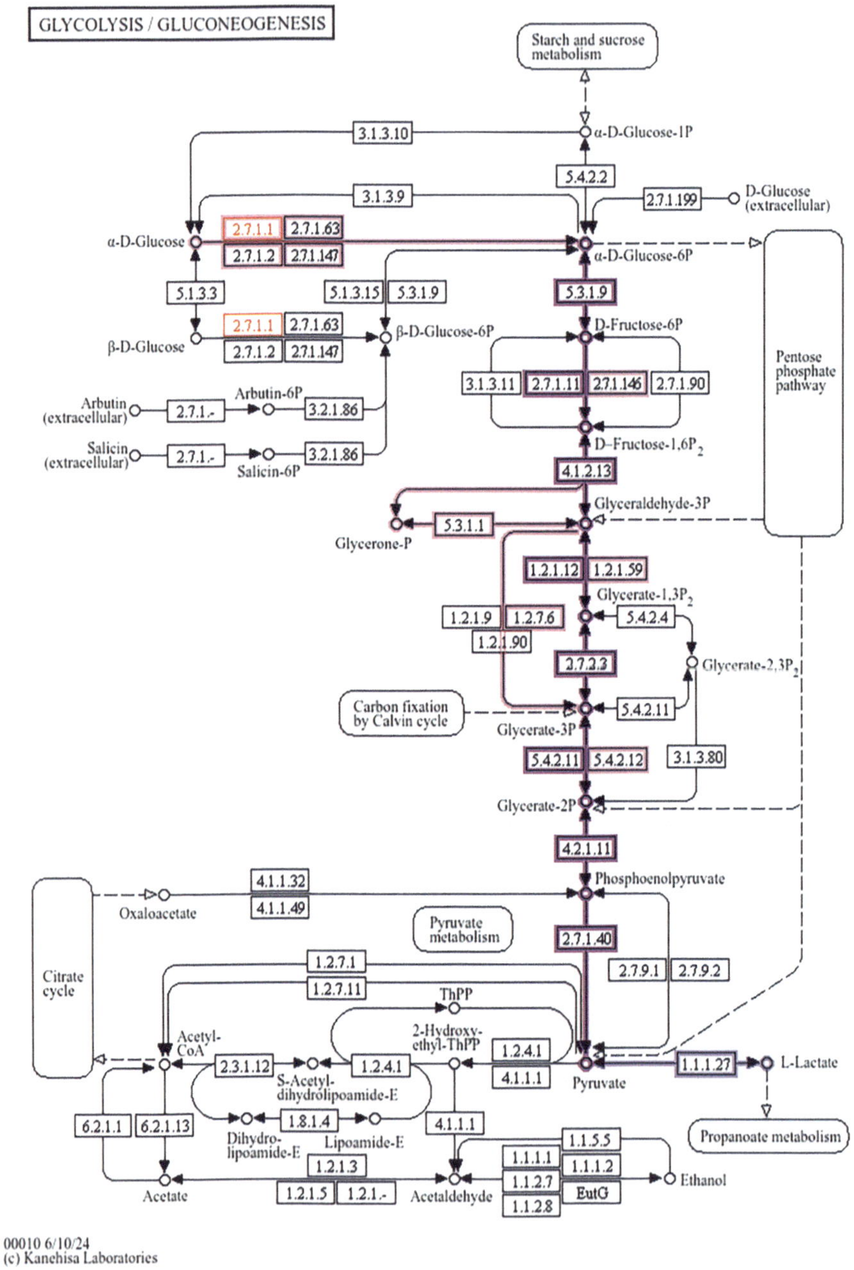

Fig. 3.1 Gluconeogenesis plant metabolic pathway from *A. marina* genomes were constructed by KEGG

References

Alsharif SM, Ismaeil M, Saeed AM, El-Sayed WS (2024) Metagenomic 16S rRNA analysis and predictive functional profiling revealed intrinsic organohalides respiration and bioremediation potential in mangrove sediment. BMC Microbiol 24(1):176

Alzubaidy H, Essack M, Malas TB, Bokhari A, Motwalli O, Kamanu FK et al (2016) Rhizosphere microbiome metagenomics of gray mangroves (*Avicennia marina*) in the Red Sea. Gene 576(2):626–636

Andreote FD, Jiménez DJ, Chaves D, Dias ACF, Luvizotto DM, Dini-Andreote F et al (2012) The microbiome of Brazilian mangrove sediments as revealed by metagenomics. PLoS One 7(6):e38600

Basak P, Pramanik A, Sengupta S, Nag S, Bhattacharyya A, Roy D et al (2016) Bacterial diversity assessment of pristine mangrove microbial community from Dhulibhashani, Sundarbans using 16S rRNA gene tag sequencing. Genom Data 7:76–78

Dai Z, Zhang N, Wang F, Li Y, Peng J, Xiang T et al (2024) Loss of microbial functional diversity following Spartina alterniflora invasion reduces the potential of carbon sequestration and nitrogen removal in mangrove sediments—from a gene perspective. J Environ Manag 365:121569

Haldar S, Nazareth SW (2018) Taxonomic diversity of bacteria from mangrove sediments of Goa: metagenomic and functional analysis. 3 Biotech 8:1–10

Liao S, Wang Y, Liu H, Fan G, Sahu SK, Jin T et al (2020) Deciphering the microbial taxonomy and functionality of two diverse mangrove ecosystems and their potential abilities to produce bioactive compounds. Msystems 5(5):10–1128

Ma D, Ding Q, Guo Z, Xu C, Liang P, Zhao Z et al (2022) The genome of a mangrove plant, *Avicennia marina*, provides insights into adaptation to coastal intertidal habitats. Planta 256(1):6

Muñoz-García A, Arbeli Z, Boyacá-Vásquez V, Vanegas J (2022) Metagenomic and genomic characterization of heavy metal tolerance and resistance genes in the rhizosphere microbiome of *Avicennia germinans* in a semi-arid mangrove forest in the tropics. Mar Pollut Bull 184:114204

Priya G, Lau NS, Furusawa G, Dinesh B, Foong SY, Amirul AAA (2018) Metagenomic insights into the phylogenetic and functional profiles of soil microbiome from a managed mangrove in Malaysia. Agri Gene 9:5–15

Shang C, Chen J, Nkoh JN, Wang J, Chen S, Hu Z, Hussain Q (2024) Biochemical and multi-omics analyses of response mechanisms of rhizobacteria to long-term copper and salt stress: effect on soil physicochemical properties and growth of *Avicennia marina*. J Hazard Mater 466:133601

Simões MF, Antunes A, Ottoni CA, Amini MS, Alam I, Alzubaidy H et al (2015) Soil and rhizosphere associated fungi in gray mangroves (*Avicennia marina*) from the Red Sea—a metagenomic approach. Genom Proteom Bioinform 13(5):310–320

Yang X, Dai Z, Yuan R, Guo Z, Xi H, He Z, Wei M (2023) Effects of salinity on assembly characteristics and function of microbial communities in the phyllosphere and rhizosphere of salt-tolerant *Avicennia marina* mangrove species. Microbiol Spectr 11(2):e03000–e03022

Impact of Fungal Endophytes on Nutrient Cycling and Related Microbes in the Rhizosphere Microbiome

4

Nissar Ahmad Wani, Naveed ul Mushtaq, Amina Manzoor, and Reiaz Ul Rehman

Abstract

The term "endophyte" was coined by Anton de Bary—a prominent figure in the field of plant pathology—in 1866. It is derived from the combination of the Greek prefix "endo", meaning within, and the Greek word "phyto", which pertains to plants. Endophytes are ubiquitously present in many plant species, and a wide range of microorganisms spanning multiple domains of life have successfully transitioned to an endophytic mode of existence. The rhizosphere microbiome plays a pivotal function in the processes of nutrient cycling, plant vitality, and soil integrity. Comprehending this intricate interplay is crucial for progress in the fields of agriculture and environmental sustainability. The diversity and activity of endophytes can be strongly influenced by factors such as soil pH, temperature, moisture levels, and nutrient availability. The populations and interactions of endophytes with plants are also influenced by human activities, such as the application of fertilisers and pesticides. The examination of fungal endophytes in the field of agriculture, particularly in the context of shifting climatic circumstances, presents several obstacles and offers novel opportunities for further investigation. Plant growth and yield are enhanced through the improvement of nutrient absorption and the provision of resistance against diseases and environmental challenges. Endophytic microorganisms have the potential to enhance soil health through their involvement in nutrient cycling and the breakdown of organic waste. There exist multiple routes by which endophytes can potentially contribute to the synthesis of metabolites, encompassing direct provision to the host and gene transfer between the endophyte and the host or vice versa. Further,

N. A. Wani · N. u. Mushtaq · A. Manzoor · R. U. Rehman (✉)
Department of Bioresources, School of Biological Sciences, University of Kashmir, Srinagar, India
e-mail: nisarwani.brct@uok.edu.in

M. Meena et al. (eds.), *Multi-omics Approach to Investigate Endophyte Diversity*, Microorganisms for Sustainability 53, https://doi.org/10.1007/978-981-96-4100-0_4

endophytic fungal communities typically demonstrate a specific preference for a single host at the level of plant species. However, it is important to note that this preference can be additionally influenced by factors such as microhabitat and microclimatic circumstances.

Keywords

Endophytes · Nutrient cycling · Fungi · Agriculture and sustainability

4.1 Introduction

The phenomenon of symbiosis characterised by the coexistence of distinct organisms has gained considerable attention since its initial conceptualisation in 1879 (Beinart et al. 2012). In the last century, extensive research has demonstrated that the majority of plants in their natural habitats establish symbiotic associations with fungi, such as mycorrhizal fungi and fungal endophytes (Rodriguez et al. 2009). These interactions exert significant influence on plant ecology, evolution, and community dynamics (Table 4.1), hence affecting the composition and variety of related species such as bacteria and insects. The fossil record indicates that this correlation has played a pivotal role in the development of life on land over a period of time. There are variations in the manner in which these mutualistic fungi engage with plants (Field and Pressel 2018). The term "endophyte" was coined by Anton de Bary—a prominent figure in the field of plant pathology—in 1866. It is derived from the combination of the Greek prefix "endo", meaning within, and the Greek word "phyto", which pertains to plants. De Bary used this phrase to refer to any organism that exhibits growth within the tissues of plants (Pandey et al. 2022). The current definition has been revised to include solely those organisms that do not induce diseases and, in many instances, may even confer benefits to their host

Table 4.1 Interaction between fungal endophytes and soil microbes

Class of fungal endophyte	Host plant	Mode of transmission	Examples of endophytes	References
Clavicipitaceous endophytes	Grasses (e.g., ryegrass, fescue)	Vertically (through seeds)	*Neotyphodium coenophialum*, *Epichloë festucae*	Clay (1989)
Non-clavicipitaceous endophytes	Various, including trees and shrubs	Horizontally (via spores or hyphal fragments)	*Phialocephala fortinii*, *Colletotrichum* spp.	Márquez et al. (2012)
Balansiaceous endophytes	Tropical and subtropical plants	Both vertically and horizontally	*Balansia* spp., *Piriformospora indica*	Schouten (2019)
Dark septate endophytes (DSE)	Wide range, including temperate and boreal plants	Primarily horizontally	*Leptodontidium orchidicola*, *Phialophora finlandia*	Jumpponen and Trappe (1998)

plants. Endophytes are ubiquitously present in many plant species, and a wide range of microorganisms spanning multiple domains of life have successfully transitioned to an endophytic mode of existence (Gouda et al. 2016). Beneficial effects of these microorganisms include promoting plant growth, improving plants' capacity to cope with environmental stress, and aiding in the defence against pests and pathogens. Fungal endophytes are a type of microorganism that resides within the tissues of plants, exhibiting a relationship by not inflicting any detrimental effects. They constitute an integral component of the plant's internal environment, exerting a significant impact on both development and overall well-being (Grabka et al. 2022). Endophytic fungi are characterised by their exclusive residence within plant tissues, including roots, stems, and leaves, with their emergence commonly coinciding with the ageing of the plant tissue. There exist two primary categories of endophytic fungi. The initial category, known as clavicipitaceous endophytes, mostly exhibits a tendency to infect grasses (Table 4.2). Non-clavicipitaceous endophytes, which comprise the second group, exhibit a broad distribution across many plant species, encompassing nonvascular plants, ferns, conifers, and flowering plants. Every group possesses unique evolutionary histories, specific host plants, and diverse ecological functions (Tidke et al. 2017). The rhizosphere is abundant with a diverse array of microscopic organisms like bacteria, fungus, and other microorganisms. The rhizosphere microbiome plays a pivotal function in the processes of nutrient cycling, plant vitality, and soil integrity. Comprehending this intricate interplay is crucial for progress in the fields of agriculture and environmental sustainability. Various categories of fungal endophytes exist, each possessing distinct attributes and roles (Amballa and Bhumi 2016). Mycorrhizal fungi, a widely recognised group, establish symbiotic associations with the roots of a majority of plant species. There are two main classifications of fungi involved in mycorrhizal associations: arbuscular mycorrhizal fungi (AMF) and ectomycorrhizal fungi (Yadav and Meena 2022; Bhupenchandra et al. 2024). AMF are characterised by their ability to enter the root cells, while ectomycorrhizal fungi build a protective sheath around the roots (Brundrett 2004; Yadav and Meena 2021). These fungal organisms facilitate the process of nutrient intake, specifically in relation to phosphate and nitrogen, and contribute to the absorption of water, so playing a crucial role in promoting plant growth. The alternative category is referred to as *endophytic Trichoderma*, which encompasses *Trichoderma* species that are widely prevalent in soil and recognised for their significant contributions to plant growth enhancement and biocontrol activities (Delian et al. 2011; Kumar et al. 2022). Endophytes are capable of colonising the roots of plants, so exhibiting the potential to promote plant vigour, stimulate root growth, and confer protection against a range of soil-borne infections. Dark septate endophytes (DSE) exhibit distinctive features such as fungal hyphae with a dark pigmentation and are frequently observed in challenging habitats such as arid or polluted soils. The utilisation of DSE (dark septate endophytes) has the potential to augment the ability of host plants to tolerate stress and play a significant role in nutrient acquisition, particularly in situations where nutrient availability is limited (Chaudhary et al. 2022). *Piriformospora indica*—an endophyte that was initially isolated from desert soils—has gained significant interest due to its capacity to

Table 4.2 Different classes of fungal endophytes and the applications of their interactions in agriculture

Fungal endophyte	Class	Role in agriculture	Interaction with rhizomicrobiome	Application of interaction	References
Arbuscular mycorrhizal fungi (AMF)	Glomeromycetes	Enhances phosphorus uptake, benefits a wide range of crops like wheat, corn, and soybeans	Forms mutualistic interactions, enhances nutrient absorption and soil structure	Used in sustainable agriculture, enhances crop yields, reduces fertiliser use	Bagyaraj et al. (2015)
Ectomycorrhizal fungi	Various (e.g., basidiomycetes)	Forms symbiotic relationships with trees, important for nutrient uptake in forests	Alters rhizosphere microbial communities, aids in nutrient cycling	Important in forestry and reforestation, supports diverse forest ecosystems	Cairney (2005)
Trichoderma spp.	Ascomycetes	Biocontrol agent, promotes plant growth and root development, enhances nutrient uptake	Competes with pathogenic microbes, promotes beneficial microbes	Widely used in biocontrol, enhances plant health and soil quality	Li et al. (2020)
Dark septate endophytes (DSE)	Ascomycetes	Improves plant nutrient uptake and stress tolerance, thrives in extreme environments	Associates with plant roots under stress, influences microbial diversity	Potential in phytoremediation, helps plants in stressed environments	Govindasamy et al. (2018)
Piriformospora indica	Basidiomycetes	Enhances plant growth, nutrient uptake, and resistance to stress, similar to mycorrhiza	Interacts similarly to mycorrhizal fungi, affects microbial dynamics	Utilised in crop stress management, potential in desert agriculture	Varma et al. (2012)
Endophytic yeasts	Various (e.g., ascomycetes)	Produces phytohormones, enhances plant stress tolerance, less studied than filamentous fungi	Impacts rhizosphere microbial balance, role in biocontrol and plant stress management	Emerging applications in drought and salt stress management in crops	Deng and Cao (2017)

augment plant development and bolster resistance against diseases and stressors. The observed characteristics have resemblance to those exhibited by mycorrhizal fungi in terms of their colonisation and functional attributes (Gill et al. 2016). Endophytic yeasts although relatively poorly recognised in comparison to filamentous fungi can exert substantial influence on the encouragement of plant growth. Certain yeast endophytes have the ability to synthesise bioactive compounds that support plant growth, hence enhancing plant resilience against adverse environmental conditions such as drought and salinity (Joubert and Doty 2018). Every distinct fungal endophyte has a distinct mode of interaction with its host plant and the adjacent soil environment. The functions of these microorganisms can encompass a wide range of activities including the absorption of nutrients, promotion of growth, suppression of diseases, and enhancement of stress tolerance. Fungal endophytes due to their diverse nature and intricate interactions with plants and soil microorganisms play a crucial role in the functioning of soil and plant ecosystems. Fungal endophytes have been found to augment the process of nutrient uptake in plants. This is accomplished by a variety of techniques. For example, several endophytes have the capability to synthesise enzymes that enhance the bioavailability of nutrients for plants. The aforementioned nutrients encompass nitrogen, phosphorus, and vital minerals (Khare et al. 2018). Endophytic microorganisms have the capacity to modify the structure of plant roots, resulting in an augmentation of the root surface area. Consequently, this alteration enhances the plant's capacity to assimilate essential nutrients. The establishment of this interaction is of utmost importance for the overall well-being of plants, particularly in soil environments that lack sufficient nutrients. Endophytes serve a broader range of functions beyond their role as nutrient sources. Plant growth and development are strongly impacted by their effect. Through the secretion of growth hormones and various biochemical compounds, these organisms are capable of inducing the growth of both roots and shoots (Watts et al. 2023). The symbiotic interaction between organisms is especially advantageous under challenging circumstances such as drought or limited nutrient availability. Plants that harbour endophytes frequently exhibit enhanced growth and durability in comparison to plants lacking such associations (Omomowo and Babalola 2019). This chapter presents a thorough examination of the influence exerted by fungal endophytes on nutrient cycling and the rhizosphere microbiome, with a specific focus on their relevance in agricultural contexts and their ability to adapt to shifting climatic circumstances.

4.2 Clavicipitaceous Endophytes

Epichloë and *Neotyphodium* species, as previously documented in scientific literature, are endophytic microorganisms that form intercellular associations within specific grass species. The endophytes in question exhibit a strict preference for colonising the aerial portions of grasses and participate in a mutually dependent symbiotic relationship (Xia et al. 2018). Every endophytic species possesses the ability to establish colonies in various grass hosts, although many endophytes demonstrate affinities for particular genera and species of grasses (Schardl and Phillips 1997). The spatial arrangement of endophytic hyphae within the plant has a distinct pattern, characterised by elevated densities in the basal areas and progressively decreasing densities towards the apical sections. This distribution remains consistent even among several organs of the grass species (Schardl et al. 2004). It can be observed that in the context of leaves, the leaf sheaths exhibit notable infection primarily at the basal ends while displaying diminished infection towards the apical ends. Furthermore, the occurrence of hyphae beyond ligules is unusual. It is worth noting that the roots of grasses growing in soil that is infected do not undergo colonisation. The progression of endophytic hyphae along the plant axis is marked by regulated growth patterns, which include branching at the base or lower crown of the plant, but a notable absence of branching in the upper stem (Scott et al. 2012). The controlled expansion is ascribed to the compatibility of the association. It is noteworthy to add that the presence of branching hyphae is exclusively found in the upper shoot during the manifestation of pathogenic expression and stromata formation in the *Epichloë* species (Scannerini et al. 2002). The regulation of controlled branching is governed by the accumulation of reactive oxygen species (ROS) in the extracellular matrix. This process is closely linked to the noxA gene, which encodes NADPH oxidase, and its regulator noxR (Sumimoto 2008). The spatial arrangement of these endophytes inside grasses corresponds to the inherent nutritional reservoirs present in these plants, such as seeds. Nevertheless, the lack of endophytes in the roots, which are organs abundant in nutrients, implies that there are additional variables that impact this particular distribution. The predominant mode of growth for endophytic hyphae is apical, although there have been indications of intercalary growth, a less frequently observed occurrence in fungal organisms. The development of endophytic hyphae is intricately connected to the life cycle of the host plant, and these hyphae continue to exhibit metabolic activity even in aged and fully developed host tissues (Vermot et al. 2021). It is noteworthy that *Epichloë* and *Neotyphodium* species exhibit a primary association with grasses that thrive in cool-season environments. Furthermore, it is worth mentioning that grasses exhibit intercellular spaces, which constitute the apoplasm, a non-living network present within the plant that is separate from the symplasm or cell-to-cell communication. The presence of intercellular spaces can be observed in the cortical tissue of roots, as well as in the parenchyma tissue of stems and leaves. The nature of these substances exhibits variability as certain specimens may contain gaseous or desiccated sediments, whereas others are comprised of aqueous solutions abundant in inorganic ions, including potassium, calcium, sulphur, phosphorus, and chlorine. Intercellular spaces within leaves are mostly located in the extrafascicular plant parenchyma, where they house oxygen, water saturated with air, and fluid. While the apoplastic pathway allows for fluid movement, the precise relationship with the symplasm is still not fully understood and presents a complex and ambiguous connection. Fungal endophytes get advantages from their inhabitation within the internal tissues of grasses owing to the sheltered ecological niche it affords, characterised by limited microbial competition and a consistent supply of nutrients (Taylor et al. 2014). The complete comprehension of the precise mechanisms involved in the availability and interchange of certain nutrients inside the apoplasm remains incomplete. However, there exists data indicating that fungal endophytes may possess hydrolytic enzymes that could potentially contribute to the pathogenic process.

4.3 Non-clavicipitaceous Endophytes

Non-clavicipitaceous endophytes can be distinguished from clavicipitaceous endophytes due to their distinct classification as a specific group of endophytic organisms that colonise a wide range of plant species. The endophytes have gained considerable interest owing to their unique attributes and interactions with their respective host plants (Deng and Cao 2017). The group of microorganisms referred to as "they" exhibits a wide variety of taxonomic classifications, including fungi and bacteria, without being limited to any particular category. The endophytes establish mutually advantageous associations with their host plants, similar to the symbiotic interactions reported in clavicipitaceous endophytes, hence enhancing the general health and vigour of the plants (Rai et al. 2014). In comparison to clavicipitaceous endophytes, which predominantly form associations with grasses, non-clavicipitaceous endophytes demonstrate a wider spectrum of host specificity, encompassing diverse plant taxa such as trees, shrubs, and herbaceous plants. The ecological roles of these organisms play a crucial part in the ecosystems they inhabit (Stachowicz 2001). They contribute to the enhancement of plant tolerance to various environmental stressors, optimise nutrient uptake processes, and provide protection against infections and herbivores. The field of scientific enquiry has explored the fascinating domain of non-clavicipitaceous endophytes, uncovering their ecological roles and possible uses in the fields of agriculture and forestry. These findings serve to enhance our understanding of these covert plant symbionts. The acknowledgement of the existence and functions of these endophytes in many ecosystems carries significant implications for efforts in environmental conservation and restoration, particularly in relation to nutrient cycling and enhancing the ability of plants to withstand challenges. Non-clavicipitaceous endophytes may be classified according to the associations they establish. An instance of this can be observed in the establishment of arbuscular mycorrhizae by *Glomus* spp., which form symbiotic associations with a diverse range of land plants, including significant agricultural crops such as maize, wheat, and soybeans (Ahanger et al. 2014). They have a crucial role in promoting the absorption of nutrients, particularly phosphorus, and improving the ability of plants to withstand environmental stressors.

Ectomycorrhizal relationships, which are established by fungi such as *Mycena* spp. and *Tricholoma* spp., are commonly observed in a wide range of tree species, including both conifers and hardwoods. These fungal organisms play a crucial role in the process of nutrient absorption, specifically within forest habitats. *Colletotrichum* spp. exhibit a wide distribution across several plant taxa, encompassing grasses, legumes, and woody plants. Moreover, these organisms frequently confer defensive benefits by offering protection against herbivores and pathogens. In addition, it is worth noting that *Cladosporium* spp. form non-pathogenic endophytic associations with a wide range of plant species, such as trees, shrubs, and herbaceous plants. These associations have been found to contribute to the enhancement of plant resilience in the face of diverse stressors.

Recent research has brought attention to the intricate interplay between fungal endophytes and soil microorganisms. The interactions between different components can exhibit either synergistic effects, leading to increased availability of nutrients, or antagonistic effects, which might impact the diversity of microorganisms present in the soil (Ramakrishnan et al. 2011). For example, certain endophytes have the ability to synthesise chemical compounds that exhibit suppressive effects on deleterious soil pathogens, thereby fostering an enhanced state of soil health. Consequently, this process facilitates the improvement of nutrient cycling and the subsequent increase in nutrient availability. Additionally, scholarly investigations have demonstrated that endophytes possess the ability to exert an influence on the microbial communities present in the rhizosphere, hence exerting a further impact on the dynamics of nutrient availability (Vishwakarma et al. 2020). The investigation of nutrient cycling within the rhizosphere is a fundamental subject of scholarly enquiry. Fungal endophytes are known to have a substantial role in this particular process. Microorganisms play a significant role in the process of organic matter breakdown, hence facilitating the release of essential nutrients back into the soil (Condron et al. 2010; Chandran et al. 2021; Chauhan et al. 2023; Meena et al. 2023). Recent literature studies have highlighted the significant role of endophytes in augmenting the accessibility of key nutrients, such as nitrogen and phosphorus. This is achieved by the process of decomposing intricate chemicals into simpler forms that can be readily assimilated by plants. The significance of this role lies in its ability to uphold the equilibrium of nutrients within the soil, hence facilitating the implementation of sustainable agricultural methodologies. An area of endophyte study that is considered pioneering is their involvement in the process of nitrogen fixation (Kennedy and Islam 2001). Certain endophytes have been discovered to contain the ability to fix nitrogen, a capability that was previously believed to be restricted to specific bacteria. The findings of this study carry substantial consequences for the field of agriculture, particularly in terms of mitigating reliance on chemical fertilisers. Endophytic microorganisms possess the ability to transform air nitrogen into biologically available forms, hence augmenting soil fertility and promoting plant growth. Current study is investigating the potential of these endophytes in many agricultural systems. Fungal endophytes are of significant importance in the process of phosphorus mobilisation, which is a critical factor in facilitating optimal plant growth by providing essential nutrients (Chhabra and Dowling 2017). Phosphatases are enzymatic proteins that facilitate the hydrolysis of organic phosphorus molecules within the soil matrix, hence enabling the release of phosphorus for plant uptake. The aforementioned procedure holds significant importance in soils that are weak in phosphorus. Numerous studies have demonstrated that plants that form associations with phosphorus-solubilising endophytes exhibit notable improvements in both growth and production. The results of this study indicate the possibility of utilising these discoveries in the development of biofertilisers that have the ability to organically augment phosphorus availability, hence mitigating the reliance on synthetic fertilisers. Endophytes have a substantial influence on the carbon cycle (Owen et al. 2015). These organisms participate in the process of organic matter breakdown, thereby releasing carbon into both the soil and the atmosphere. Certain endophytic microorganisms have the ability to augment the process of plant root exudation, hence leading to an amplified influx of carbon into the rhizosphere. This particular method has the potential to enhance microbial activity, hence exerting a subsequent impact on the cycling and availability of nutrients (Rashid et al. 2016). Recent research findings imply that endophytes have the capacity to enhance carbon sequestration in plants, hence indicating a potential avenue for mitigating the impacts of climate change. In conjunction with nutrient enrichment, fungal endophytes play a role in the control of diseases and the enhancement of plant stress tolerance. These organisms synthesise a diverse array of bioactive substances that possess the ability to impede the growth and development of plant pathogens, hence mitigating the occurrence of diseases (Manganyi and Ateba 2020). In addition, endophytes play a crucial role in assisting plants in adapting to abiotic stressors, such as drought and salinity, through the facilitation of nutrient absorption and improvement of water retention capabilities. The potential of endophytes to enhance crop resilience in the face of shifting climatic conditions has garnered significant interest.

4.4 Fungal Endophytes and Environmental Factors Affecting Endophyte Function

The process of isolating and analysing fungal endophytes encompasses a variety of sophisticated methodologies. In the first stage, endophytes are extracted from plant tissues employing aseptic techniques in order to limit the risk of contamination (Quambusch and Winkelmann 2018). In the standard procedure, tissue samples are subjected to surface sterilisation and thereafter placed on selected media in order to facilitate the development of fungi. Recent scientific developments have involved the utilisation of molecular methodologies, such as polymerase chain reaction (PCR) and DNA sequencing, for the purpose of identification and categorisation of endophytes. The utilisation of metagenomic methodologies has become more prevalent in the examination of the entire endophyte community, yielding valuable information regarding their variety and roles within the plant host (Fadiji and Babalola 2020). However, it is imperative to acknowledge that environmental conditions exert a pivotal influence on the operational dynamics of fungal endophytes. The diversity and activity of endophytes can be strongly influenced by factors such as soil pH, temperature, moisture levels, and nutrient availability. The populations and interactions of endophytes with plants are also influenced by human activities, such as the application of fertilisers and pesticides (Proença et al. 2017). Comprehending these environmental factors is crucial for maximising the efficacy of endophytes in the realms of agriculture and environmental stewardship.

4.5 Applications in Agriculture and Sustainability

The examination of fungal endophytes in the field of agriculture, particularly in the context of shifting climatic circumstances, presents several obstacles and offers novel opportunities for further investigation (Baron and Rigobelo 2022). A significant obstacle lies in comprehending the intricate interplay among endophytes, plants, and environmental elements among fluctuating climatic circumstances. The advantages of fungal endophytes in the field of agriculture are significant (Fig. 4.1). Plant growth and yield are enhanced through the improvement of nutrient absorption and the provision of resistance against diseases and environmental challenges. Endophytic microorganisms have the potential to enhance soil health through their involvement in nutrient cycling and the breakdown of organic waste (Khatoon et al. 2020). The potential of sustainable agricultural practices is substantial as they provide a viable alternative to artificial fertilisers and pesticides. Moreover, endophytes exhibit potential in the field of phytoremediation, offering assistance in the process of soil remediation for the purpose of eliminating contaminants. The potential of

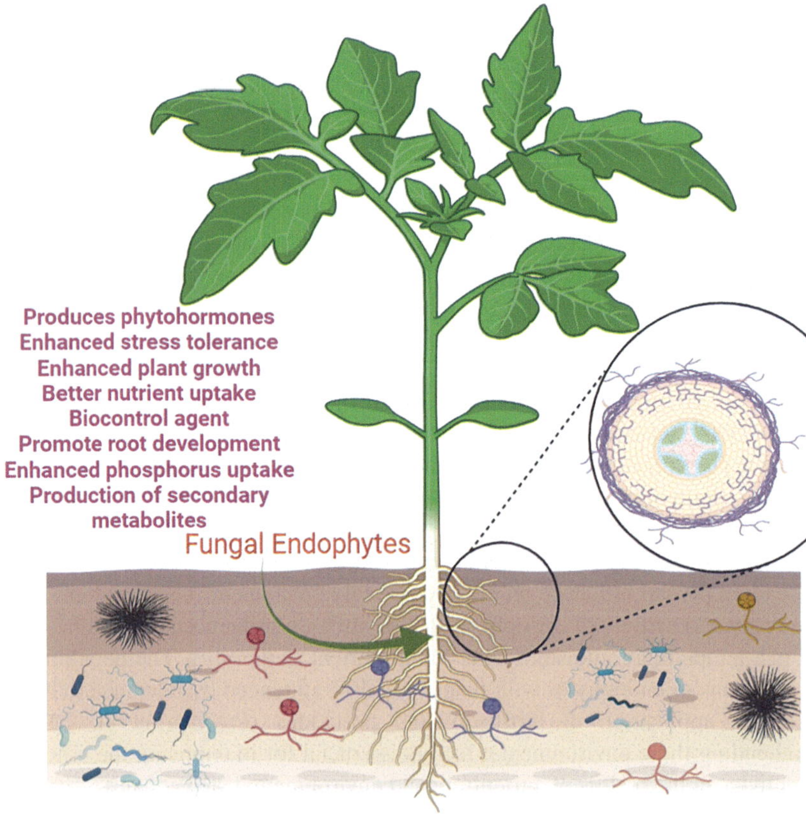

Fig. 4.1 Benefits of fungal endophytes for agriculture and sustainability

fungal endophytes in climate-resilient agriculture is becoming more acknowledged. One potential benefit of their application is the facilitation of crop adaptation to dynamic environmental conditions through the augmentation of nutrient absorption and stress resilience (Prasad et al. 2020). One illustrative instance involves the potential of endophytes to enhance water use efficiency in plants, a crucial characteristic particularly relevant in regions susceptible to drought conditions. Additionally, they contribute to the amelioration of the effects of severe temperatures on the growth of plants. The incorporation of endophytes into agricultural practices has the potential to serve as a pivotal approach in the advancement of sustainable and resilient food systems amidst the challenges posed by climate change. The implementation of sustainable soil management practices plays a critical role in both adapting to and mitigating the impacts of climate change (Delgado et al. 2011). Fungal endophytes play a significant role in promoting soil health through their ability to enhance nutrient cycling and facilitate the breakdown of organic waste. The use of certain practices has the potential to enhance the structure and fertility of soil, hence increasing its resistance to erosion and deterioration. Endophytic microorganisms possess the capacity to mitigate the dependence on chemical fertilisers, hence mitigating the associated greenhouse gas emissions. The promotion of endophytes in agricultural practices could potentially contribute to both climate change adaptation and the mitigation of agriculture's carbon emissions. Fungal endophytes have the potential to greatly enhance agricultural productivity, even across a wide range of climatic circumstances (Omomowo and Babalola 2019). The desirable attributes of these organisms, namely, their capacity to augment nutrient absorption and withstand adverse conditions, render them highly advantageous in agricultural practices in both temperate and tropical regions. Numerous studies have demonstrated that crops associated with endophytes frequently exhibit enhanced growth and production under many environmental conditions. The ability to adapt is of utmost importance in guaranteeing food security in light of the escalating unpredictability of climate patterns. Pollinators play a crucial role in promoting the vitality and adaptability of plant groups, hence fostering the richness and variety of ecosystems (Kremsa 2021). Endophytes play a crucial role in the preservation of biodiversity within agricultural settings through their facilitation of a diverse array of advantageous soil organisms. The significance of this matter is particularly pronounced within the framework of monoculture agriculture, a practice that frequently leads to a decline in biodiversity. Endophytic microorganisms present a potential avenue for achieving a harmonious equilibrium between agricultural output and ecological preservation (Fig. 4.1).

4.6 The Enigmatic Nature of Host Preference

The phenomena of host preference continue to be a complex and puzzling aspect of both antagonistic and mutualistic symbiotic relationships. It is a subject that lies at the crossroads of ecology and evolution, yet its full understanding remains elusive. In order to achieve host specificity, a significant level of adaptability between the host plant and its fungal counterpart is necessary (Naranjo-Ortiz and Gabaldón 2019). This implies a reciprocal impact originating from a longstanding cohabitation and simultaneous development. Over the course of time, this affiliation gets ingrained throughout the genetic composition of both individuals, resulting in the development of mutually supportive genetic networks. This phenomenon is apparent through the discovery that endosymbionts residing within a certain host frequently exhibit a shared evolutionary lineage, known as co-cladogenesis (Moya et al. 2008). Additional indications of co-evolution can be observed through the lack of plant defence mechanisms against microbial endophytes and the capacity of endophytes to synthesise identical bioactive chemicals as their host plants. Additional evidence is provided by the synchronisation of fungal and host reproductive systems, exemplified by the vertical transport of fungal spores with seeds. This mutually advantageous association contributes to the effective reproduction of the host organism.

The initial discovery of the endophytic fungus capable of manufacturing paclitaxel occurred exclusively on the indigenous yew tree located within Glacier National Park, Montana (Aly et al. 2011). It is worth mentioning that certain prevalent tree endophytes, such as *Trichoderma* spp., have been observed in yew trees, whereas endophytic *Penicillium* spp., which are commonly linked with pine trees, are infrequently discovered in yew trees. Yews located in subtropical areas, such as the lower slopes of the Himalayas, the Philippines, Florida, and South China, are known to harbour a diverse group of fungal species belonging to the *Pestalotia* and *Pestalotiopsis* genera. In the context of conifers, there is a notable prevalence of specific fungal lineages in plants belonging to particular families (Nisa et al. 2015). For instance, *Lophodermium* spp. is commonly observed in Pinaceae, but its presence is missing in other conifers or angiosperms. While it is possible for certain tree species to harbour numerous endophyte species within a certain tissue type, the majority of communities are mostly composed of a limited number of species that are exclusive to the host. It is noteworthy that within plant families, fungal communities in host species frequently exhibit a prevalence of closely related endophyte species. The degree of relatedness among dominant endophytes reduces in proportion to the taxonomic dissimilarity of host plants.

Endophytic fungal communities typically demonstrate a specific preference for a single host at the level of plant species. However, it is important to note that this preference can be additionally influenced by factors such as microhabitat and microclimatic circumstances (Ziesche and Roth 2008). In contrast, numerous endophytes, like those found within the genera *Phomopsis*, *Phoma*, *Colletotrichum*, and *Phyllosticta*, exhibit a wide range of hosts and successfully establish themselves in taxonomically unrelated plant species, thereby demonstrating their capacity to adapt to diverse host defence mechanisms. The influence of tissue specificity of endophytic fungi may also demonstrate selectivity towards particular organs and tissues as a result of their adaptation to diverse physiological conditions inside plant hosts (Schulz and Boyle 2005). The endophytes residing in a certain plant tissue type are influenced by the selective forces exerted on it, leading to notable variations compared to endophytes found in other tissues. As a result, several fungal species exhibit dominance in diverse above-ground anatomical structures, thereby establishing discrete and tissue-specific communities. The analysis of foliar and root endophyte populations indicates a restricted degree of similarity. On the other hand, research indicates that there may be a degree of overlap in the species composition of endophyte communities found in leaves and wood (Arnold 2007). However, it has been shown that leaves that are fully infected later in the season tend to display greater diversity and harbour unique endophytic organisms. As an example, a research investigation on fungal communities in *Cupressus arizonica* revealed the presence of 53 distinct isolates, with 21 originating from foliage and 32 obtained from twigs. It is worth noting that a significant proportion of genotypes, specifically 83%, exhibited exclusivity to either twigs or leaves, hence emphasising the phenomenon of tissue-specific colonisation.

The age of host plant tissues can also have an impact on the species composition of endophyte communities (Vesterlund et al. 2011). Moreover, it is probable that the bulk of cultivable endophytes found in the foliage of woody plants are derived from external sources through horizontal transmission. However, there are instances where endophytes recovered from seeds are transmitted vertically. It is worth noting that prevalent foliar endophytes are frequently not present in seeds, which instead contain a smaller number of fungi that are not commonly seen in leaves. The aforementioned distinction highlights the significance of employing culture-independent techniques, such as environmental polymerase chain reaction (PCR), to effectively capture vertically transmitted fungi that may provide difficulties in cultivation (Carper et al. 2018). Enhancements in cultivation techniques may potentially facilitate the identification and capture of fungi that are only identifiable through cloning methodologies.

The phenomenon of tissue specificity in mangrove trees demonstrates a certain degree of tissue specialisation, albeit with a lesser degree of specificity towards their host taxon. A research investigation conducted in Hong Kong examined the mangrove species *Kandelia candel* and discovered consistent communities of endophytic fungi in samples of bark and wood (Li et al. 2016). In contrast, leaf samples harboured a different and separate population of fungi. The dominant fungi that undergo sporulation in this study encompass worldwide species, namely, *Phomopsis* sp., *Pestalotiopsis* sp., *Guignardia* sp., and *Xylaria* sp. In order to obtain a thorough understanding of endophyte variety, it is imperative to conduct screenings across many tissues within a single host plant, as opposed to just examining homologous tissues across different plant species.

4.7 Production of Bioactive Secondary Metabolites

The production of several bioactive secondary metabolites by endophytic fungus has been documented, revealing the existence of compounds that were previously exclusive to plants (Gakuubi et al. 2021). The identification of endophytes as potential alternate sources for these important chemicals has presented a promising opportunity. There exist multiple routes by which endophytes can potentially contribute to the synthesis of these metabolites, encompassing direct provision to the host and gene transfer between the endophyte and the host or vice versa. An example that is widely recognised in the scientific community is the identification of plant-derived chemicals in endophytic fungi, as demonstrated by the discovery of *Taxomyces andreanae*, an endophytic fungus capable of making paclitaxel (commonly known as Taxol) (Daley and Cordell 2021). This particular fungus was initially isolated from the Pacific yew tree, *Taxus brevifolia*. Paclitaxel is an important pharmacological agent used in the treatment of cancer, exerting its therapeutic effects through the stabilisation of microtubules and the inhibition of mitotic processes. Subsequent investigations conducted with the objective of identifying alternate sources for paclitaxel production have uncovered the ability of many endophytic fungi, including *Seimatoantlerium tepuiense*, *Seimatoantlerium nepalense*, *Tubercularia* sp. strain TF5, and *Metarhizium anisopliae*, to synthesise paclitaxel. Furthermore, the presence of paclitaxel has been identified in endophytic microorganisms inhabiting plant species other than those belonging to the *Taxus* genus. Vincristine, a significant pharmaceutical agent with anticancer properties, was initially derived from *Catharanthus roseus* (Alam et al. 2017). However, subsequent research revealed its presence in cultures of the endophytic fungus *Fusarium oxysporum*. The mechanism of action of vincristine involves its binding to tubulin dimers, so impeding their ability to assemble into microtubule assemblies. The *Podophyllum* genus is known for its production of podophyllotoxin, a compound that serves as the precursor for the synthesis of anticancer medications such as etoposide and teniposide (Shah et al. 2021). The presence of endophytes, specifically *Trametes hirsuta* and *Phialocephala fortinii*, in these plants resulted in the production of podophyllotoxin. Furthermore, a deoxy derivative of podophyllotoxin was documented to be derived from the endophytic fungus *Aspergillus fumigatus*, which was found in *Juniperus communis*. The compound known as camptothecin, which is a naturally occurring inhibitor of topoisomerase I, was initially identified in the plant species *Camptotheca acuminata*. Endophytic fungi, such as *Entrophosporain frequens* and *Fusarium solani*, have been documented to be associated with diverse plant hosts (Sharma et al. 2022). Endophytic microorganisms residing within grasses synthesise alkaloids that possess deterring and poisonous properties, thereby serving as a defence mechanism against herbivorous organisms, ultimately safeguarding the grasses from their detrimental impact. For instance, the presence of loline alkaloids was observed in *Festuca pratensis*, a grass species, as well as in the root hemiparasitic plant *Rhinanthus serotinus* (Schardl et al. 2007). It is noteworthy that although endophytic fungi possess the ability to generate secondary metabolites, they may not be the exclusive providers. The phenomenon of horizontal gene transfer occurring between fungus and their host plants has the potential to contribute to the manufacture of various substances. This process may provide insights into the uneven distribution of specific natural products among the plant kingdom. Endophytic fungus exhibits a high capacity for synthesising bioactive secondary metabolites that exhibit distinct structural properties compared to those present in plants (Ancheeva et al. 2020). These chemicals exhibit promising prospects for application in the fields of medicine and agriculture. In recent years, extensive investigation into fungal endophytes has resulted in the identification of more than 100 novel natural products. This significant finding underscores the increasing significance of these organisms in the field of bioprospecting. These bioactive chemicals possess a range of capabilities, such as antibacterial, antiparasitic, cytotoxic, neuroprotective, antioxidant, insulin mimetic, or immunosuppressant actions (Shukla et al. 2014). As an illustration, pestalotheols A–D were obtained from the endophytic fungus *Pestalotiopsis theae* and have shown inhibitory properties against HIV-1LAI replication. Enfumafungin, which was derived from an endophytic strain of *Hormonema* sp., has shown significant efficacy as an antifungal agent through its ability to impede glucan production (Vicente et al. 2003). Consequently, this compound holds considerable potential as a therapeutic option for the treatment of fungal infections. As a consequence, the emergence of MK-3118 ensued, which is an orally administrable inhibitor of β-1,3-glucan synthase, designed to address the therapeutic needs associated with fungal infections.

4.8 Conclusion

Fungal endophytes possess significant promise for facilitating the transformation of agricultural practices towards enhanced sustainability and climate resilience. The invaluable contributions of these organisms in agriculture stem from their capacity to augment nutrient cycling, boost plant growth and stress resilience, and foster soil health and biodiversity. The appropriate utilisation of these microscopic allies has the potential to significantly alter the future of farming. The comprehensive exploration of fungal endophytes in conjunction with the implementation of favourable legislation and educational initiatives will play a pivotal role in maximising the sustainable prospects they provide for the future.

References

Ahanger MA, Hashem A, Abd-Allah EF, Ahmad P (2014) Arbuscular mycorrhiza in crop improvement under environmental stress. In: Emerging technologies and management of crop stress tolerance. Academic Press, Cambridge, MA, pp 69–95

Alam MM, Naeem M, Khan MMA, Uddin M (2017) Vincristine and vinblastine anticancer catharanthus alkaloids: pharmacological applications and strategies for yield improvement. In: Catharanthus roseus: current research and future prospects. Springer, pp 277–307

Aly AH, Debbab A, Proksch P (2011) Fungal endophytes: unique plant inhabitants with great promises. Appl Microbiol Biotechnol 90:1829–1845

Amballa H, Bhumi NR (2016) Significance of arbuscular mycorrhizal fungi and rhizosphere microflora in plant growth and nutrition. In: Plant-microbe interaction: an approach to sustainable agriculture. Springer, Singapore, pp 417–452

Ancheeva E, Daletos G, Proksch P (2020) Bioactive secondary metabolites from endophytic fungi. Curr Med Chem 27(11):1836–1854

Arnold AE (2007) Understanding the diversity of foliar endophytic fungi: progress, challenges, and frontiers. Fungal Biol Rev 21(2–3):51–66

Bagyaraj DJ, Sharma MP, Maiti D (2015) Phosphorus nutrition of crops through arbuscular mycorrhizal fungi. Curr Sci 108:1288–1293

Baron NC, Rigobelo EC (2022) Endophytic fungi: a tool for plant growth promotion and sustainable agriculture. Mycology 13(1):39–55

Beinart RA, Sanders JG, Faure B, Sylva SP, Lee RW, Becker EL et al (2012) Evidence for the role of endosymbionts in regional-scale habitat partitioning by hydrothermal vent symbioses. Proc Natl Acad Sci 109(47):E3241–E3250

Bhupenchandra I, Chongtham SK, Devi AG, Dutta P, Sahoo MR, Mohanty S, Kumar S, Choudhary AK, Devi EL, Sinyorita S, Devi SH, Mahanta M, Kumari A, Devi HL, Josmee RK, Pusparani A, Pathaw N, Gupta S, Meena M, Janmeda P, Seth CS, Arora A, Sahoo A, Swapnil P (2024) Unlocking the potential of arbuscular mycorrhizal fungi: exploring role in plant growth promotion, nutrient uptake mechanisms, biotic stress alleviation, and sustaining agricultural production systems. J Plant Growth Regul 43(9):1–39

Brundrett M (2004) Diversity and classification of mycorrhizal associations. Biol Rev 79(3):473–495

Cairney JW (2005) Basidiomycete mycelia in forest soils: dimensions, dynamics and roles in nutrient distribution. Mycol Res 109(1):7–20

Carper DL, Carrell AA, Kueppers LM, Frank AC (2018) Bacterial endophyte communities in *Pinus flexilis* are structured by host age, tissue type, and environmental factors. Plant Soil 428:335–352

Chandran H, Meena M, Swapnil P (2021) Plant growth-promoting rhizobacteria as a green alternative for sustainable agriculture. Sustain For 13(19):10986

Chaudhary P, Agri U, Chaudhary A, Kumar A, Kumar G (2022) Endophytes and their potential in biotic stress management and crop production. Front Microbiol 13:933017

Chauhan P, Sharma N, Tapwal A, Kumar A, Verma GS, Meena M, Seth CS, Swapnil P (2023) Soil microbiome: diversity, benefits and interactions with plants. Sustain For 15:14643

Chhabra S, Dowling DN (2017) Endophyte-promoted nutrient acquisition: phosphorus and iron. In: Functional importance of the plant microbiome: implications for agriculture, forestry and bioenergy. Springer, Cham, pp 21–42

Clay K (1989) Clavicipitaceous endophytes of grasses: their potential as biocontrol agents. Mycol Res 92(1):1–12

Condron L, Stark C, O'Callaghan M, Clinton P, Huang Z (2010) The role of microbial communities in the formation and decomposition of soil organic matter. In: Soil microbiology and sustainable crop production. Springer, Dordrecht, pp 81–118

Daley SK, Cordell GA (2021) Biologically significant and recently isolated alkaloids from endophytic fungi. J Nat Prod 84(3):871–897

Delgado JA, Groffman PM, Nearing MA, Goddard T, Reicosky D, Lal R et al (2011) Conservation practices to mitigate and adapt to climate change. J Soil Water Conserv 66(4):118A–129A

Delian E, Chira A, Chira L, Savulescu E (2011) Arbuscular mycorrhizae: an overview. South Western J HortBiol Env 2:167–192

Deng Z, Cao L (2017) Fungal endophytes and their interactions with plants in phytoremediation: a review. Chemosphere 168:1100–1106

Fadiji AE, Babalola OO (2020) Metagenomics methods for the study of plant-associated microbial communities: a review. J Microbiol Methods 170:105860

Field KJ, Pressel S (2018) Unity in diversity: structural and functional insights into the ancient partnerships between plants and fungi. New Phytol 220(4):996–1011

Gakuubi MM, Munusamy M, Liang ZX, Ng SB (2021) Fungal endophytes: a promising frontier for discovery of novel bioactive compounds. J Fungi 7(10):786

Gill SS, Gill R, Trivedi DK, Anjum NA, Sharma KK, Ansari MW et al (2016) Piriformospora indica: potential and significance in plant stress tolerance. Front Microbiol 7:332

Gouda S, Das G, Sen SK, Shin HS, Patra JK (2016) Endophytes: a treasure house of bioactive compounds of medicinal importance. Front Microbiol 7:1538

Govindasamy V, George P, Raina SK, Kumar M, Rane J, Annapurna K (2018) Plant-associated microbial interactions in the soil environment: role of endophytes in imparting abiotic stress

tolerance to crops. In: Advances in crop environment interaction. Springer, Singapore, pp 245–284
Grabka R, d'Entremont TW, Adams SJ, Walker AK, Tanney JB, Abbasi PA, Ali S (2022) Fungal endophytes and their role in agricultural plant protection against pests and pathogens. Plan Theory 11(3):384
Joubert PM, Doty SL (2018) Endophytic yeasts: biology, ecology and applications. In: Endophytes of forest trees: biology and applications. Springer, Cham, pp 3–14
Jumpponen ARI, Trappe JM (1998) Dark septate endophytes: a review of facultative biotrophic root-colonizing fungi. New Phytol 140(2):295–310
Kennedy IR, Islam NJAJ (2001) The current and potential contribution of asymbiotic nitrogen fixation to nitrogen requirements on farms: a review. Aust J Exp Agric 41(3):447–457
Khare E, Mishra J, Arora NK (2018) Multifaceted interactions between endophytes and plant: developments and prospects. Front Microbiol 9:2732
Khatoon Z, Huang S, Rafique M, Fakhar A, Kamran MA, Santoyo G (2020) Unlocking the potential of plant growth-promoting rhizobacteria on soil health and the sustainability of agricultural systems. J Environ Manag 273:111118
Kremsa VŠ (2021) Sustainable management of agricultural resources (agricultural crops and animals). In: Sustainable resource management. Elsevier, pp 99–145
Kumar S, Chandra R, Keswani C, Minkina T, Mandzhieva S, Voloshina M, Meena M (2022) *Trichoderma viride*—mediated modulation of oxidative stress network in potato challenged with *Alternaria solani*. J Plant Growth Regul 42:1919–1936
Li JL, Sun X, Chen L, Guo LD (2016) Community structure of endophytic fungi of four mangrove species in Southern China. Mycology 7(4):180–190
Li Y, Guo Q, He F, Li Y, Xue Q, Lai H (2020) Biocontrol of root diseases and growth promotion of the tuberous plant *Aconitum carmichaelii* induced by Actinomycetes are related to shifts in the rhizosphere microbiota. Microb Ecol 79:134–147
Manganyi MC, Ateba CN (2020) Untapped potentials of endophytic fungi: a review of novel bioactive compounds with biological applications. Microorganisms 8(12):1934
Márquez SS, Bills GF, Herrero N, Zabalgogeazcoa I (2012) Non-systemic fungal endophytes of grasses. Fungal Ecol 5(3):289–297
Meena M, Yadav G, Sonigra P, Nagda A, Mehta T, Swapnil P, Harish Marwal A, Kumar S (2023) Multifarious responses of forest soil microbial community toward climate change. Microb Ecol 86:49–74
Moya A, Peretó J, Gil R, Latorre A (2008) Learning how to live together: genomic insights into prokaryote–animal symbioses. Nat Rev Genet 9(3):218–229
Naranjo-Ortiz MA, Gabaldón T (2019) Fungal evolution: major ecological adaptations and evolutionary transitions. Biol Rev 94(4):1443–1476
Nisa H, Kamili AN, Nawchoo IA, Shafi S, Shameem N, Bandh SA (2015) Fungal endophytes as prolific source of phytochemicals and other bioactive natural products: a review. Microb Pathog 82:50–59
Omomowo OI, Babalola OO (2019) Bacterial and fungal endophytes: tiny giants with immense beneficial potential for plant growth and sustainable agricultural productivity. Microorganisms 7(11):481
Owen D, Williams AP, Griffith GW, Withers PJ (2015) Use of commercial bio-inoculants to increase agricultural production through improved phosphrous acquisition. Appl Soil Ecol 86:41–54
Pandey SS, Jain R, Bhardwaj P, Thakur A, Kumari M, Bhushan S, Kumar S (2022) Plant Probiotics–Endophytes pivotal to plant health. Microbiol Res 127148:127148
Prasad M, Srinivasan R, Chaudhary M, Mahawer SK, Jat LK (2020) Endophytic bacteria: role in sustainable agriculture. In: Microbial endophytes. Woodhead Publishing, pp 37–60
Proença DN, Schwab S, Baldani JI, Morais PV (2017) Diversity and function of endophytic microbial community of plants with economical potential. In: Diversity and benefits of microorganisms from the tropics. Springer, Cham, pp 209–243

Quambusch M, Winkelmann T (2018) Bacterial endophytes in plant tissue culture: mode of action, detection, and control. In: Plant cell culture protocols. Humana, New York, pp 69–88
Rai M, Rathod D, Agarkar G, Dar M, Brestic M, Pastore GM, Junior MRM (2014) Fungal growth promotor endophytes: a pragmatic approach towards sustainable food and agriculture. Symbiosis 62:63–79
Ramakrishnan B, Megharaj M, Venkateswarlu K, Sethunathan N, Naidu R (2011) Mixtures of environmental pollutants: effects on microorganisms and their activities in soils. Springer, New York, pp 63–120
Rashid MI, Mujawar LH, Shahzad T, Almeelbi T, Ismail IM, Oves M (2016) Bacteria and fungi can contribute to nutrients bioavailability and aggregate formation in degraded soils. Microbiol Res 183:26–41
Rodriguez RJ, White JF Jr, Arnold AE, Redman ARA (2009) Fungal endophytes: diversity and functional roles. New Phytol 182(2):314–330
Scannerini S, Fusconi A, Mucciarelli M (2002) The effect of endophytic fungi on host plant morphogenesis. In: Symbiosis: mechanisms and model systems. Springer, Dordrecht, pp 425–447
Schardl CL, Phillips TD (1997) Protective grass endophytes: where are they from and where are they going? Plant Dis 81(5):430–438
Schardl CL, Leuchtmann A, Spiering MJ (2004) Symbioses of grasses with seedborne fungal endophytes. Annu Rev Plant Biol 55:315–340
Schardl CL, Grossman RB, Nagabhyru P, Faulkner JR, Mallik UP (2007) Loline alkaloids: currencies of mutualism. Phytochemistry 68(7):980–996
Schouten A (2019) Endophytic fungi: definitions, diversity, distribution and their significance in plant life. In: Endophyte biotechnology: potential for agriculture and pharmacology. CABI, Wallingford UK, pp 6–31
Schulz B, Boyle C (2005) The endophytic continuum. Mycol Res 109(6):661–686
Scott B, Becker Y, Becker M, Cartwright G (2012) Morphogenesis, growth, and development of the grass symbiont Epichlöe festucae. In: Morphogenesis and pathogenicity in fungi. Springer, Berlin/Heidelberg, pp 243–264
Shah Z, Gohar UF, Jamshed I, Mushtaq A, Mukhtar H, Zia-UI-Haq M et al (2021) Podophyllotoxin: history, recent advances and future prospects. Biomol Ther 11(4):603
Sharma R, Shukla AC, Tangjang S (2022) Fungal endophytes: a potential source of low-cost entrepreneurship. In: Applied mycology: entrepreneurship with fungi. Springer International Publishing, Cham, pp 39–73
Shukla ST, Habbu PV, Kulkarni VH, Jagadish KS, Pandey AR, Sutariya VN (2014) Endophytic microbes: a novel source for biologically/pharmacologically active secondary metabolites. Asian J PharmacolToxicol 2(3):1–6
Stachowicz JJ (2001) Mutualism, facilitation, and the structure of ecological communities: positive interactions play a critical, but underappreciated, role in ecological communities by reducing physical or biotic stresses in existing habitats and by creating new habitats on which many species depend. Bioscience 51(3):235–246
Sumimoto H (2008) Structure, regulation and evolution of Nox-family NADPH oxidases that produce reactive oxygen species. FEBS J 275(13):3249–3277
Taylor TN, Krings M, Taylor EL (2014) Fossil fungi. Academic Press, pp 1–323
Tidke SA, Kumar RK, Ramakrishna D, Kiran S, Kosturkova G, Gokare RA (2017) Current understanding of endophytes: their relevance, importance, and industrial potentials. J Biotechnol Biochem 3(3):43–59
Varma A, Bakshi M, Lou B, Hartmann A, Oelmueller R (2012) Piriformospora indica: a novel plant growth-promoting mycorrhizal fungus. Agric Res 1:117–131
Vermot A, Petit-Härtlein I, Smith SM, Fieschi F (2021) NADPH oxidases (NOX): an overview from discovery, molecular mechanisms to physiology and pathology. Antioxidants 10(6):890
Vesterlund SR, Helander M, Faeth SH, Hyvönen T, Saikkonen K (2011) Environmental conditions and host plant origin override endophyte effects on invertebrate communities. Fungal Divers 47:109–118

Vicente MF, Basilio A, Cabello A, Peláez F (2003) Microbial natural products as a source of antifungals. Clin Microbiol Infect 9(1):15–32
Vishwakarma K, Kumar N, Shandilya C, Mohapatra S, Bhayana S, Varma A (2020) Revisiting plant–microbe interactions and microbial consortia application for enhancing sustainable agriculture: a review. Front Microbiol 11:560406
Watts D, Palombo EA, Jaimes Castillo A, Zaferanloo B (2023) Endophytes in agriculture: potential to improve yields and tolerances of agricultural crops. Microorganisms 11(5):1276
Xia C, Li N, Zhang Y, Li C, Zhang X, Nan Z (2018) Role of Epichloë endophytes in defense responses of cool-season grasses to pathogens: a review. Plant Dis 102(11):2061–2073
Yadav G, Meena M (2021) Bioprospecting of endophytes in medicinal plants of Thar Desert: an attractive resource for biopharmaceuticals. Biotechnol Rep 30:e00629
Yadav G, Meena M (2022) Biological control of plant diseases by endophytes. In: Shah MP, Deka D (eds) Endophytic Association: what, why and how (volume: developments in applied microbiology and biotechnology). Academic Press, Elsevier Inc., pp 119–135
Ziesche TM, Roth M (2008) Influence of environmental parameters on small-scale distribution of soil-dwelling spiders in forests: what makes the difference, tree species or microhabitat? For Ecol Manag 255(3–4):738–752

Endophytes and Microbiome: Modern Techniques in Sustainable Crop Production

5

K. Greeshma, R. Arutselvan, T. Ramya Teja, Disha Baishnabi, and Shivani Chaudhary

Abstract

Endophytes have emerged as critical components in modern sustainable crop production. These symbiotic microorganisms residing within plant tissues enhance plant growth and nutrient acquisition, improve stress tolerance, and increase resistance to pests and diseases. Modern techniques such as metagenomics, high-throughput sequencing, and genome editing are revolutionizing our understanding of endophyte-plant interactions enabling the development of innovative strategies for crop improvement. Farmers can improve crop yields and sustainability, promoting healthier plants without relying on harmful agrochemicals. The integration of microbiome management into farming practices represents a promising shift toward environmentally friendly and efficient agriculture. This approach not only supports environmental sustainability but also contributes to food security by improving crop yields and quality. As research advances, integrating endophytes and the broader plant microbiome into agricultural practices holds great potential for transforming crop production systems globally.

K. Greeshma (✉) · T. R. Teja
Department of Plant Pathology, School of Agricultural Sciences, Malla Reddy University, Hyderabad, Telangana, India

R. Arutselvan
Regional Station, ICAR- Central Tuber Crops Research Institute, Bhubaneswar, Odisha, India

D. Baishnabi
Department of Plant Pathology, School of Agricultural Sciences, Malla Reddy University, Hyderabad, Telangana, India

Department of Vegetable Science, College of Agriculture, OUAT, Bhubaneswar, Odisha, India

S. Chaudhary
Department of Engineering and Technology of Chemical Processes, Faculty of Chemistry, Wroclaw University of Science and Technology, Wroclaw, Poland

M. Meena et al. (eds.), *Multi-omics Approach to Investigate Endophyte Diversity*, Microorganisms for Sustainability 53, https://doi.org/10.1007/978-981-96-4100-0_5

Keywords

Diversity · Endophytes · Plant growth · Stress tolerance · Sustainability

5.1 Introduction

Endophytes are a diverse group of microorganisms including bacteria, fungi, and actinomycetes that reside within the internal tissues of plants without inducing visible symptoms of any diseases. These microorganisms have emerged as a pivotal component of the plant microbiome, playing significant roles in the health and productivity of their host plants. The concept of endophytism was first introduced in the early twentieth century by the German scientist Heinrich Anton de Bary, who is often credited with discovering the first endophytes in 1866 while studying the symbiotic relationship between fungi and plants.

The plant microbiome in general and its endophytic component in particular influence a wide array of plant physiological processes (Brundrett 2006) starting from nutrient uptake to growth regulation and stress response. Understanding the complex interactions between plants and their associated endophytes offers promising avenues for developing sustainable agricultural practices that reduce dependency on chemical inputs and improve crop resilience and productivity (Ryan et al. 2008; Schulz and Boyle 2005). This chapter introduces the concept of endophytes, explores their diversity and functional roles, and sets the stage for a detailed discussion on their potential applications in sustainable crop production.

Over the past few decades, research into endophytes has gained a new momentum, driven by their potential applications in sustainable agriculture. These microorganisms have been shown to contribute to plant growth promotion through various mechanisms, such as nitrogen fixation, phosphate solubilization, and production of phytohormones like indole-3-acetic acid (IAA) (Compant et al. 2005; Seth et al. 2023). Additionally, endophytes enhance plant resilience to both biotic and abiotic stresses, including drought, salinity, and pathogen attacks, by modulating the plant's physiological and biochemical responses (Rodriguez et al. 2009). For instance, bacterial endophytes like *Azospirillum brasilense* have been extensively studied for their ability to fix atmospheric nitrogen, thus reducing the reliance on chemical nitrogen fertilizers (Bashan and de-Bashan 2010). Similarly, fungal endophytes such as *Piriformospora indica* have been shown to enhance nutrient uptake and increase plant tolerance to drought and salinity by influencing root architecture and water use efficiency (Varma et al. 1999). The beneficial traits of endophytes make them a promising alternative to traditional agricultural inputs, which often involve the use of synthetic fertilizers and pesticides with significant negative impact on the environment. As a result, the integration of endophytes into agricultural practices is being explored as a means to promote sustainable crop production, improve soil health, and reduce the ecological footprint of agriculture (Saikkonen et al. 2004).

The exploration of endophytes and their roles within the plant microbiome is increasingly recognized as a critical area of research for sustainable agriculture. As the global agricultural sector faces mounting pressures from climate change, soil degradation, and the overuse of chemical inputs, the need for innovative and eco-friendly solutions has never been more urgent. Endophytes, with their ability to enhance plant growth, bolster stress tolerance, and suppress diseases, represent a promising alternative to conventional agricultural practices. Their application can potentially reduce the environmental footprint of farming while improving crop resilience and productivity (Santoyo et al. 2016; Lata et al. 2018).

Despite the significant progress made in understanding the diversity and functional capabilities of endophytes, much remains to be explored. Future research should focus on uncovering the molecular mechanisms underlying plant-endophyte interactions, which could lead to the development of more targeted bioinoculants. Additionally, long-term field studies are needed to assess the efficacy and stability of endophytic applications under diverse environmental conditions. There is also a pressing need to explore the potential of endophytes in improving the resilience of crops to emerging threats, such as new pests and pathogens exacerbated by climate change (Ameen et al. 2024). The integration of endophytes into sustainable agricultural practices offers a promising pathway toward enhancing crop productivity and resilience in an environmentally responsible manner (Fadiji and Babalola 2020).

5.2 Diversity and Role of Endophytes in Crop Plants

5.2.1 Diversity of Endophytes

Endophytes are highly diverse, with different species colonizing a wide range of plant tissues, including roots, stems, leaves, and seeds. The diversity of endophytic communities is influenced by several factors, including plant species, genotype, environmental conditions, and agricultural practices (Hardoim et al. 2008; Bhupenchandra et al. 2024). Among endophytic microbes, *Actinobacteria*, *Proteobacteria*, and *Firmicutes* are notably prominent. Commonly identified genera include *Bacillus*, *Burkholderia*, *Enterobacter*, *Herbaspirillum*, *Pantoea*, *Pseudomonas*, *Rhizobium*, and *Streptomyces*, which are prevalent in a variety of host plants. Additionally, numerous niche-specific genera have been observed, reflecting the unique microbial communities associated with different host plants (Suman et al. 2016). This rich microbial diversity highlights the potential for leveraging endophytes in sustainable agricultural practices, where they can contribute to plant health and environmental management.

For instance, *Pseudomonas fluorescens*, a well-studied bacterial endophyte, has been found in various crops and is known for its ability to promote plant growth by producing phytohormones such as indole-3-acetic acid (IAA) and by solubilizing phosphate, thereby enhancing nutrient availability (Vessey 2003). Another example is the fungal endophyte *Piriformospora indica*, which has been identified in rice (*Oryza sativa*) and other crops. This fungus not only enhances nutrient uptake but also significantly improves the plant's tolerance to drought stress by promoting root growth and increasing water uptake efficiency (Waller et al. 2005).

5.2.2 Role in Nutrient Uptake and Growth Promotion

Endophytes play a crucial role in enhancing nutrient uptake and promoting plant growth. Endophytic fungi, isolated from tomato plants, have been reported to show promising results in enhancing their health and productivity by providing a range of benefits. These include effective control of pests and pathogens, promotion of plant growth, improved availability and uptake of essential nutrients, and increased tolerance to abiotic stress (Sinno et al. 2020). These multifaceted contributions underscore the significant impact of endophytic fungi on advancing tomato plant physiology and resilience.

Azospirillum brasilense, a nitrogen-fixing bacterium, found in cereal crops such as maize and wheat, colonizes the root surfaces and rhizosphere, where it fixes atmospheric nitrogen, making it available to the plant. This process not only reduces the need for chemical nitrogen fertilizers but also promotes plant growth and increases crop yields (Hungria et al. 2010). Some endophytes are involved in addition to nitrogen fixation and phosphate solubilization, making the plant nutrients more accessible to plants. For instance, *Bacillus subtilis*, an endophyte commonly found in the rhizosphere, solubilizes phosphate and produces siderophores, which chelate iron and make it more available to the plant (Li et al. 2023). These activities contribute to enhanced plant growth and resilience. Bioformulations based on *Bacillus velezensis* have been shown to increase rice yield and provide protection against fungal pathogens, demonstrating the potential of endophytes to improve crop productivity in a sustainable manner (Cawoy et al. 2011). Similarly, plant growth-promoting rhizobacteria were found to enhance the tolerance of *Solanum tuberosum* to salinity by inducing the expression of enzymes that scavenge reactive oxygen species (ROS), such as superoxide dismutase (SOD), catalase (CAT), and peroxidase (POD), thus reducing oxidative stress (Gururani et al. 2013). Endophytes associated with tropical tuber crops were examined for their potential and benefits, revealing their role in enhancing resistance to cold shock, drought, and salinity, as well as their ability to antagonize pathogens and fix nitrogen (Nayak et al. 2017). Endophytes are also known to enhance plant tolerance to various biotic and abiotic stresses. They can either activate the plant's inherent stress response systems or produce anti-stress metabolites, such as plant growth regulators, 1-aminocyclopropane-1-carboxylate (ACC) deaminase, and various volatile and bioactive compounds (Rehman et al. 2022). The approach taken by endophytes to alleviate abiotic stress depends on the specific type of stress encountered. Common mechanisms include the synthesis and accumulation of compatible solutes, modulation of ion transport, regulation of stress-responsive genes, and generation of reactive oxygen species (ROS) scavengers (Phurailatpam and Mishra 2020).

5.3 Isolation, Identification, and Characterization of Endophytic Microbiomes

5.3.1 Isolation of Endophytes

Isolating endophytes from plant tissues requires meticulous procedures to ensure that the internal microbiome is accurately represented. The process typically begins with surface sterilization of plant tissues to remove any epiphytic (surface-dwelling) microorganisms followed by either tissue maceration and streaking in growing media or placing small, equal sterilized segments onto the growing media (Hallmann et al. 2006). For example, endophytes isolated from maize roots are often cultured on nutrient agar or potato dextrose agar to encourage the growth of specific bacterial or fungal communities (Rosenblueth and Martínez-Romero 2006). Similarly, isolation of bacterial endophytes from *Amaranthus spinosus* was performed by first collecting and washing the plant samples to remove soil particles. After air-drying, the roots, stem, and leaves were separated, weighed, and subjected to surface sterilization using a series of ethanol and sodium hypochlorite treatments before being rinsed in sterile water. The sterilized plant parts were macerated, and serial dilutions were prepared for plating on nutrient agar and potato dextrose agar to isolate bacterial and fungal endophytes, respectively. The plates were incubated at appropriate temperatures for growth, with sterility checks performed to ensure the validity of the isolation process (Sharma and Roy 2015).

5.3.2 Identification of Endophytes

Accurate identification of endophytes is critical for understanding their roles and potential applications. The isolation and characterization of endophytes have been extensively studied across various plant species, revealing their significant roles in plant health and growth promotion. Molecular techniques, particularly PCR amplification of 16S rRNA genes for bacteria and ITS regions for fungi, followed by sequencing, are widely used for this purpose (Sarsaiya et al. 2020). This method allows precise identification of endophytic strains and their classification within the broader microbial taxonomy. For instance, Gotz et al. (2006) investigated the fungal endophytes present in the roots of both transgenic potatoes and their parental lines, utilizing biomalt agar for isolation and techniques such as BOX-PCR genomic fingerprinting and ARDRA for characterization. *Bacillus pumilus*, isolated from soybean roots, was identified using 16S rRNA sequencing, which confirmed its potential as a plant growth-promoting bacterium (McInroy and Kloepper 1995).

5.3.3 Characterization of Endophytes

Characterizing endophytes involves assessing their functional capabilities, such as their ability to produce phytohormones, solubilize nutrients, or suppress pathogens.

Advanced techniques like genomics, proteomics, and metabolomics are increasingly being used to explore the metabolic pathways and gene expression profiles of endophytes. For example, *Klebsiella pneumoniae*, isolated from sugarcane, was characterized for its nitrogen-fixing ability, which was confirmed through genomic analysis revealing key genes involved in nitrogen fixation and growth promotion (Rosenblueth and Martínez-Romero 2006). Bacterial endophytes from sweet potato plants were characterized based on their abilities to fix nitrogen, produce IAA, and tolerate stress (Khan and Doty 2009). Similarly, bacterial endophytes isolated from surface-sterilized tomato seeds grown in soil samples collected from various Canadian locations were characterized based on ACC deaminase activity, IAA production, siderophore synthesis, and salt tolerance, with identification confirmed through Biolog plates and 16S rRNA gene sequencing, analyzed using the BLAST program against the National Center for Biotechnology Information database (Rashid et al. 2012).

Advancements in molecular biology and genomic technologies have revolutionized the study of endophytes and the plant microbiome. Techniques such as metagenomics and high-throughput sequencing allow researchers to analyze the microbial communities associated with plants at an unprecedented scale and depth. This has led to the identification of novel endophytes with potential applications in agriculture. Genome editing tools, such as CRISPR-Cas9, are also being used to enhance the beneficial traits of endophytes, such as improving their ability to colonize plants or increasing their production of bioactive compounds. Furthermore, synthetic biology approaches enable the engineering of microbial consortia that can be introduced into crops to provide tailored benefits, such as enhanced nutrient uptake or stress resistance (Sarsaiya et al. 2020).

5.4 Endophytes in Enhancing Biotic and Abiotic Stress Tolerance

Plant growth and development is hindered by a range of extreme conditions, including biotic and abiotic stresses. Biotic stresses include pathogens such as fungi, bacteria, viruses, and nematodes that are responsible for causing a range of plant diseases (rots, wilts, spots) (Lamichhane and Venturi 2015; Martins et al. 2018; Nazarov et al. 2020; Yadav and Meena 2021). Moreover, insects also feed on plant tissues and compete for nutrients (Oukala et al. 2021) directly causing damage to the plants, while some also serve as potential vectors for plant pathogens.

Abiotic stresses include drought, salinity, extreme temperatures, flooding, and ultraviolet radiation, which significantly restrict the overall growth and yield of crop plants (Hartman and Tringe 2019; Zhu 2016). Drought stress in particular affects nutrient diffusion and disrupts the relationship between plants and water, leading to changes in plant morphology and physiology (Vurukonda et al. 2016). Salinity stress hampers plant growth and productivity by causing specific ion toxicity and osmotic imbalances, which result in nutritional deficiencies (Singh and Roy 2016). Extreme temperature stress impairs enzyme function, disrupts cell division, and leads to the excessive denaturation of membrane proteins ultimately causing cell death (Nievola et al. 2017).

Biotic and abiotic stresses negatively impact agricultural productivity, making it essential to develop sustainable microbe-based strategies to address these challenges and improve crop yield. In this context, endophytic microorganisms offer a promising alternative providing protection against both biotic and abiotic stresses improving nutrient supply and enhancing the plant health (Omomowo and Babalola 2019; Patle et al. 2018).

5.4.1 Biotic Stress Tolerance

Endophytes are microorganisms that live within host plants and are capable of colonizing plant roots without causing any harm to the host plants. They promote plant growth and serve as natural biocontrol agents against pests and disease (Chaudhary et al. 2022a, b). Further, they enhance the plant's tolerance to various biotic and abiotic stresses. Endophytes can produce several growth hormones such as IAA and ACC deaminase, improve potassium ion uptake in plants, and lower ethylene levels offering alternative mechanisms to mitigate stress conditions in plants (Agri et al. 2022).

Both bacterial and fungal endophytes show great potential as biocontrol agents. These endophytes exhibit antagonistic activity against disease-causing phytopathogens, reducing the damage they inflict on plants. They produce a variety of bioactive antimicrobial and antiviral compounds along with antioxidants that suppress pathogens (Gouda et al. 2016). Additionally, a wide range of fungal species, particularly entomopathogenic fungi, are known to provide long-term preventive control of insect populations (Litwin et al. 2020). Several fungal strains, such as *Beauveria bassiana*, *M. robertsii*, *B. metarhizium*, *Acremonium* spp., and *Chaetomium globosum*, have also been effective in plant protection (Grabka et al. 2022).

Biocontrol using endophytes can be achieved either by directly inhibiting pathogens or by indirectly enhancing the plant's systemic resistance (Santoyo et al. 2016). Additional mechanisms include competing for ecological niches and resources, producing cell wall-degrading enzymes, activating induced systemic resistance (ISR), and disrupting pathogen quorum sensing (Rajesh and Rai 2014).

5.4.2 Abiotic Stress Tolerance

Plants can tolerate abiotic stress through two mechanisms: (i) they can mitigate the negative effects of stress by activating response systems immediately upon exposure to stress and (ii) endophytes synthesized biochemical compounds that function as anti-stress agents (Godara and Ramakrishna 2023; Schulz et al. 2002).

Stress conditions lead to a reduction in phytohormone levels, which negatively impacts plants. Hormones produced by endophytes can influence plant physiology or assist in host colonization by interacting with plant defense systems. Additionally, some microbial hormones can serve as nutrient sources or possess antibacterial properties, directly affecting nearby microbial communities (Eichmann et al. 2021). Under stress conditions, these endophytic microbes produce growth-promoting hormones like IAA and gibberellic acid, which promote plant growth and confer stress tolerance (Gupta et al. 2022). Auxins produced by microbes alter the root architecture, resulting in increased root hair production and development. IAA also plays an important role in biofilm production and growth and development of bacteria under stress conditions (Kudoyarova et al. 2019). Antioxidants produced by microbes and ABA accumulation in plants reduced the ROS levels allowing the plants to grow under stress conditions. Further, these endophytes can also influence the expression of plant genes involved in ABA biosynthesis, such as ABA2, ABA3, and NCED3 (Vargas et al. 2014). Endophytic plant growth-promoting bacteria and fungi with high level ACC (1-aminocyclopropane-1-carboxylic acid) deaminase activity are commonly found in the rhizosphere of plants under stress (Guo et al. 2020; Sagar et al. 2021).

5.5 Mechanisms of Plant-Endophyte Interactions

The interaction between plant endophytes and their host plants is a multifaceted relationship that significantly enhances plant resilience to both biotic and abiotic stresses. Endophytes, residing within plant tissues without causing harm, contribute to biotic stress resistance by producing antimicrobial compounds that inhibit or kill pathogens, such as bacteria and fungi. In terms of abiotic stress, endophytes support plants by producing osmoprotectants that maintain cellular hydration under drought conditions, regulating ion homeostasis to counteract salinity and heavy metal toxicity, and enhancing antioxidant defenses to mitigate oxidative damage. Overall, endophytes play a crucial role in improving plant health and stress tolerance through their diverse mechanisms of action (Fig. 5.1). A detailed exploration of these mechanisms and their effects is provided in the sections that follow.

5.5.1 Mechanisms of Endophyte-Mediated Tolerance to Biotic Stress

Endophytes play a critical role in enhancing plant tolerance to biotic stresses, primarily through mechanisms that involve direct antagonism of pathogens, modulation of plant defense pathways, and priming of the plant immune system. These interactions not only help plants defend against a variety of pathogens but also improve overall plant health and productivity.

I. *Induced Systemic Resistance (ISR)*

Induced systemic resistance (ISR) is a state of enhanced defensive capacity developed by plants in response to certain endophyte, which is activated by specific elicitors produced by endophytes. These pathways lead to the priming of

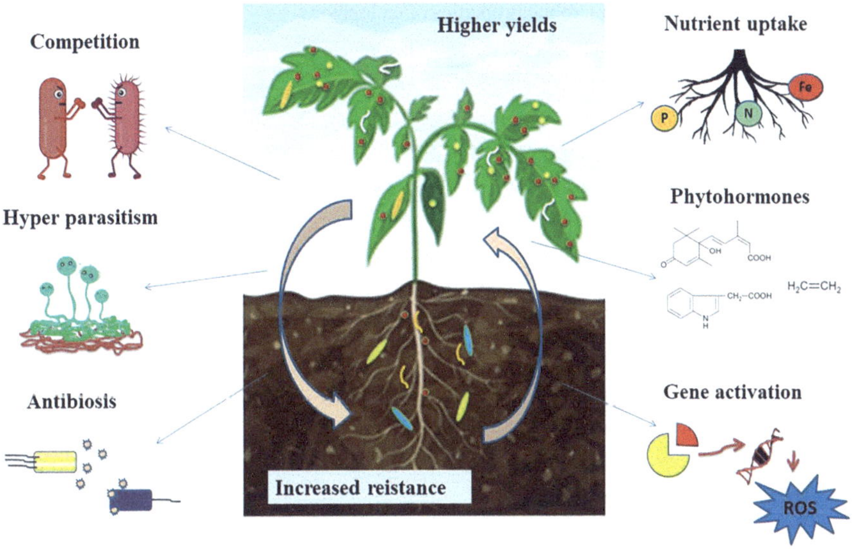

Fig. 5.1 Mechanisms of plant-endophyte interaction

the plant's immune system, allowing for a faster and stronger defensive response upon pathogen attack. The endophyte *Trichoderma harzianum* is known to trigger ISR in various plants, including tomato and cucumber, by activating the JA and ET pathways. This priming enhances the plant's resistance against pathogens like *Botrytis cinerea* and *Pseudomonas syringae* (Pieterse et al. 2014). It also provides cross-protection against a broad spectrum of pathogens, including bacteria, fungi, viruses, and nematodes. This broad-spectrum resistance is crucial for plants growing in environments with diverse pathogenic threats. *Pseudomonas fluorescens* is an endophyte that induces ISR in *Arabidopsis*, providing protection against a wide range of pathogens, including *Alternaria brassicicola* (fungus) and *Erwinia carotovora* (bacterium) (Van Wees et al. 2008).

II. *Direct Antagonism of Pathogens*

(a) *Production of Antimicrobial Compounds*

Many endophytes produce secondary metabolites with antimicrobial properties, which directly inhibit the growth of pathogens (Gunatilaka 2006). These compounds include antibiotics, lytic enzymes, and volatile organic compounds (VOCs). Ongoing research aims to discover and harness these metabolites for commercial applications. Various bioactive compounds have been investigated for their effectiveness against phytopathogens (Suryanarayanan 2013; Daguerre et al. 2016). Among the recently reviewed antimicrobial metabolites identified in endophytes are flavonoids, peptides, quinones, alkaloids, phenols, steroids, terpenoids, and polyketides (Mousa and Raizada 2013; Lugtenberg et al. 2016; Yadav and Meena 2022). Table 5.1 provides an overview of research on the antimicrobial properties of endophytes.

Table 5.1 Summary of antimicrobial properties of endophytes

Endophytes	Compounds produced	Host plant	Activity	References
Endophytic bacteria				
B. subtilis	Bacilysocin	–	Antifungal	Tamehiro et al. (2002)
B. subtilis	Terpene	*Allamanda cathartica*	Antifungal	Nithya and Muthumary (2011)
B. subtilis	Subtilin	–	Antibacterial	Singh et al. (2017)
Bacillus atrophaeus, *Bacillus mojavensis*	1,2-Benzenedicarboxylic acid and methyl ester	*Glycyrrhiza uralensis*	Antifungal	Mohamad et al. (2018)
B. licheniformis, *B. subtilis* subsp. *inaquosorum*	–	*Moringa peregrina*	Antibacterial and antifungal	Aljuraifani et al. (2019)
Endophytic fungi				
Phoma sp.	5-Hydroxyramulsin	*Cinnamomum mollissimum*	Antifungal	Santiago et al. (2012)
Phomopsis sp.	Gavodermside and clavasterols	*Aconitum carmichaelii*	Antifungal	Wu et al. (2013)
Pestalotiopsis mangiferae	4-(2,4,7-trioxabicyclo (4.1.0)heptan-3-yl)phenol	*Mangifera indica*	Antibacterial	Subban et al. (2013)
Aspergillus sp.	Fumigaclavine C and Pseurotin C	*Bauhinia guianensis*	Antibacterial	Pinheiro et al. (2013)
Endophytic actinomycetes				
Streptomyces rimosus	Tetracycline	–	Antifungal	Nelson (2001)
Streptomyces sp.	Kakadumycin A Echinomycin	*Grevillea pteridifolia*	Antibacterial	Castillo et al. (2003)
Streptomyces sp. TP-A0595	6-Prenylindole	*Allium tuberosum*	Antifungal	Singh and Dubey (2018)
Streptomyces sp. neau-D50	3-Acetonylidene-7-prenylindolin-2-one	*Glycine max*	Antifungal	Zhang et al. (2014)
Actinosynnema pretiosum	Ansamitocin	*Maytenus serrata*	Antibacterial	Siyu-Mao et al. (2013)
Streptomyces sp. TP-A0456	Cedarmycins A and B	*Aucuba japonica*	Antibacterial	Sasaki et al. (2001)

Antibiotics Endophytes like *Bacillus subtilis* produce lipopeptides (e.g., surfactin, iturin, and fengycin) that disrupt the cell membranes of pathogens, leading to their death. *Bacillus subtilis* has been shown to produce these lipopeptides that effectively inhibit the growth of fungal pathogens such as *Fusarium oxysporum*, which causes root rot in a variety of crops (Ongena and Jacques 2008).

Lytic Enzymes Endophytes can secrete enzymes like chitinases, glucanases, and proteases that degrade the cell walls of fungal pathogens, weakening them and preventing infection. The endophyte *Trichoderma viride* produces chitinase, which degrades the chitin in the cell walls of *Rhizoctonia solani*, a pathogen responsible for damping off disease in seedlings (Harman et al. 2004).

Volatile Organic Compounds (VOCs) Some endophytes produce VOCs that have antimicrobial properties, creating an inhospitable environment for pathogens. *Bacillus amyloliquefaciens* emits VOCs such as 2,3-butanediol and acetoin, which inhibit the growth of pathogens like *Sclerotinia sclerotiorum* on soybean plants (Ryu et al. 2003).

(b) *Competition for Resources*

Endophytes can outcompete pathogens for essential nutrients and ecological niches within the plant. By colonizing plant tissues, endophytes occupy the space and utilize the nutrients that would otherwise be available to pathogenic microbes, effectively reducing the likelihood of infection. *Pseudomonas putida* competes with pathogenic bacteria for iron by producing siderophores—molecules that bind iron tightly and sequester it away from pathogens—thereby inhibiting their growth (Loper and Henkels 1999).

III. *Interference with Pathogen Signaling*

(a) *Quorum Sensing Inhibition*

Quorum sensing is a communication mechanism used by many bacterial pathogens to coordinate the expression of virulence factors. Some endophytes can interfere with quorum sensing by degrading the signaling molecules (e.g., N-acyl homoserine lactones (AHLs)) or by producing molecules that block these signals, thereby preventing the pathogen from becoming virulent. The endophyte *Pseudomonas aeruginosa* produces enzymes that degrade AHLs, disrupting the quorum sensing system of *Erwinia carotovora*, a pathogen responsible for soft rot in plants (Haas and Keel 2003).

(b) *Manipulation of Plant Hormone Levels*

Pathogens often manipulate plant hormone levels to suppress plant defenses and facilitate infection. Some endophytes counteract this by modulating the same hormone pathways, thereby restoring the plant's natural defenses. The endophyte *Azospirillum brasilense* modulates the levels of auxins and cytokinins in the plant, which can counteract the effects of pathogen-produced hormones like gibberellins that promote susceptibility (Spaepen et al. 2007).

IV. *Epigenetic and Transcriptomic Changes*

Endophytes can induce significant changes in the plant's transcriptome, particularly in genes involved in defense responses. These changes often include the upregulation of genes related to secondary metabolism, which produces compounds that directly inhibit pathogens or strengthen cell walls to prevent pathogen entry. RNA-seq studies on plants treated with *Pseudomonas fluorescens* have shown an upregulation of genes involved in lignin biosynthesis, which enhances cell wall strength and provides a physical barrier against pathogen invasion (Doornbos et al. 2012).

Endophytes can also induce epigenetic modifications, such as DNA methylation and histone acetylation, which alter the expression of defense-related genes. These changes can lead to long-term resistance to pathogens. Research has shown that endophytes like *Serratia marcescens* can induce DNA methylation changes in rice, leading to the enhanced expression of defense genes and improved resistance to bacterial blight caused by *Xanthomonas oryzae* (Gómez-Gómez et al. 2019).

5.5.2 Mechanisms of Endophyte-Mediated Tolerance to Abiotic Stress

I. *Drought Tolerance*

(a) *Osmoprotectant Production*

Endophytes can improve plant drought tolerance by producing osmoprotectants—compounds that help maintain cell turgor and prevent dehydration. These include proline, glycine betaine, and trehalose, which accumulate in plant cells to protect cellular structures and proteins from damage caused by water deficit. The endophyte *Piriformospora indica* has been shown to enhance drought tolerance in plants like wheat and barley by increasing the accumulation of osmoprotectants and enhancing root growth, which improves water uptake (Varma et al. 2012).

(b) *Enhanced Root Growth and Hydraulic Conductivity*

Endophytes often stimulate root growth, leading to a more extensive root system that can access deeper soil layers where moisture is more available. This increased root biomass is crucial for maintaining water uptake during periods of drought. Endophytes such as *Azospirillum brasilense* promote root elongation and branching, thereby increasing the surface area for water absorption and improving the plant's overall water status under drought conditions (Spaepen et al. 2007).

(c) *Abscisic Acid (ABA) Signaling*

ABA is a plant hormone that plays a central role in the plant's response to drought by regulating stomatal closure to reduce water loss. Some endophytes can modulate ABA levels or enhance ABA signaling pathways, leading to more effective water conservation. The endophyte *Trichoderma harzianum* has been found to influence ABA levels in maize, leading to improved drought tolerance by optimizing stomatal closure and reducing water loss through transpiration (Vinale et al. 2008).

II. *Salinity Tolerance*

(a) *Ion Homeostasis*

Salinity stress is primarily caused by the accumulation of sodium (Na+) and chloride (Cl−) ions in the soil, which disrupts the balance of essential nutrients and leads to ion toxicity. Endophytes can help plants maintain ion homeostasis by regulating the uptake and transport of ions, particularly by reducing Na+ uptake and enhancing potassium (K+) levels, which are crucial for maintaining cellular functions. Endophytic *Bacillus* spp. have been shown to enhance the expression of genes involved in ion transport, such as those encoding salt overly sensitive (SOS) proteins, which help in the sequestration of Na+ into vacuoles, thereby reducing its toxic effects on the cytoplasm (Zhang et al. 2016).

(b) *Reactive Oxygen Species (ROS) Scavenging*

Salinity stress often leads to the overproduction of ROS, which can cause oxidative damage to cellular components like lipids, proteins, and DNA. Endophytes mitigate this by enhancing the plant's antioxidant defense system, which includes enzymes such as superoxide dismutase (SOD), catalase, and glutathione peroxidase. The endophyte *Pseudomonas fluorescens* produces enzymes that neutralize ROS, thereby protecting rice plants from oxidative stress caused by high salinity (Jha et al. 2011).

III. *Heavy Metal Tolerance*

(a) *Metal Chelation and Sequestration*

Endophytes can produce metal-chelating compounds, such as siderophores and organic acids, which bind heavy metals and reduce their bioavailability, thereby protecting the plant from toxic effects. These compounds can also facilitate the sequestration of metals into less harmful forms or into specific compartments within the plant, such as vacuoles. *Pseudomonas putida* produces siderophores that chelate cadmium, reducing its toxicity and allowing the plant to grow in contaminated soils (Rajkumar et al. 2009).

(b) *Phytochelatin and Metallothionein Synthesis*

Some endophytes induce the production of phytochelatins and metallothioneins in plants, which are small, cysteine-rich proteins that bind heavy metals and aid in their detoxification and sequestration within the plant tissues. *Serratia marcescens* endophytes have been shown to increase the synthesis of phytochelatins in *Brassica* plants, improving their tolerance to heavy metal stress (Singh et al. 2010).

IV. *Modulation of Plant Hormones in Stress Tolerance*

(a) *Auxin Production*

Endophytes can produce auxins, particularly indole-3-acetic acid (IAA), which promotes root elongation and enhances root architecture. This improved root system allows for better water and nutrient uptake. *Azospirillum* spp. are known to produce auxins that significantly enhance root growth and development in host plants, contributing to improved water and nutrient acquisition under stress (Spaepen et al. 2007).

(b) *Cytokinin and Gibberellin Modulation*

Cytokinins and gibberellins are plant hormones involved in cell division, growth, and development. Endophytes can modulate the levels of these hormones to help plants maintain growth under stress conditions. *Bacillus licheniformis* endophytes have been shown to increase cytokinin levels in host plants, promoting cell division and growth even under suboptimal environmental conditions (Kohler et al. 2012).

5.6 Integrating Endophytes into Sustainable Agricultural Practices

The practical application of endophytes in agriculture is gaining momentum as research continues to uncover their benefits. On inoculating the seed or potting mix with endophytic bacteria has reported substantial reduction of disease causing microorganisms (Kloepper and Ryu 2006). Fungal endophytes on the other hand are reported to have an ability to produce a wide range of secondary metabolites and bioactive compounds, offering a viable alternative source for secondary plant metabolites and natural pharmaceuticals. Over the past 20 years, bioremediation has emerged as an effective approach for managing large-scale pollution. Additionally, endophytic fungi that generate ligninolytic enzymes are of great interest for their potential applications in lignocellulosic biorefineries (Rana et al. 2019). Endophytes in wheat have been reported to enhance growth and provide resistance to various stresses, including salinity and drought. They contribute to improved physiological performance and stress tolerance (Ahlawat et al. 2022; Kumar et al. 2022). Not just in cereals but endophytes present in cucumber plants have shown to possess the ability to enhance growth and improve resistance to fungal pathogens, thereby boosting overall plant health and yield (Sathe et al. 2019; Chandran et al. 2021).

5.6.1 Holistic Approach

Current industrial methods for combating stressors rely heavily on fertilizers, herbicides, and pesticides, which raise concerns about their impact on consumer health and long-term soil quality (Le Cocq et al. 2017). In contrast, natural approaches offer a safer and more consumer-friendly alternative. Endophytes, including both bacteria and fungi, have gained recognition as powerful natural resources. Their use in inoculation has been shown to enhance plant stress tolerance and reduce vulnerability to diseases and pests (Watts et al. 2023). However, when exposed to uncontrollable variables, the protective effects of endophytes observed in laboratory conditions may become less effective in field environments (Le Cocq et al. 2017; Serfling et al. 2007).

Introducing beneficial endophytes in field applications is more challenging than theoretical models suggested. The intricate relationships between soil biosystems, including rhizospheric microbes, actinomycetes, and plants, create significant obstacles for successfully introducing any new microorganisms (Khaksar et al. 2016).

Inoculation methods are crucial for plant-microbe interactions that lack vertical transmission, which is particularly relevant for many crop plants like wheat and grains. Since these crops do not naturally transmit microbes, implementing large-scale, farm-based applications of endophyte-assisted processes presents significant challenges (Robinson et al. 2016). The most obvious approach involves applying inoculants directly to the soil or using them as seed dressings. This method has shown success in crops like sugarcane (Silva et al. 2012). Seedling colonization is a complex and competitive process, often hindered by other microorganisms and the presence of inhibitors within the seedlings (Gaiero et al. 2013). This competition is especially intense during the first 24 hours after sowing, a critical period where early colonization by endophytes significantly enhances their long-term presence in mature plants (Nelson 2004; Robinson et al. 2016).

While each approach has its pros and cons, endophytic treatment of plants holds promising potential as a future strategy for enhancing food nutrition. In theory, bio-prospecting for endophytes could lead to innovative methods that naturally boost food nutrition from within, eliminating the need for potentially harmful chemical treatments or GMOs. Despite the success of some well-known endophyte-plant relationships (Hardoim et al. 2015), the use of endophytes to address threats to plant health remains uncommon in conventional agriculture. As the agrochemicals continue to dominate as the preferred solution, overshadowing these alternative approaches (Le Cocq et al. 2017). However, before these methods can be considered safe for human consumption, significant knowledge gaps must be addressed, and the effects of endophytes should be thoroughly understood.

5.6.2 Conclusion

Endophytes offer a powerful tool for advancing sustainable agricultural practices, and the integration of endophytes and the broader plant microbiome into sustainable crop production represents a promising shift toward more resilient and environmentally friendly agricultural practices. Modern techniques of genome editing have greatly enhanced our understanding of these microorganisms and their interactions with plants, paving the way for innovative crop improvement strategies. By exploiting the natural benefits of endophytes, we can reduce the dependency on chemical inputs, improve soil health, and boost crop productivity. As research continues to advance, the adoption of these biological approaches in agriculture will be critical in addressing the challenges of food security and environmental sustainability in the face of a growing global population. However, realizing their full potential will require overcoming challenges related to field performance, regulation, and scalability. With continued research and innovation, endophytes could play a central role in the future of sustainable agriculture.

References

Agri U, Chaudhary P, Sharma A, Kukreti B (2022) Physiological response of maize plants and its rhizospheric microbiome under the influence of potential bioinoculants and nanochitosan. Plant Soil 474:451–468. https://doi.org/10.1007/s11104-022-05351-2

Ahlawat OP, Yadav D, Kashyap PL, Khippal A, Singh G (2022) Wheat endophytes and their potential role in managing abiotic stress under changing climate. J Appl Microbiol 132(4):2501–2520. https://doi.org/10.1111/jam.15375

Aljuraifani A, Aldosary S, Ababutain I (2019) *In vitro* antimicrobial activity of endophytes, isolated from *Moringa peregrina* growing in Eastern region of Saudi Arabia. Nat Acad Sci Lett 42:75–80. https://doi.org/10.1007/s40009-018-0739-6

Ameen M, Mahmood A, Sahkoor A, Zia MA, Ullah MS (2024) The role of endophytes to combat abiotic stress in plants. Plant Stress 12:100435. https://doi.org/10.1016/j.stress.2024.100435

Bashan Y, de-Bashan LE (2010) How the plant growth-promoting bacterium *Azospirillum* promotes plant growth—a critical assessment. Adv Agron 108:77–136. https://doi.org/10.1016/S0065-2113(10)08002-8

Bhupenchandra I, Chongtham SK, Devi AG, Dutta P, Sahoo MR, Mohanty S, Kumar S, Choudhary AK, Devi EL, Sinyorita S, Devi SH, Mahanta M, Kumari A, Devi HL, Josmee RK, Pusparani A, Pathaw N, Gupta S, Meena M, Janmeda P, Seth CS, Arora A, Sahoo A, Swapnil P (2024) Unlocking the potential of arbuscular mycorrhizal fungi: exploring role in plant growth promotion, nutrient uptake mechanisms, biotic stress alleviation, and sustaining agricultural production systems. J Plant Growth Regul 43(9):1–39. https://doi.org/10.1007/s00344-024-11467-9

Brundrett M (2006) Understanding the roles of multifunctional Mycorrhizal and Endophytic fungi. In: Microbial root endophytes. Soil biology. https://doi.org/10.1007/3-540-33526-9_16

Castillo U, Harper JK, Strobel GA, Sears J, Alesi K, Ford E (2003) Kakadumycins, novel antibiotics from *Streptomyces* sp. NRRL 30566, an endophyte of *Grevillea pteridifolia*. FEMS Microbiol Lett 224:183–190

Cawoy H, Bettiol W, Fickers P, Ongena M (2011) Bacillus-based biological control of plant diseases. In: Pesticides in the modern world - pesticides use and management

Chandran H, Meena M, Swapnil P (2021) Plant growth-promoting rhizobacteria as a green alternative for sustainable agriculture. Sustain For 13(19):10986. https://doi.org/10.3390/su131910986

Chaudhary P, Agri U, Chaudhary A, Kumar A, Kumar G (2022a) Endophytes and their potential in biotic stress management and crop production. Front Microbiol 13:933017. https://doi.org/10.3389/fmicb.2022.933017

Chaudhary P, Chaudhary A, Bhatt P, Kumar G, Khatoon H, Rani A (2022b) Assessment of soil health indicators under the influence of nanocompounds and *Bacillus* spp. in field condition. Front Environ Sci 9:769871. https://doi.org/10.3389/fenvs.2021.769871

Compant S, Duffy B, Nowak J, Clement C, Barka EA (2005) Use of plant growth-promoting bacteria for biocontrol of plant diseases: principles, mechanisms of action, and future prospects. Appl Environ Microbiol 71(9):4951–4959. https://doi.org/10.1128/AEM.71.9.4951-4959.2005

Daguerre Y, Edel-Hermann V, Steinberg C (2016) Fungal genes and metabolites associated with the biocontrol of soil-borne plant pathogenic fungi. In: Fungal Metabolites. Springer International Publishing, p 1001. https://doi.org/10.1007/978-3-319-19456-1_27-1

Doornbos RF et al (2012) Impact of root exudates and plant defense signaling on bacterial communities in the rhizosphere. A review. Agron Sustain Dev 32(1):227–243

Eichmann R, Richards L, Schafer P (2021) Hormones as go-betweens in plant microbiome assembly. Plant J 105:518–541

Fadiji AE, Babalola OO (2020) Exploring the potentialities of beneficial endophytes for improved plant growth. Saudi J Biol Sci 27(12):3622–3633. https://doi.org/10.1016/j.sjbs.2020.08.002

Gaiero JR, McCall CA, Thompson KA, Day NJ, Best AS, Dunfield KE (2013) Inside the root microbiome: bacterial root endophytes and plant growth promotion. Am J Bot 100:1738–1750

Godara H, Ramakrishna W (2023) Endophytes as nature's gift to plants to combat abiotic stresses. Lett Appl Microbiol 76(2):ovac067

Gómez-Gómez E et al (2019) Epigenetic changes in rice plants treated with a plant growth-promoting rhizobacterium. Plant Physiol Biochem 139:1–10

Gotz M, Nirenberg H, Krause S, Wolters H, Draeger S, Buchner A, Lottman J, Berg G, Smalla K (2006) Fungal endophytes in potato roots studied by traditional isolation and cultivation-independent DNA-based methods. FEMS Microbiol Ecol 58(3):404–413. https://doi.org/10.1111/j.1574-6941.2006.00163.x

Gouda S, Das G, Sen SK, Shin HS, Patra JK (2016) Endophytes: a treasure house of bioactive compounds of medicinal importance. Front Microbiol 7:1538. https://doi.org/10.3389/fmicb.2016.01538

Grabka R, d'Entremont TW, Adams SJ, Walker AK, Tanney JB, Abbasi PA (2022) Fungal endophytes and their role in agricultural plant protection against pests and pathogens. Plan Theory 11:384

Gunatilaka AL (2006) Natural products from plant-associated microorganisms: distribution, structural diversity, bioactivity, and implications of their occurrence. J Nat Prod 69:509–526

Guo DJ, Singh RK, Singh P (2020) Complete genome sequence of *Enterobacter roggenkampii ED5*, a nitrogen fixing plant growth promoting endophytic bacterium with biocontrol and stress tolerance properties, isolated from sugarcane root. Front Microbiol 11:2270

Gupta A, Mishra R, Rai S, Bano A, Pathak N, Fujita M, ... & Hasanuzzaman M (2022) Mechanistic insights of plant growth promoting bacteria mediated drought and salt stress tolerance in plants for sustainable agriculture. International Journal of Molecular Sciences, 23(7):3741

Gururani MA, Upadhyaya CP, Baskar V, Venkatesh J, Nookaraju A, Park SW (2013) Plant growth-promoting rhizobacteria enhance abiotic stress tolerance in *Solanum tuberosum* through inducing changes in the expression of ROS-scavenging enzymes and improved photosynthetic performance. J Plant Growth Regul 32(2):245–258

Haas D, Keel C (2003) Regulation of antibiotic production in root-colonizing *Pseudomonas* spp. and relevance for biological control of plant disease. Annu Rev Phytopathol 41(1):117–153

Hallmann J, Berg G, Schulz B (2006) Isolation procedures for endophytic microorganisms. In Microbial root endophytes (pp. 299–319). Berlin, Heidelberg: Springer Berlin Heidelberg.

Hardoim PR, van Overbeek LS, van Elsas JD (2008) Properties of bacterial endophytes and their proposed role in plant growth. Trends Microbiol 16(10):463–471. https://doi.org/10.1016/j.tim.2008.07.008

Hardoim PR, Van Overbeek LS, Berg G, Pirttila AM, Compant S, Campisano A, Döring M, Sessitsch A (2015) The hidden world within plants: ecological and evolutionary considerations for defining functioning of microbial endophytes. Microbiol Mol Biol Rev 79:293–320

Harman GE, Howell CR, Viterbo A, Chet I, Lorito M (2004) Trichoderma species—opportunistic, avirulent plant symbionts. Nat Rev Microbiol 2(1):43–56. https://doi.org/10.1038/nrmicro797

Hartman K, Tringe SG (2019) Interactions between plants and soil shaping the root microbiome under abiotic stress. Biochem J 476(19):2705–2724

Hungria M, Campo RJ, Souza EM, Pedrosa FO (2010) Inoculation with selected strains of *Azospirillum brasilense* and *A. lipoferum* improves yields of maize and wheat in Brazil. Plant Soil 331(1–2):413–425. https://doi.org/10.1007/s11104-009-0262-0

Jha Y et al (2011) Salinity stress tolerance mechanisms in plants: an overview. Plant Signal Behav 6(10):1467–1480

Khaksar G, Treesubsuntorn C, Thiravetyan P (2016) Effect of endophytic *Bacillus cereus* ERBP inoculation into non-native host: potentials and challenges for airborne formaldehyde removal. Plant Physiol Biochem 107:326–336

Khan Z, Doty SL (2009) Characterization of bacterial endophytes of sweet potato plants. Plant Soil 322(1):197–207

Kloepper JW, Ryu CM (2006) Bacterial endophytes as elicitors of induced systemic resistance. In: Schulz BJE, Boyle CJC, Sieber TN (eds) Microbial root endophytes. Soil biology, vol 9. Springer, Berlin, Heidelberg. https://doi.org/10.1007/3-540-33526-9_3

Kohler A et al (2012) Mycorrhizal plant interaction: the role of the fungal partners. Front Plant Sci 3:182
Kudoyarova G, Arkhipova T, Korshunova T (2019) Phytohormone mediation of interactions between plants and non-symbiotic growth promoting bacteria under edaphic stresses. Front Plant Sci 10:1638
Kumar R, Swapnil P, Meena M, Selpair S, Yadav BG (2022) Plant growth-promoting rhizobacteria (PGPR): approaches to alleviate abiotic stresses for enhancement of growth and development of medicinal plants. Sustain For 14(23):15514. https://doi.org/10.3390/su142315514
Lamichhane JR, Venturi V (2015) Synergisms between plant disease complexes: a growing trend. Front Plant Sci 6:385
Lata R, Chowdhury S, Gond SK, White JF (2018) Induction of abiotic stress tolerance in plants by endophytic microbes. Lett Appl Microbiol 66(4):268–276. https://doi.org/10.1111/lam.12855
Le Cocq K, Gurr SJ, Hirsch PR, Mauchline TH (2017) Exploitation of endophytes for sustainable agricultural intensification. Mol Plant Pathol 18(3):469–473
Li H, He Y, Yan Z, Yang Z, Tian F, Liu X, Wu Z (2023) Insight into microbial mechanisms for the improvement of spent mushroom substrate composting efficiency driven by phosphate-solubilizing *Bacillus subtilis*. J Environ Manag 336:117561
Litwin A, Nowak M, Różalska S (2020) Entomopathogenic fungi: unconventional applications. Rev Environ Sci Biotechnol 19:23–42. https://doi.org/10.1007/s11157-020-09525-1
Loper JE, Henkels MD (1999) Rhizosphere biology: interactions between plants and beneficial microorganisms. Plant Soil 212(1):71–85
Lugtenberg BJ, Caradus JR, Johnson LJ (2016) Fungal endophytes for sustainable crop production. FEMS Microbiol Ecol 92:1–17. https://doi.org/10.1093/femsec/fiw194
Martins PM, Merfa MV, Takita MA, De Souza AA (2018) Persistence in phytopathogenic bacteria: do we know enough? Front Microbiol 9:1099
McInroy JA, Kloepper JW (1995) Survey of indigenous bacterial endophytes from cotton and sweet corn. Plant Soil 173(2):337–342. https://doi.org/10.1007/BF00011472
Mohamad A, Abdalla O, Li L, Ma J, Hatab SR, Xu L (2018) Evaluation of the antimicrobial activity of endophytic bacterial populations from Chinese traditional medicinal plant licorice and characterization of the bioactive secondary metabolites produced by *Bacillus atrophaeus* against *Verticillium dahliae*. Front Microbiol 9:924. https://doi.org/10.3389/fmicb.2018.00924
Mousa WK, Raizada MN (2013) The diversity of anti-microbial secondary metabolites produced by fungal endophytes: an interdisciplinary perspective. Frontiers in Microbiology 4:65
Nayak S, Mukherjee A, Samanta S (2017) Endophytic microorganisms of tropical tuber crops: potential and perspectives. J Appl Nat Sci 9(2):860–865
Nazarov PA, Baleev DN, Ivanova MI, Sokolova LM, Karakozova MV (2020) Infectious plant diseases: etiology, current status, problems and prospects in plant protection. Acta Nat 12(3):46
Nelson ML (2001) The chemistry and cellular biology of the tetracyclines. In: Nelson M, Hillen W, Greenwald AA (eds) Tetracyclines in biology, chemistry and medicine. Birkhäuser, Basel, pp 3–63
Nelson EB (2004) Microbial dynamics and interactions in the spermosphere. Annu Rev Phytopathol 42:271–309
Nievola CC, Carvalho CP, Carvalho V, Rodrigues E (2017) Rapid responses of plants to temperature changes. Temperature 4(4):371–405
Nithya K, Muthumary J (2011) Bioactive metabolite produced by *Phomopsis* sp., an endophytic fungus in *Allamanda cathartica* Linn. Recent Res Sci Technol 3:44–48
Omomowo OI, Babalola OO (2019) Bacterial and fungal endophytes: tiny giants with immense beneficial potential for plant growth and sustainable agricultural productivity. Microorganisms 7(11):481
Ongena M, Jacques P (2008) Bacillus lipopeptides: versatile weapons for plant disease biocontrol. Trends Microbiol 16(3):115–125
Oukala N, Aissat K, Pastor V (2021) Bacterial endophytes: the hidden actor in plant immune responses against biotic stress. Plan Theory 10(5):1012

Patle PN, Navnage NP, Ramteke PR (2018) Endophytes in plant system: roles in growth promotion, mechanism and their potentiality in achieving agriculture sustainability. Int J Chem Stud 6:270–274

Phurailatpam L, Mishra S (2020) Role of plant endophytes in conferring abiotic stress tolerance. In: Hasanuzzaman M (ed) Plant ecophysiology and adaptation under climate change: mechanisms and perspectives II. Springer, Singapore. https://doi.org/10.1007/978-981-15-2172-0_22

Pieterse CM et al (2014) Induced systemic resistance by beneficial microbes. Annu Rev Phytopathol 52:347–375

Pinheiro EAA, Carvalho JM, Dos Santos DCP, Feitosa ADO, Marinho PSB, Guilhon GMSP (2013) Antibacterial activity of alkaloids produced by endophytic fungus *Aspergillus* sp. EJC08 isolated from medical plant *Bauhinia guianensis*. Nat Prod Res 27:1633–1638. https://doi.org/10.1080/14786419.2012.750316

Rajesh PS, Rai RV (2014) Quorum quenching activity in cell-free lysate of endophytic bacteria isolated from *Pterocarpus santalinus* Linn., and its effect on quorum sensing regulated biofilm in *Pseudomonas aeruginosa* PAO1. Microbiol Res 169:561–569. https://doi.org/10.1016/j.micres.2013.10.005

Rajkumar M et al (2009) Metals in environmental context: soil microbial–metal interactions and impact on plant growth and productivity. Ecotoxicol Environ Saf 72(1):179–189

Rana KL et al (2019) Endophytic fungi: biodiversity, ecological significance, and potential industrial applications. In: Yadav A, Mishra S, Singh S, Gupta A (eds) Recent advancement in white biotechnology through fungi. Fungal biology. Springer, Cham. https://doi.org/10.1007/978-3-030-10480-1_1

Rashid S, Charles TC, Glick BR (2012) Isolation and characterization of new plant growth-promoting bacterial endophytes. Appl Soil Ecol 61:217–224

Rehman B, Javed J, Rauf M, Khan SA, Arif M, Hamayun M, Gul H, Khilji SA, Sajid ZA, Kim W-C, Lee I-J (2022) ACC deaminase-producing endophytic fungal consortia promote drought stress tolerance in *Moringa oleifera* by mitigating ethylene and H2O2. Front Plant Sci 13:967672. https://doi.org/10.3389/fpls.2022.967672

Robinson RJ, Fraaije BA, Clark IM, Jackson RW, Hirsch PR, Mauchline TH (2016) Wheat seed embryo excision enables the creation of axenic seedlings and Koch's postulates testing of putative bacterial endophytes. Sci Rep 6:25581

Rodriguez RJ, White Jr, JF, Arnold AE, & Redman ARA (2009) Fungal endophytes: diversity and functional roles. New phytologist, 182(2):314–330

Rosenblueth M, Martínez-Romero E (2006) Bacterial endophytes and their interactions with hosts. Mol Plant-Microbe Interact 19(8):827–837. https://doi.org/10.1094/MPMI-19-0827

Ryan RP, Germaine K, Franks A, Ryan DJ, Dowling DN (2008) Bacterial endophytes: recent developments and applications. FEMS Microbiol Lett 278:1–9

Ryu CM et al (2003) Bacterial volatiles promote growth in Arabidopsis. Proc Natl Acad Sci 100(8):4927–4932

Sagar A, Rathore P, Ramteke PW (2021) Plant growth promoting rhizobacteria, arbuscular mycorrhizal fungi and their synergistic interactions to counteract the negative effects of saline soil on agriculture: key macromolecules and mechanisms. Microorganisms 9:1491

Saikkonen K, Faeth SH, Helander M, Sullivan TJ (2004) Fungal endophytes: a continuum of interactions with host plants. Annu Rev Ecol Evol Syst 35(1):319–343

Santiago C, Fitchett C, Munro MH, Jalil J, Santhanam J (2012) Cytotoxic and antifungal activities of 5-hydroxyramulosin, a compound produced by an endophytic fungus isolated from *Cinnamomum mollisimum*. Evid Based Compl Altern Med Stud 2012:689310. https://doi.org/10.1155/2012/689310

Santoyo G, Moreno-Hagelsieb G, Del Carmen Orozco-Mosqueda M, Glick BR (2016) Plant growth-promoting bacterial endophytes. Microbiol Res 183:92–99. https://doi.org/10.1016/j.micres.2015.11.008

Sarsaiya S, Jain A, Jia Q, Fan X, Shu F, Chen Z, Zhou Q, Shi J, Chen J (2020) Molecular identification of endophytic fungi and their pathogenicity evaluation against *Dendrobium nobile* and *Dendrobium officinale*. Int J Mol Sci 21(1):316. https://doi.org/10.3390/ijms21010316

Sasaki T, Igarashi Y, Saito N, Furumai T (2001) Cedarmycins A and B, new antimicrobial antibiotics from *Streptomyces* sp. TP-A0456. J Antibiot 54:567–572. https://doi.org/10.7164/antibiotics.54.567

Sathe P, Krishnamoorthy S, Paramandham L, Kalaiselvi T (2019) Exploration of bacterial endophytes in cucumber (*Cucumis sativus* L.). Madras Agric J 106:406. https://doi.org/10.29321/MAJ.2019.000283

Schulz B, Boyle C (2005) The endophytic continuum. Mycol Res 109(6):661–686. https://doi.org/10.1017/S095375620500273X

Schulz B, Boyle C, Draeger S, Reommert AK, Krohn K (2002) Endophytic fungi: a source of novel biologically active secondary metabolites. Mycol Res 109:996–1004

Serfling A, Wirsel SG, Lind V, Deising HB (2007) Performance of the biocontrol fungus *Piriformospora indica* on wheat under greenhouse and field conditions. Phytopathology 97:523–531

Seth K, Vyas P, Deora S, Gupta AK, Meena M, Swapnil P, Harish. (2023) Understanding plant-plant growth-promoting rhizobacteria (PGPR) interactions for inducing plant defense. In: Swapnil P, Meena M, Harish MA, Vijayalakshmi S, Zehra A (eds) Plant-microbe interaction—recent advances in molecular and biochemical approaches, vol 2. Academic Press, Elsevier, pp 201–226. https://doi.org/10.1016/B978-0-323-91876-3.00010-5

Sharma S, Roy S (2015) Isolation and identification of a novel endophyte from a plant *Amaranthus spinosus*. Int J Curr Microbiol App Sci 4(2):785–798

Silva MF, Antônio CS, Oliveira PJ, Xavier GR, Rumjanek NG, Soares LHB, Reis VM (2012) Survival of endophytic bacteria in polymer-based inoculants and efficiency of their application to sugarcane. Plant Soil 356:231–243

Singh R, Dubey AK (2018) Diversity and applications of endophytic actinobacteria of plants in special and other ecological niches. Front Microbiol 9:1767. https://doi.org/10.3389/fmicb.2018.01767

Singh D, Roy BK (2016) Salt stress affects mitotic activity and modulates antioxidant systems in onion roots. Rev Bras Bot 39:67–76

Singh RP et al (2010) Phytochelatin synthesis and the role of endophytes in heavy metal tolerance. J Hazard Mater 181(1–3):384–391

Singh M, Kumar A, Singh R, Pandey KD (2017) Endophytic bacteria: a new source of bioactive compounds. 3 Biotech 7:315. https://doi.org/10.1007/s13205-017-0942-z

Sinno M, Ranesi M, Gioia L, d'Errico G, Woo SL (2020) Endophytic fungi of tomato and their potential applications for crop improvement. Agriculture 10:587. https://doi.org/10.3390/agriculture10120587

Siyu-Mao H-C, Li-Chen C-W, Wei-Jia, Xiaoming-Chen (2013) Two novel ansamitocin analogs from *Actinosynnema pretiosum*. Nat Prod Res 27:1532–1536. https://doi.org/10.1080/14786419.2012.733388

Spaepen S, Vanderleyden J, & Remans R (2007) Indole-3-acetic acid in microbial and microorganism-plant signaling. FEMS microbiology reviews 31(4):425–448

Subban K, Subramani R, Johnpaul M (2013) A novel antibacterial and antifungal phenolic compound from the endophytic fungus *Pestalotiopsis mangiferae*. Nat Prod Res 27:1445–1449

Suman A, Yadav AN, Verma P (2016) Endophytic microbes in crops: diversity and beneficial impact for sustainable agriculture. In: Singh D, Singh H, Prabha R (eds) Microbial inoculants in sustainable agricultural productivity. Springer, New Delhi. https://doi.org/10.1007/978-81-322-2647-5_7

Suryanarayanan TS (2013) Endophyte research: going beyond isolation and metabolite documentation. Fungal Ecol 6:561–568. https://doi.org/10.1016/j.funeco.2013.09.007

Tamehiro N, Okamoto-Hosoya Y, Okamoto S, Ubukata M, Hamada M, Naganawa H et al (2002) Bacilysocin, a novel phospholipid antibiotic produced by *Bacillus subtilis* 168. Antimicrob Agents Chemother 46:315–320. https://doi.org/10.1128/AAC.46.2.315-320.2002

Van Wees SC et al (2008) Differential activation of systemic acquired resistance and induced systemic resistance in Arabidopsis. Mol Plant-Microbe Interact 21(2):153–160

Vargas L, Brigida ABS, Mota Filho JP (2014) Drought tolerance conferred to sugarcane by association with *Gluconacetobacter diazotrophicus*: a transcriptomic view of hormone pathways. PLoS One 9:e114744

Varma A, Verma S, Sudha SN, Bütehorn B, Franken P (1999) *Piriformospora indica*, a cultivable plant-growth-promoting root endophyte with similarities to arbuscular mycorrhizal fungi. Appl Environ Microbiol 65(6):2741–2744. https://doi.org/10.1128/AEM.65.6.2741-2744.1999

Varma A et al (2012) Piriformospora indica: a cultivable plant-growth-promoting root endophyte with substantial impact on growth, yield, and disease resistance in crop plants. Physiol Mol Biol Plants 18(3):207–220

Vessey JK (2003) Plant Growth Promoting Rhizobacteria as Biofertilizers. Plant and Soil, 255, 571–586. https://doi.org/10.1023/A:1026037216893

Vinale F, Sivasithamparam K, Ghisalberti EL, Marra R, Woo SL, & Lorito M (2008) Trichoderma–plant–pathogen interactions. Soil Biology and Biochemistry, 40(1):1–10

Vurukonda SSKP, Vardharajula S, Shrivastava M, Ali SKZ (2016) Enhancement of drought stress tolerance in crops by plant growth promoting rhizobacteria. Microbiol Res 184:13–24

Waller F, Achatz B, Baltruschat H, Fodor J, Becker K, Fischer M, Kogel KH (2005) The endophytic fungus *Piriformospora indica* reprograms barley to salt-stress tolerance, disease resistance, and higher yield. Proc Natl Acad Sci 102(38):13386–13391. https://doi.org/10.1073/pnas.0504423102

Watts D, Palombo EA, Jaimes Castillo A, Zaferanloo B (2023) Endophytes in agriculture: potential to improve yields and tolerances of agricultural crops. Microorganisms 11(5):1276

Wu SH, Huang R, Miao CP, Chen YW (2013) Two new steroids from an endophytic fungus *Phomopsis* sp. Chem Biodivers 10:1276–1283. https://doi.org/10.1002/cbdv.201200415

Yadav G, Meena M (2021) Bioprospecting of endophytes in medicinal plants of Thar Desert: an attractive resource for biopharmaceuticals. Biotechnol Rep 30:e00629. https://doi.org/10.1016/j.btre.2021.e00629

Yadav G, Meena M (2022) Biological control of plant diseases by endophytes. In: Shah MP, Deka D (eds) Endophytic association: what, why and how (volume: developments in applied microbiology and biotechnology). Academic Press, Elsevier Inc., pp 119–135. https://doi.org/10.1016/B978-0-323-91245-7.00007-9

Zhang J, Wang J-D, Liu C-X, Yuan J-H, Wang X-J, Xiang W-S (2014) A new prenylated indole derivative from endophytic actinobacteria *Streptomyces* sp. neau-D50. Nat Prod Res 28:431–437. https://doi.org/10.1080/14786419.2013.871546

Zhang H et al (2016) Plant growth-promoting bacteria alleviate abiotic stress in crops. Front Microbiol 7:1487

Zhu JK (2016) Abiotic stress signaling and responses in plants. Cell 167:313–324

Unlocking Endophytic Mysteries by Leveraging Multi-omics Technologies

6

Sonia Kumari Shishodia, Raman Thakur, Jata Shankar, Ravinder Singh, and P. De LosRios-Escalante

Abstract

Endophytes serve as valuable reservoirs of bioactive compounds with a wide-range of applications. Investigating these compounds not only enhances our comprehension of ecological dynamics but also holds immense potential for developing innovative products with agricultural, medicinal, and industrial relevance. The ongoing exploration of endophyte diversity, coupled with understanding the ecological functions of bioactive compounds, presents avenues for novel discoveries and applications. Omics methodologies, which include genomics, transcriptomics, proteomics, and metabolomics, are crucial in revealing the characteristics and potential applications of endophytes. By incorporating omics techniques, a broader range of molecular information can be captured. Multi-omics methodologies are essential for investigating the symbiotic relationship. These omics approaches are instrumental in deciphering the intricate interactions between endophytes and their host plants, offering valuable insights into the molecular mechanisms underpinning these associations and the possible applica-

S. K. Shishodia (✉) · R. Singh
University Institute of Biotechnology, Chandigarh University, Mohali, Punjab, India

R. Thakur
Department of Medical Laboratory Science, Lovely Professional University, Jalandhar, Punjab, India

J. Shankar
Genomic Laboratory, Department of Biotechnology and Bioinformatics, Jaypee University of Information Technology, Waknaghat Solan, Himachal Pradesh, India

P. De LosRios-Escalante
Departamento de Ciencias Biológicas y Químicas, Facultad de Recursos Naturales, Universidad Católica de Temuco, Casilla, Temuco, Chile

Núcleo de Estudios Ambientales UC Temuco, Casilla, Temuco, Chile

M. Meena et al. (eds.), *Multi-omics Approach to Investigate Endophyte Diversity*, Microorganisms for Sustainability 53, https://doi.org/10.1007/978-981-96-4100-0_6

tions. This book chapter offers a comprehensive overview on how major "omics" techniques contribute to the advancement of our knowledge about endophytes e.g., bacteria (*Pseudomonas, Bacillus, Burkholderia, Stenotrophomonas, Micrococcus, Pantoea, Microbacterium,* etc) and fungi (*Aspergillus Alternaria, Fusarium, Trichoderma, Penicillium, Piriformospora,* etc) and their various roles e.g., stress-tolerance genes to withstand plant-induced oxidative stress; release of bioactive secondary metabolites (e.g., antibiotics, antifungals); and activation of secretion systems (T3SS, T6SS) for host colonization and immune evasion . Moreover, the advancement of meta-omics technologies is revolutionizing our understanding of the microbial communities linked to plants, uncovering the mutual evolution of host plants and endophytes.

Keywords

Transcriptomics · Meta-omics · Host-pathogen interaction · Next-generation sequencing

6.1 Introduction

Plants' microbial communities, including endophytes, play a crucial role in promoting growth and protection against various stresses. Endophytes, including fungi and bacteria, have mobility, cell wall degradation, and reactive oxygen species (Godara and Ramakrishna 2023). Common bacterial genera found as endophytes include *Pseudomonas*, *Bacillus*, *Burkholderia*, *Stenotrophomonas*, *Micrococcus*, *Pantoea*, and *Microbacterium*, while fungal endophytes include *Aspergillus*, *Alternaria*, *Bipolaris*, *Cladosporium*, *Diaporthe*, *Fusarium*, *Mucor*, *Nigrospora*, *Penicillium*, *Piriformospora*, *Trichoderma*, *Ulocladium*, and *Yarrowia.* These microbes aid in plant growth promotion by synthesizing plant hormones and facilitating growth in diverse rhizospheric zone (Kumar et al. 2023). Abiotic stresses like high temperatures, salinity, and drought can significantly impact plant growth and productivity (Godara and Ramakrishna 2023; Qin et al. 2016). These symbionts reside inside plant tissues, synthesize advantageous compounds that enhance both plant growth and defence mechanisms, and also contribute to sustainable agriculture through their ability to fix nitrogen biologically, uptake nutrients, suppress diseases, and clean up the environment (Jha et al. 2023). Endophytes employ diverse strategies like mutualism, latent pathogenicity, and adaptation to thrive within host plants, offering benefits such as growth promotion, improved nutrient uptake, and pathogen resistance (Sonali Jaiswal et al. 2023). To achieve successful implementation in agriculture and other fields, it is crucial to have a comprehensive understanding of endophyte interactions, colonization, diversity, and functions (Burragoni and Jeon 2021). Additional investigation into endophytic relationships has the potential to enhance medicine, agriculture, and biotechnology, underscoring the importance of implementing sustainable practices (Qin et al. 2016; Sonali Jaiswal et al. 2023). Considering that the conditions of the host plant have a significant impact on

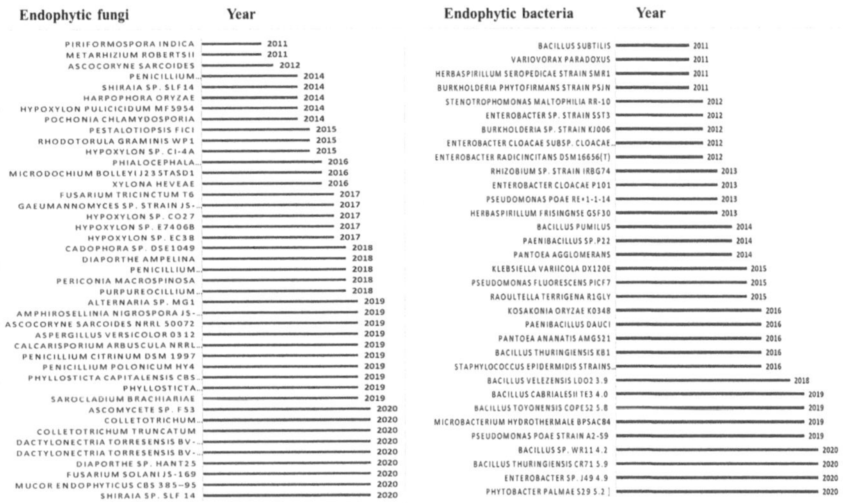

Fig. 6.1 Advantageous mechanisms of interactions between plants and microbial endophytes

endophytic associations, this field of study is extremely interesting. Agriculture, disease control, and the development of biopesticides are some of the potential biotechnological applications that endophytes could have (Quach et al. 2023; Slama et al. 2021). Figure 6.1 shows the details of endophytic influence on plant and their benefits conferred by endophytes. For sustainable agriculture, environmental conservation, and biotechnological innovations, it is essential to have a solid understanding of these strategies (Adeleke and Babalola 2022; Burragoni and Jeon 2021). The analyses of the genome and transcriptome are essential tools for gaining an understanding of endophytic organisms (Burragoni and Jeon 2021). These analyses provide insights into the biology of these organisms, as well as their interactions with host plants and potential applications. The results of these analyses contribute to the evaluation of genetic diversity; the identification of genes responsible for characteristics such as stress tolerance, nutrient acquisition, and biocontrol capabilities; and the comprehension of molecular interactions between endophytes and host plants (Bogas et al. 2022; Kaul et al. 2016). There are a number of biotechnological applications of genome, transcriptome, and proteome analyses, including the development of biofertilizers and biopesticides, the improvement of strains, and the comprehension of ecological environments (Quach et al. 2023; Slama et al. 2021). For the purposes of selective breeding or genetic engineering, genomic information is essential for the enhancement of desirable traits in endophytic organisms, which can lead to increased agricultural productivity or improvement in the effectiveness of biocontrol (Al-Amrani et al. 2021; Kumar et al. 2022). The identification of genes that are involved in nutrient cycling, plant growth promotion, and pathogen inhibition can be facilitated through the use of transcriptome analysis. The omics technologies are crucial for comprehending the interactions between endophytes and plants, as well as for exploring the potential applications of these interactions, and by employing diverse technologies, including meta-omics, metabolomics, genomics, transcriptomics, and proteomics, one can gain a comprehensive understanding of the intricate connections between endophytic fungi and their host plants (Al-Amrani et al. 2021; Balotf et al. 2022; Hilário and Gonçalves 2023; Khoomrung et al. 2017; Pun et al. 2022; Slama et al. 2021; Yuan et al. 2019a). This chapter discusses the use of genomic and transcriptomic integration with other emerging omics methods in endophytic research to discover new metabolites and identify new sources for retrieval. Currently, the use of metabolomics is crucial for discovering new metabolites from endophytic fungi, and metabolic screening can increase the likelihood of acquiring targeted metabolites without harming natural sources. Additionally, rapid development of computational tools and parallel technologies has revolutionized plant-microbe interaction sciences. Integrating omics datasets through robust statistical testing provides holistic system knowledge, ensuring synergistic interaction between biological components (Graw et al. 2021). Furthermore, they can help to uncover concealed gene clusters in endophytic fungi/bacteria that activate dormant metabolites and increase metabolite production by employing genetic manipulation and metabolomic interventions. Additionally, it is possible to explore the structural diversity of endophytes and their functions using omics approaches. This sets the stage for the idea of genetic manipulation, which has the capacity to lead to higher agricultural output and the creation of new biomolecules that can be used in various fields such as medicine, agriculture, industry, and the environment. Emerging omics technologies can help us understand how plant-dwelling endophytes can improve plant growth and interact with host plants (Mishra et al. 2021). These technologies also allow the study of medicinal plant microbial communities (Pun et al. 2022), the chemical and functional diversity of endophytic fungi-derived natural products (Bogas et al. 2022), how endophytic microbes affect host plant responses (Mishra et al. 2022), the molecular mechanisms of endophytes like *Diaporthe* infection (Hilário and Gonçalves 2023), and plant-endophyte interactions from colonization to defence (Chen et al. 2022). Thus, the omics technologies improve understanding of host plant-endophyte coevolution, identify new bioactive compounds, improve growth conditions, and establish sustainable agricultural practices.

6.2 Exploring Endophyte Function Through Genomics

6.2.1 Genomic Analysis in Decoding Endophytic Commodities

The field of genomics has been revolutionized by next-generation sequencing (NGS), which has significantly improved our understanding of genome structure, function, and variability. Next-generation sequencing (NGS), due to its capacity to rapidly and effectively handle substantial volumes of data at a reasonable cost, has become an indispensable instrument for researchers across various disciplines. Advanced NGS platforms like Illumina, Pacific Biosciences, and Oxford Nanopore have revolutionized genomics by enabling parallel sequencing of millions to billions of DNA fragments. This ability provides opportunities to better understand genetic variation, gene expression, epigenetic modifications, and microbial diversity (Levy and Myers 2016; Satam et al. 2023). The comparative genomics elucidates the molecular mechanisms underlying mutualistic bacterial communities, with a specific emphasis on endophytic and rhizospheric bacteria in comparison to phytopathogens. The study emphasized the significance of varied microbiomes for the development of plants and the intricate mechanisms by which endophytes transport nutrients (Hardoim and Hardoim 2017). Recently, a literature-based analysis showed the significance of next-generation sequencing technologies for elucidating biosynthetic pathways in endophytic fungi and host plants, almost 26 strains that have been used in experiments using parallel genome and transcriptome sequencing methods (Bielecka et al. 2022). Whole-genome analysis of endophytes can identify genes involved in plant growth-promoting activities, providing insights into their molecular and functional mechanisms (Guo et al. 2020; Kamada et al. 2015). Several studies have revealed that genome sequencing has provided insights into the genetic mechanisms employed by endophytes to colonize plants, interact with hosts, and potentially generate bioactive substances such as antibiotics, antiviral agents, and anticancer agents (Udayan and Gnanadoss Joel 2023; Franco-Sierra et al. 2020; Jha et al. 2023). Genomic methods help to explain the complex relationship between endophytes and host plants, revealing stress tolerance and growth promotion mechanisms (Pun et al. 2022). These technologies enable the study of plant-associated microbial communities, revealing endophytes' diversity and roles (Adeleke and Babalola 2022). Also, endophytic microorganisms play a crucial role in the nutrition of plants by aiding in the absorption and production of nutrients. Figure 6.2

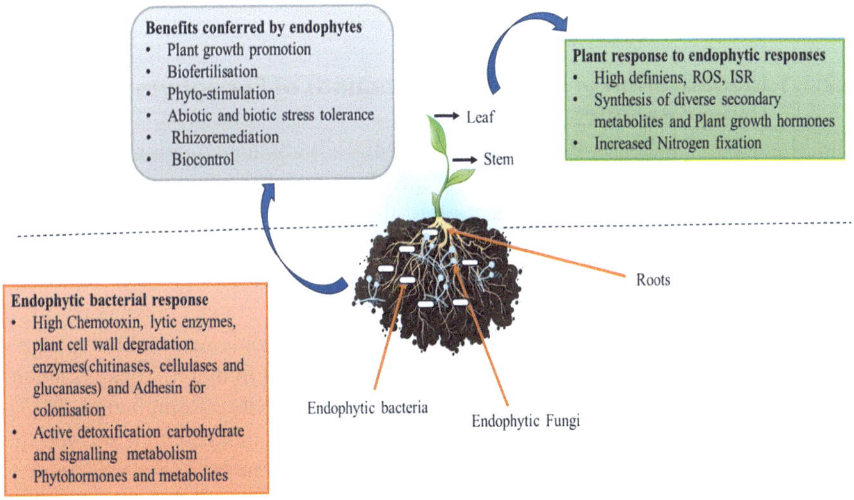

Fig. 6.2 Diagram illustrating the network connections between endophytes and host plants by integrating omics tools. The symbiotic or mutualistic behaviour of endophytes relies on intricate protein, genetic, and metabolic interactions. Genetic factors play a crucial role in controlling the biosynthetic routes, with transcriptional control being particularly important

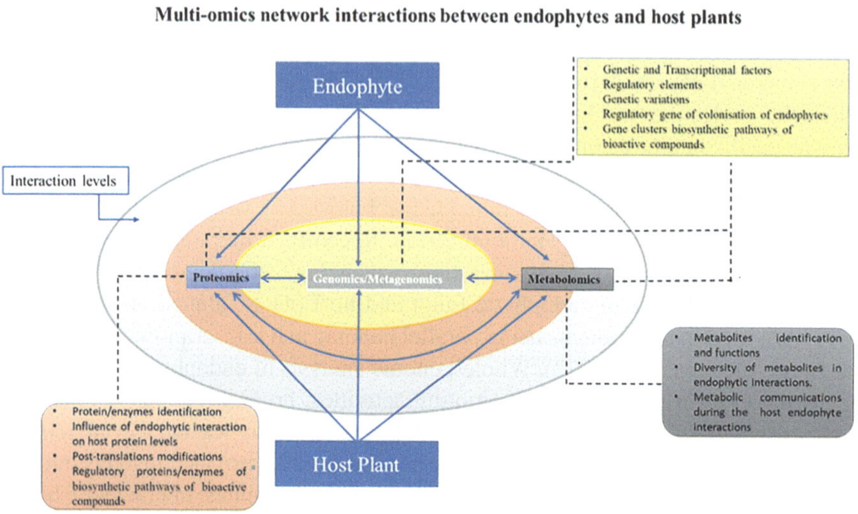

Fig. 6.3 Summary of year-by-year progress in genome sequencing of endophytic bacteria and fungi (del Carmen Orozco-Mosqueda and Santoyo 2021; Kaul et al. 2016; Sagita et al. 2021; Verma et al. 2023)

shows the application of genomics in endophytic interaction with host plant. A large number of bacterial and fungal endophytes have had their whole genomes sequenced, presented in Fig. 6.3, and this number is rising. Accessible endophyte genomes are the gold standard systems for studying plant-microbe interactions. Individual genome sequences improve meta-omics (metagenomics, transcriptomics, and proteomics) data analysis of plant-associated microbes.

6.2.1.1 Identification of Bioactive Compounds of Endophytes Using Genomics

The Malaysian Genomics Resource Centre (MGRC) conducted genomic sequencing of *S. kebangsaanensis* strain, which involved fragmentation, end-repair, and enrichment processes. The quantification of the sequencing library was performed using a KAPA kit, and the validation of the library size was conducted using an Agilent Bioanalyzer High Sensitivity DNA chip. The entire genome was sequenced utilizing the Illumina Genome Analyzer. The study found 24 gene clusters involved in polyketides, peptides, terpenes, bacteriocins, and siderophores, including a predicted phenazine biosynthesis gene and additionally confirm that S. kebangsaanensis produces a variety of antibiotics and secondary metabolites (Remali et al. 2017). Further, with the advancements in NGS technologies, another research showed the capacity of *Streptomyces* sp. VITGV100, an endophyte from *Lycopersicon esculentum*, to generate secondary metabolites with antimicrobial properties. The Illumina NextSeq 500 platform was used to study the whole genome of *Streptomyces* sp. VITGV100, and the results showed 8.08 megabases that comprised 35 biosynthetic gene clusters, some of which may be involved in the production of bioactive secondary metabolites. (Veilumuthu et al. 2022). Similarly, the metabolic pathways of *Bacillus subtilis* EA-CB0575 and *Bacillus toyonensis* BAC3151 have been analysed, revealing potential for developing antimicrobial agents for controlling crop diseases (Guo et al. 2020). By gaining an understanding of the genomic basis of endophytic contributions, it is possible to improve the utilization of these contributions for the purpose of promoting sustainable plant growth and health as well as the creation of novel natural products (Bogas et al. 2022; Dhamodaran et al. 2023). Genomic screening and in silico screening identified specific characteristics for plant stimulation and biocontrol endophytic bacteria from *Alkanna tinctoria* Tausch with biocontrol properties against *Fusarium graminearum*. Also, the selected strains demonstrated antifungal activity, with *Pseudomonas* sp. R-71838 being a promising candidate for biocontrol applications. This research serves as a foundation for developing a biocontrol strategy (Naranjo et al. 2023). Endophytic bacterial consortium in the tissues of *Glehnia littoralis* using high-throughput sequencing techniques was investigated, and results indicate that there is a greater variety and abundance of microorganisms in the root tissues, with *Actinobacteria* and *Proteobacteria* being the most prevalent. The study shows *Glehnia littoralis* tissue endophytic bacteria diversity and abundance (Huo et al. 2020). Recently, 60 *Streptomyces* strains were screened for their antifungal activity against various fungal pathogens in Vietnam. *Streptomyces albus* endophytic strain RC2 was effective against *S. gossypii* Co1 and *F. fujikuroi* L3. The genome of strain RC2 showed 15 biosynthetic gene clusters (BGCs) with ≥45% similarity to antiSMASH database reference BGCs. Eight previously unidentified secondary metabolites were found by UPLC-HRMS, and this study found RC2 exhibited biocontrol properties against phytopathogenic fungi (Quach et al. 2023).

6.2.1.2 Identification of Endophytic Bacteria Interaction with Host Using Genomics

To overcome the limitation of old genetic techniques, evolutionary histories of various organisms have been studied using next-generation sequencing technologies, enabling the creation of genome models that reveal their natural populations and evolution (Sharma et al. 2011). Integration of genome mining and metabolic profiling discovered a novel linear lipopeptide named holrhizin A in *Rhizopus* endosymbionts, which reduces surface tension and influences bacterial colonization and invasion. This study found a functional link between the orphan NRPS genes and holrhizin A, promoting bacterial invasion into the fungal host, providing insight into the unique *Burkholderia Rhizopus* symbiosis (Niehs et al. 2018). The interaction between endophytic bacteria and arbuscular mycorrhizal fungi (AMF) has a significant impact on plant symbiotic signalling pathways, root colonization patterns, and phytohormone synthesis, which has been recently reviewed (Vandana et al. 2021). Comparative genome analysis of nine *Proteobacteria* genomes to identify endophytic genes sheds light on plant-endophyte molecular interactions. Furthermore, investigating the essential genetic determinants that enable bacteria to survive as endophytes facilitates a deeper comprehension of the beneficial and detrimental impacts bacteria exert on plants (Mushtaq et al. 2023).

6.2.2 Comparative Genomic Analysis in Decoding Endophytic Lifestyle

Distinguishing between endophytes and free-living bacteria is challenging due to their varying relationships with plants. Some endophytes form long-term connections, while others form temporary or optional associations. Rhizospheric bacteria can enter plant tissues through mechanisms like wounds or tissue ruptures (Santoyo et al. 2016). Rhizobia and leguminous plants establish a simple connection through nodules, oxygen-free nitrogen fixation, and mutual chemical communication. Rhizobia and leguminous plants release flavonoids, facilitating a symbiotic relationship through nodulation factors and plant release (Poole et al. 2018).

Valuable genomic data was obtained by sequencing the genomes of *Bordetella bronchiseptica* RB50, *Bordetella parapertussis* 12,822, and *Bordetella pertussis* Tohama I in the study. The study uncovered significant gene loss and inactivation in *B. parapertussis* and *B. pertussis*, which could potentially affect their ability to survive outside of their host. The study identified the precise gene functions that have been lost in these bacteria (Parkhill et al. 2003). First report by Ali et al. (2014) on comparative genomic studies of endophytic and non-endophytic bacteria. In this study endophytic strains *Burkholderia phytofirmans* PsJN, *Burkholderia* spp. JK006, *Azospirillum lipoferum* 4B, *Enterobacter cloacae* ENHKU01, *Klebsiella pneumoniae* 342, *Pseudomonas putida* W619, *Enterobacter* spp. 638, *Azoarcus* spp. BH72, and *Serratia proteamaculans* 568 were compared genome-wise. In addition to *rhizospheric* bacterium *Burkholderia xenovorans* LB400 and the non-plant host pathogen *Burkholderia cenocepacia* AU1054. These genomic comparisons identified genes involved in secretion, transport, plant polymer degradation or modification, detoxification, and redox potential (Ali et al. 2014). The above functions may be involved in endophytic behaviour. A comparison of *Burkholderia phytofirmans* PsJN genome with nine other genomes has shown that detoxification activities, cell wall degradation by lytic enzymes, and motility are all very important to its endophytic lifestyle (Mitter et al. 2013). More recently, researchers found intriguing features in the genome of the plant-growing bacterium *Kosakonia radicincitans*. Multiple copies of a gene cluster contain two flagellar systems and three type VI secretion systems. The bacterium needs these traits to use plant carbon sources and colonize the rhizosphere and possibly plants. A megaplasmid contained several of these functions. Comparing the core genome of *Kosakonia radicincitans* to closely related species confirmed its unique traits (Becker et al. 2018). Brader and colleagues studied the ecological characteristics and genomic indicators of plant pathogenic and non-pathogenic endophytes, highlighting the importance of effector proteins like T3SSs and T6SSs in activating the plant immune system and suppressing it (Brader et al. 2014). They suggested that certain genes may not be present in endophytic bacteria or are not fully developed, highlighting their role in plant interaction (Brader et al. 2014). According to an excellent study by Knapp et al., microbial endophytes share these behaviours because they can colonize plants and get nutrients from inside their cell walls. Their study have found that endophytic genomes contain active carbohydrate and degrading enzymes when comparing the genomes of 2 dark septate fungi, *Cadophora* sp. and *Periconia macrospinosa*, to 32 other ascomycetes (Knapp et al. 2018). Also in the case of some fungal endophytes, comparative genomic analysis of *Trichoderma reesei* reveals significant genes interacting with plants, including those related to mycoparasitism, chitinolytic enzymes, β-1,3-glucanases, secondary metabolites, NRPS, PKS, and pore-forming cytolytic peptides (Kubicek et al. 2011). Comparative mitochondrial genome sequencing analysis revealed in terms of gene content, tRNA lengths, gene order, and orientation, the study discovered that *Leptosphaerulina chartarum* and *Curvularia trifolii*, shared some similarities and had some differences in their mitochondrial genomes (Yuan et al. 2019b). *Aspergillus* endophytes AS31, AS33, and AS42 colonize rice roots and boost growth. A recent comparative genomic study found that endophytes increased phosphorus accumulation and transport in rice. Also endophytes had more carbohydrate-active enzyme and small secreted protein genes than the pathogens (Jing et al. 2022). Whole-genome comparisons revealed genomic differences between grapevine endophytic test bacteria and reference strains, shedding light on endophyte virulence. This study examined the entire genome of 7 Italian grapevine bacterial endophytes and 12 reference genomes from the same group but with different lifestyles. We used MAUVE and BLAST to align and search the genome (Lòpez-Fernàndez et al. 2015). An interesting report aims to shed light on the endophytic lifestyle of *Trichoderma* fungi by analysing their genomes. Genes linked to mycoparasitism, chitinase, and degradative and biosynthetic pathways are among the many new topics uncovered by the investigation (Scott et al. 2023). Emerging meta-omics techniques uncover the most prevalent bacterial and fungal phyla in plant species, offering valuable information about their structural variety. Endophytic microbes in different plant species, including sunflower, maize, rice, and wheat, have been studied using meta-omics approaches. These studies have identified dominant bacterial phyla such as *Proteobacteria*, *Firmicutes*, *Actinobacteria*, *Bacteroidetes*, and *Chloroflexi*, as well as fungal phyla like *Ascomycota*, *Basidiomycota*, and *Zygomycota* (Adeleke and Babalola 2022). This means that endophytic genomes exhibit high rates of nucleotide substitution, considerable preservation, and a progressive reduction of the genome. These characteristics shed light on the dynamics and molecular evolution of gene loss. Therefore, the aforementioned research shows that comparative genomics is crucial for comprehending the life strategies and evolutionary adaptations of endophytic microbes.

6.3 Exploring Endophyte Function Through Transcriptomics

6.3.1 Transcriptomic Analysis in Decoding Endophytic Commodities

Omics technologies enable a comprehensive analysis of biological material on a large scale. The development of Sanger sequencing method made a groundbreaking advancement in omics technology. This method was only appropriate and precise for analysing brief sequences. Subsequently, advancements have been made in genomics that allow the human genome to be sequenced. Nevertheless, significant amount of expenses and time period taken to complete Sanger sequencing led to its hindering widespread adoption and use in many research projects. Next-generation sequencing (NGS) enables a high-throughput and cost-effective method, allowing detailed study of a species' transcriptomes and genome. Further, the advancement of next-generation sequencing methods made it possible to read long sequencing, which represents a significant progression in addressing this issue (van Dijk et al. 2018). Advancements in sequencing technology aid in uncovering novel genes, therefore enabling the establishment of different regulatory networks for established genes. Other omics techniques have been created and are now used in other research disciplines.

The single transcriptomics offers a perspective on different biotic processes restricted at the level of single cell. The study primarily focuses on correlations, often resulting in observed variations rather than causal relationships. We must now have worldwide knowledge that is combined via several individual omics technologies, known as the multi-omics method. An interactome, created via genomes and transcriptomics, is crucial for investigating pathways involved in plant and microorganism interactions. Also sequencing technology provides essential genetic information, and transcriptome findings identify various specific conditions linked to genes (Qin et al. 2016). Thus, multi-omics studies combining genetic, protein, and metabolite data can help to identify various molecules during plant-microbe interaction such as endophytic involvement. This approach enhances our understanding of how microbes help plants cope with stress. Figure 6.4 shows the details.

The interaction between plants and their endophytes at the level of transcriptomics can be analysed using sequencing of mRNA (Sheibani-Tezerji et al. 2015). Transcriptomic studies on endophytes during interaction with plant and without plants will help to identify the influence of host response.

Transcriptomic profiling may be conducted by using different approaches ranging from microarray to RNA sequencing (RNA-seq). RNA-seq is advantageous for doing differential gene expression analysis under varying settings, detecting microbe-host gene expression simultaneously, and identifying new RNA entrants. Technologies such as SOLiD-SAGE and microarray are valuable for doing a comparative gene expression investigation, whereas next-generation sequencing allows for a comprehensive transcriptome analysis, enabling the investigation of cell-specific transcriptional responses. Transcriptomic analysis allows us to reveal the changing aspects and control of active genes, providing significant benefits compared to genomic studies alone (Levy et al. 2018). Genome-based researches are essential for conducting effective transcriptomic investigations. Conducting genome and transcriptome analysis together is more beneficial for understanding plant endophytic relations (Kaul et al. 2016).

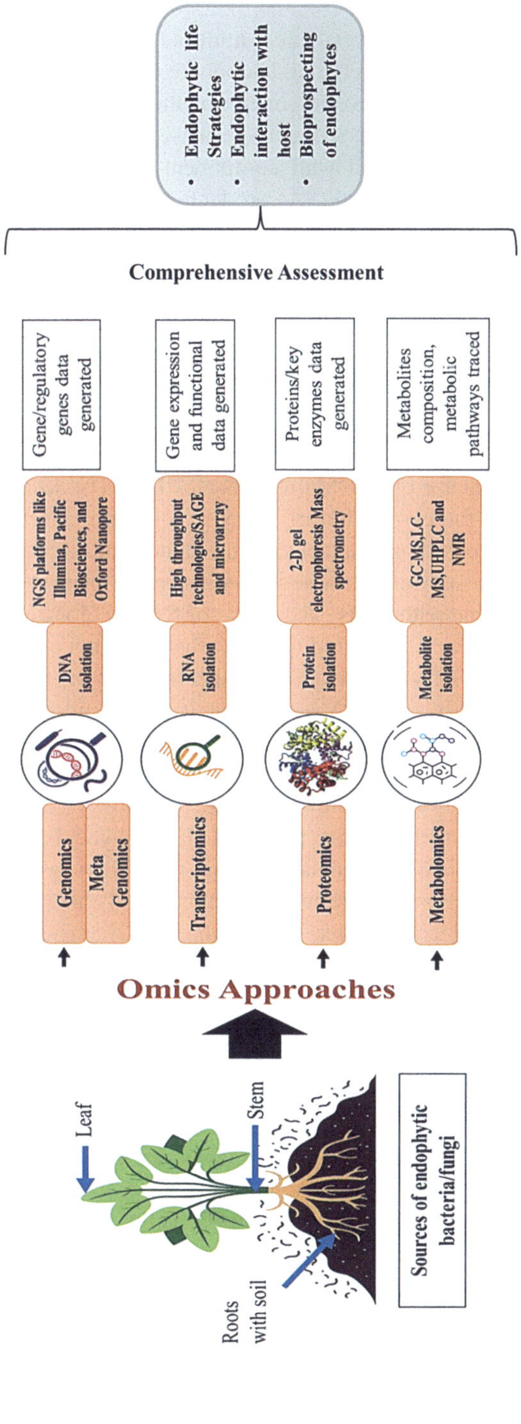

Fig. 6.4 An outline of the omics technologies utilized in endophytic researches

6.3.2 Insights Gained from Transcriptomic Analysis

Transcriptomic analyses have investigated multiple aspects of the connection between plants and endophytes, including endophyte colonization and stress tolerance. Endophytic bacteria have several contact methods with hosts to avoid host defences before colonization in the recognition phase. For instance, certain fungi-based endophytes conceal the cell wall component chitin to prevent defensive responses, inhibit plant salicylic acid (SA) production, and enhance jasmonic acid (JA) biosynthesis. Certain endophytes use miRNA-targeted regulatory pathways to defend them from plant immunity (Plett and Martin 2018). During the recognition stage, the different chemicals released by endophytes function as messengers' molecules when they enter the host, stimulating plant growth and triggering stress and defence responses towards pathogens. When *Arabidopsis* was exposed to *S. indica*, it was observed that the JA content was less as compared to SA, which was the opposite of the initial state before the exposure (Vahabi et al. 2015a, b). The study by Malinich et al. (2019) observed that *Trichoderma virens* decreased the production of salicylic acid in maize plants (Malinich et al. 2019). The decrease in salicylic acid creation inhibited the conversion of salicylic acid into catechol, making the plant more vulnerable to pathogen attacks. At the same time, genes responsible for jasmonic acid biosynthesis were activated to enhance the plant's defence mechanism against necrotrophic pathogens. The levels of SA in colonized plants did not show a significant difference compared to plants that were not colonized by endophytes (Ambrose et al. 2015). The physical contact between the two different symbionts may not necessarily raise SA levels, but it could still trigger a systemic defence mechanism (Hillmer et al. 2017; Tsuda et al. 2009).

Endophytes may use several techniques to avoid being detected by the host's defence mechanisms after successfully outcompeting other microorganisms and overcoming physical obstacles inside the plant. Endophytic bacteria synthesize their own microbe-associated molecular patterns (MAMPs) or release enzymes upon encountering the host's immune system, while fungal endophytes secrete chitin deacetylase or fungal lectins. Unlike harmful bacteria that trigger resistance via SA signalling, endophytes control via JA and ethylene (ET) signalling to stimulate induced systemic resistance, which improves the resistance of plants to different biotic stress (Bastias et al. 2017; Constantin et al. 2019; De Palma et al. 2019). Moreover, they also restrict their colonization area through avoiding excessive dominance over the plant during the crucial colonization period. During the colonization phase, an endosymbiotic association that benefits both species is established. The plants sustain different endophytic microorganisms in their specialized tissues by providing essential nutrients for their survival. During the initial stages of colonization, hosts support endophytes by stimulating the activation of genes associated with motility, chemotaxis, and adhesion and supported by secretions of various proteins and involvement of ABC transporters (Coutinho et al. 2015; Pankievicz et al. 2016). The host's plant usually offered those tissue locations that were appropriate for endophytic development (Utturkar et al. 2016). This phenomenon suggests that when an endophyte is co-cultured, it shows a higher number of upregulated genes that may enhance the growth fitness of both organisms compared to those that are not co-cultured (Ambrose and Belanger 2012; Trovero et al. 2018). Symbiotic mechanisms between rapeseed and the virus-mediated endophytic fungus *Sclerotinia sclerotiorum* strain DT-8 were explained by upregulated genes encoding amylase/glucoamylase, glucose transporter genes, and the SWEET11 gene, suggesting potential interactions between the fungus and the host plant in using sugars as nutrients or signalling molecules. The host increases pectate lyase and pectin esterase activity to fight the fungus, while the fungus fights back (Qu et al. 2021).

Transcriptomics has partly analysed the interrelated components, pathways, chemicals, and different level of reorganization and colonization relationship with plants and endophytes. Transcriptome-based research is having limitation on experimental design, technology, analytic expenses, and data storage. Moreover, gene annotation can be challenging when high-quality reference genomes are lacking, particularly for co-expressed genes of hosts and endophytes that are hard to detect in plants due to factors such as the low levels of bacterial mRNA (Levy et al. 2018). Research on the molecular underpinnings of plant-endophyte interactions is in its early stages and requires additional exploration to enhance our comprehension.

6.4 Investigating Host-Endophyte Interactions Through the Lens of Proteomics

Proteomics is defined as the analysis of different proteins or set of proteins in an organism or different tissues at particular time point (Al-Amrani et al. 2021). This technology can be applied to detect endophytic protein set including those of fungi and bacteria. Therefore, it can be used to identify, describe, and measure protein relations in various pathways. Further, the use of 2D gel electrophoresis along with mass spectrometry enhanced the detection of individual proteins and helped to decipher the relationship between endophytes and plants (Al-Amrani et al. 2021). Also, the combination of these technologies could be helpful for identifying the protein set in fungal and bacterial endophytes alone. Moreover, this technique provides understanding into their relationships with hosts, which includes biological pathways and posttranslational modifications (Balotf et al. 2022). Yuan et al. performed research on the whole protein set of *Atractylodes lancea* that was infected with or without the endophytic fungus strain *Gilmaniella* spp. AL12. The study revealed an increase in host proteins that are related to CO_2 fixation, carbohydrate, and energy metabolism, indicating that the endophyte *Gilmaniella* spp. AL12 may enhance the biomass of *Atractylodes lancea* (Yuan et al. 2019a). Further, mass spectrometry was used in *Piriformospora indica* to analyse the proteome and identify proteins related to ROS defence, photosynthesis, and energy transfer under drought conditions. Yet, the restricted data gathered just from proteomics-based research highlights the intricate nature of plant-fungal endophyte interactions. By integrating various omics approaches, a wider variety of molecular information may be captured, surpassing this constraint.

6.5 Exploring Endophyte Function Through Meta-omics

Meta-omics is a multidisciplinary approach combining data from various fields to understand complex biological processes. Multi-omics research, a subset of meta-omics, offers a holistic view of biological problems, generating detailed information for research purposes. Researchers can use meta-omics approaches, such as metagenomics, metatranscriptomics, and metabolomics, to study the intricate interactions between plants and their associated endophytes. This allows for a more comprehensive understanding of the coevolutionary relationship between host plants and endophytes [4]. Research involving multiple omics techniques has demonstrated that endophytes play a significant role in promoting plant growth, enhancing metabolism, and increasing resistance to stress. These findings emphasize the complex mechanisms involved in host-endophyte interactions.

6.5.1 Insight Gained Through Metagenomic Analysis in Endophytic Research

Metagenomics is a molecular biology technique that examines complete genomes in environmental samples, offering valuable information. Transcriptomics and proteomics technologies are used to analyse the gene expression and proteomes of endophytic fungal strains, whereas metagenomics enables researchers to identify strong isolates and bioactive compounds, as well as gain insight into the intricate interactions between the fungi and their host plants (Verma et al. 2023). Metagenomics, employed by environmental microbiologists for the examination of endospheric microbiomes, has revolutionized the understanding of interactions between plants and microbes, revealing insights into functions, metabolic pathways, and previously unknown genes (Prasannakumar et al. 2020). The fast development of metagenomic analysis has thus rendered the genomic reference library obsolete when it comes to interpreting full metagenome sequenced data reads, which can include hundreds of thousands to millions of endophytic bacteria and fungi (Almeida and De Martinis 2019). Metagenomics and computational tools are employed to identify functional genes in fungal endophytes derived from host plants. In their study, Zhang et al. (2020) investigated *Ephedra sinica* and identified the presence of fungal strains OTU30589 and OTU48 in both the roots and stem. Additionally, they found the presence of *Phyllosticta*, *Aspergillus*, *Talaromyces*, and *Aporospora* in the roots (Zhang et al. 2020).

6.5.2 Insight Gained Through Metabolomic Analysis in Endophytic Research

Metabolomic analysis is a powerful approach used to comprehensively study the metabolites present in biological systems. This approach helps in identification as well as in quantification of minor molecules or metabolites, within cells, tissues, or organisms. In the context of endophytic research, metabolomic analysis helps in understanding the metabolic interactions between host plants and endophytic organisms, as well as elucidating the functional roles of endophytic metabolites.

In recent years, understanding of plant-microbe interactions has significantly advanced, but the chemical communication that triggers priming is still not completely understood. Metabolomics, using modern bioanalytical techniques along with chemo-bioinformatic tools, allows the analysis of changes in the metabolomes of interacting species, serving as a valuable addition to other "omics" methods. Metabolomics may help enhance our comprehension of different metabolites during interaction of plant and its endophyte and also metabolic communication that supports the interactome (Jorge et al. 2016). Various methods may be used to analyse metabolites obtained from endophytic interaction with plants such as chromatography-based approaches and mass spectrometry-based methods (Khoomrung et al. 2017; Mhlongo et al. 2018; Maheshwari 2017; Kumar et al. 2023). When conventional method such as MS fails to precisely identify the distribution of metabolites, spatial metabolomics can be investigated using desorption electrospray ionization-imaging mass spectrometry (DESI-IMS) and matrix-assisted laser desorption/ionization-IMS (MALDI-IMS) (He et al. 2018).

At present bioinformatics tools are employed in endophytic research, which specifically investigates the genomes and the genetic makeup of endophytic microbes, as well as the analysis of their functions, gene expression, and protein composition. It also helps in understanding the metabolic pathways, the evolutionary relationships between species, the study of microbial communities, the discovery of new drug molecules, and the application of biotechnology (Bogas et al. 2022). Bioinformatic tools are indispensable for the process of sequencing and assembling the genomes of endophytes. These tools enable the analysis of the combined genomes of endophyte communities present within a plant. Additionally, it assists in the prediction and annotation of genes, specifically identifying distinct genes and pathways that contribute to endophytic lifestyles and their advantageous impacts on host plants. It also aids in the examination of RNA-seq analysis and proteomic data, determining and measuring the proteins expressed by endophytes, and comprehending their functional roles in plant-endophyte interactions. Integrating all omics data with in silico tools has the potential to significantly advance endophytic research and uncover the mysteries of endophytic life (Bogas et al. 2022).

6.6 Conclusion

The rapid advancement of omics techniques has significantly contributed to our understanding on the mechanisms underlying endophytes and their interactions with host plants. Integrating genetics, metagenomics, and other omics tools is essential for unravelling the intricate relationships between endophytes and their host plants. Through omics techniques, we gain extensive insights into the metabolic strategies, the gene cascades, and the accumulation of metabolites, enzymes, and proteins within endophytic communities. Understanding the regulatory systems of differentially expressed genes is crucial for deciphering the complexity of these interactions. Multi-omics approaches play a pivotal role in analysing the mutualistic relationships between endophytes and their host plant communities. Endophytic genome exploration has revealed the genetic mechanisms that control colonization, interaction, and bioactive compound synthesis for plant growth and defence. Multifaceted communication and metabolic pathways have been revealed by transcriptomic analyses of host-endophyte gene expression patterns. By understanding the chemical exchange between plants and endophytes, metabolomics has shown the importance of endophytic metabolites in plant health and resilience. While transcriptomic approaches provide valuable information, they may encounter limitations in sequencing data coverage. Nonetheless, re-sequencing and analytical techniques present opportunities for improved assemblies. By identifying key proteins involved in plant-endophyte interactions, proteomic research has illuminated biological pathways and posttranslational modifications. Environmental omics have made endophytic ecology more understandable by highlighting the complex relationship between endophytes, plants, and their environment. Thus, establishing well-curated integrated databases from multi-omics research endeavours can greatly enhance our comprehension of endophytic microbial species. System biology approaches, combined with integrated omics data, hold promise for prognostic modelling and dynamic assessment of fungal endophytes within broader microbial communities.

References

Adeleke BS, Babalola OO (2022) Meta-omics of endophytic microbes in agricultural biotechnology biocatalysis and agricultural. Biotechnology 42:102332

Al-Amrani S, Al-Jabri Z, Al-Zaabi A, Alshekaili J, Al-Khabori M (2021) Proteomics: concepts and applications in human medicine. World J Biol Chem 12:57–69. https://doi.org/10.4331/wjbc.v12.i5.57

Ali S, Duan J, Charles TC, Glick BR (2014) A bioinformatics approach to the determination of genes involved in endophytic behavior in Burkholderia spp. J Theor Biol 343:193–198

Almeida OG, De Martinis EC (2019) Bioinformatics tools to assess metagenomic data for applied microbiology. Appl Microbiol Biotechnol 103:69–82

Ambrose KV, Belanger FC (2012) SOLiD-SAGE of endophyte-infected red fescue reveals numerous effects on host transcriptome and an abundance of highly expressed fungal secreted proteins. PloS One 7:e53214. https://doi.org/10.1371/journal.pone.0053214

Ambrose KV, Tian Z, Wang Y, Smith J, Zylstra G, Huang B, Belanger FC (2015) Functional characterization of salicylate hydroxylase from the fungal endophyte Epichloë festucae. Sci Rep 5:10939. https://doi.org/10.1038/srep10939

Balotf S, Wilson R, Tegg RS, Nichols DS, Wilson CR (2022) Shotgun proteomics as a powerful tool for the study of the proteomes of plants, their pathogens, and plant-pathogen interactions. Proteomes 10:5. https://doi.org/10.3390/proteomes10010005

Bastias DA, Martínez-Ghersa MA, Ballaré CL, Gundel PE (2017) Epichloë fungal endophytes and plant defenses: not just alkaloids. Trends Plant Sci 22:939–948. https://doi.org/10.1016/j.tplants.2017.08.005

Becker M et al (2018) Comparative genomics reveal a flagellar system, a type VI secretion system and plant growth-promoting gene clusters unique to the endophytic bacterium Kosakonia radicincitans. Front Microbiol 9:355324

Bielecka M, Pencakowski B, Nicoletti R (2022) Using next-generation sequencing technology to explore genetic pathways in endophytic fungi in the syntheses of plant bioactive metabolites. Agriculture 12:187

Bogas A, Cruz F, Lacava P, Sousa C (2022) Endophytic fungi: an overview on biotechnological and agronomic potential. Braz J Biol 84:e258557

Brader G, Compant S, Mitter B, Trognitz F, Sessitsch A (2014) Metabolic potential of endophytic bacteria. Curr Opin Biotechnol 27:30–37

Burragoni SG, Jeon J (2021) Applications of endophytic microbes in agriculture, biotechnology, medicine, and beyond. Microbiol Res 245:126691. https://doi.org/10.1016/j.micres.2020.126691

Chen X-l, Sun M-C, Chong S-L, Si J-P, Wu L-S (2022) Transcriptomic and metabolomic approaches deepen our knowledge of plant–endophyte interactions. Front Plant Sci 12:700200

Constantin ME, de Lamo FJ, Vlieger BV, Rep M, Takken FLW (2019) Endophyte-mediated resistance in tomato to Fusarium oxysporum is independent of ET, JA, and SA. Front Plant Sci 10:979. https://doi.org/10.3389/fpls.2019.00979

Coutinho BG, Licastro D, Mendonça-Previato L, Cámara M, Venturi V (2015) Plant-influenced gene expression in the rice endophyte Burkholderia kururiensis M130 molecular plant-microbe interactions. MPMI 28:10–21. https://doi.org/10.1094/mpmi-07-14-0225-r

De Palma M et al (2019) Transcriptome reprogramming, epigenetic modifications and alternative splicing orchestrate the tomato root response to the beneficial fungus Trichoderma harzianum. Hortic Res 6:5. https://doi.org/10.1038/s41438-018-0079-1

del Carmen Orozco-Mosqueda M, Santoyo G (2021) Plant-microbial endophytes interactions: scrutinizing their beneficial mechanisms from genomic explorations. Curr Plant Biol 25:100189

Dhamodaran N, Konappa N, Chowdappa S, Jogaiah S (2023) Endophytic fungi: application in combating plant pathogens and sustainable agriculture. In: Fungal resources for sustainable economy: current status and future perspectives. Springer, pp 251–273

Franco-Sierra ND, Posada LF, Santa-María G, Romero-Tabarez M, Villegas-Escobar V, Álvarez JC (2020) Bacillus subtilis EA-CB0575 genome reveals clues for plant growth promotion and potential for sustainable agriculture. Funct Integr Genomics 20:575–589

Godara H, Ramakrishna W (2023) Endophytes as nature's gift to plants to combat abiotic stresses letters in. Appl Microbiol 76:ovac067

Graw S, Chappell K, Washam CL, Gies A, Bird J, Robeson MS, Byrum SD (2021) Multi-omics data integration considerations and study design for biological systems and disease. Mol Omics 17:170–185

Guo D-J et al (2020) Complete genome sequence of Enterobacter roggenkampii ED5, a nitrogen fixing plant growth promoting endophytic bacterium with biocontrol and stress tolerance properties, isolated from sugarcane root. Front Microbiol 11:580081

Hardoim PR, Hardoim CCP (2017) Genomic features of mutualistic plant bacteria. In: Endophytes: biology and biotechnology, vol 1. Springer, pp 99–125

He J et al (2018) A sensitive and wide coverage ambient mass spectrometry imaging method for functional metabolites based molecular histology. Adv Sci. (Weinheim, Baden-Wurttemberg, Germany) 5:1800250. https://doi.org/10.1002/advs.201800250

Hilário S, Gonçalves MF (2023) Mechanisms underlying the pathogenic and endophytic lifestyles in diaporthe: an omics-based approach. Horticulturae 9:423

Hillmer RA et al (2017) The highly buffered Arabidopsis immune signaling network conceals the functions of its components. PLoS Genet 13:e1006639. https://doi.org/10.1371/journal.pgen.1006639

Huo X, Wang Y, Zhang D, Gao T, Liu M (2020) Characteristics and diversity of endophytic bacteria in endangered Chinese herb based on Illumina sequencing. Pol J Microbiol 69:283–291

Jha P, Kaur T, Chhabra I, Panja A, Paul S, Kumar V, Malik T (2023) Endophytic fungi: hidden treasure chest of antimicrobial metabolites interrelationship of endophytes and metabolites. Front Microbiol 14:1227830

Jing M, Xu X, Peng J, Li C, Zhang H, Lian C (2022) Comparative genomics of three aspergillus strains reveals insights into endophytic lifestyle and endophyte-induced plant growth promotion. J Fungi 8(7):690. https://doi.org/10.3390/jof8070690

Jorge TF, Rodrigues JA, Caldana C, Schmidt R, van Dongen JT, Thomas-Oates J, António C (2016) Mass spectrometry-based plant metabolomics: metabolite responses to abiotic stress. Mass Spectrom Rev 35:620–649. https://doi.org/10.1002/mas.21449

Kamada M, Hase S, Fujii K, Miyake M, Sato K, Kimura K, Sakakibara Y (2015) Whole-genome sequencing and comparative genome analysis of Bacillus subtilis strains isolated from non-salted fermented soybean foods. PloS One 10:e0141369

Kaul S, Sharma TK, Dhar M (2016) "Omics" tools for better understanding the plant–endophyte interactions. Front Plant Sci 7:955. https://doi.org/10.3389/fpls.2016.00955

Khoomrung S, Wanichthanarak K, Nookaew I, Thamsermsang O, Seubnooch P, Laohapand T, Akarasereenont P (2017) Metabolomics and integrative omics for the development of Thai traditional medicine. Front Pharmacol 8:474. https://doi.org/10.3389/fphar.2017.00474

Knapp DG et al (2018) Comparative genomics provides insights into the lifestyle and reveals functional heterogeneity of dark septate endophytic fungi. Sci Rep 8:6321

Kubicek CP et al (2011) Comparative genome sequence analysis underscores mycoparasitism as the ancestral life style of Trichoderma. Genome Biol 12:1–15

Kumar P, Rani S, Dahiya P, Kumar A, Dang AS, Suneja P (2022) Whole genome analysis for plant growth promotion profiling of Pantoea agglomerans CPHN2, a non-rhizobial nodule endophyte. Front Microbiol 13:998821

Kumar VN, SeonJu P, Kandasamy S, Anbazhagan S, Myeong-Hyeon W (2023) Ultraperformance liquid chromatography-based metabolite profiling and cytotoxic activity of the ethyl acetate extract of two endophytic Penicillium sp.,Process Biochemistry, 130:366–378. https://doi.org/10.1016/j.procbio.2023.04.027

Kumar A, Maurya VK, Susmita C, Chuarasiya U, Maurya DK, Singh SK (2023) Environmental factors and plant–microbes (endophytes) interaction: an overview and future outlook. In: Microbial endophytes and plant growth, pp 245–257

Levy SE, Myers RM (2016) Advancements in next-generation sequencing. Annu Rev Genomics Hum Genet 17:95–115

Levy A, Conway JM, Dangl JL, Woyke T (2018) Elucidating bacterial gene functions in the plant microbiome. Cell Host Microbe 24:475–485. https://doi.org/10.1016/j.chom.2018.09.005

Lòpez-Fernàndez S, Sonego P, Moretto M, Pancher M, Engelen K, Pertot I, Campisano A (2015) Whole-genome comparative analysis of virulence genes unveils similarities and differences between endophytes and other symbiotic bacteria. Front Microbiol 6:419

Maheshwari D, Annapurna K (2017) Endophytes: Crop Productivity and Protection: Volume 2. https://doi.org/10.1007/978-3-319-66544-3

Malinich EA, Wang K, Mukherjee PK, Kolomiets M, Kenerley CM (2019) Differential expression analysis of Trichoderma virens RNA reveals a dynamic transcriptome during colonization of Zea mays roots. BMC Genomics 20:280. https://doi.org/10.1186/s12864-019-5651-z

Mhlongo MI, Piater LA, Madala NE, Labuschagne N, Dubery IA (2018) The chemistry of plant-microbe interactions in the rhizosphere and the potential for metabolomics to reveal signaling related to defense priming and induced systemic resistance. Front Plant Sci 9:112. https://doi.org/10.3389/fpls.2018.00112

Mishra RC, Kalra R, Dwivedi N, Goel M (2021) Exploring endophytes using "Omics": an approach for sustainable production of bioactive metabolites. In: Mycoremediation and environmental sustainability, vol 3, pp 349–376

Mishra S, Priyanka, Sharma S (2022) Metabolomic insights into endophyte-derived bioactive compounds. Front Microbiol 13:835931

Mitter B et al (2013) Comparative genome analysis of Burkholderia phytofirmans PsJN reveals a wide spectrum of endophytic lifestyles based on interaction strategies with host plants. Front Plant Sci 4:120

Mushtaq S et al (2023) Interaction between bacterial endophytes and host plants. Front Plant Sci 13:1092105

Naranjo HD, Rat A, De Zutter N, De Ridder E, Lebbe L, Audenaert K, Willems A (2023) Uncovering genomic features and biosynthetic gene clusters in endophytic bacteria from roots of the medicinal plant Alkanna Tinctoria Tausch as a strategy to identify novel biocontrol bacteria. Microbiol Spectr 11:e00747–e00723

Niehs SP, Scherlach K, Hertweck C (2018) Genomics-driven discovery of a linear lipopeptide promoting host colonization by endofungal bacteria. Org Biomol Chem 16:8345–8352

Pankievicz VC et al (2016) RNA-seq transcriptional profiling of Herbaspirillum seropedicae colonizing wheat (Triticum aestivum) roots. Plant Mol Biol 90:589–603. https://doi.org/10.1007/s11103-016-0430-6

Parkhill J et al (2003) Comparative analysis of the genome sequences of Bordetella pertussis, Bordetella parapertussis and Bordetella bronchiseptica. Nat Genet 35:32–40

Plett JM, Martin FM (2018) Know your enemy, embrace your friend: using omics to understand how plants respond differently to pathogenic and mutualistic microorganisms. Plant J 93:729–746. https://doi.org/10.1111/tpj.13802

Poole P, Ramachandran V, Terpolilli J (2018) Rhizobia: from saprophytes to endosymbionts. Nat Rev Microbiol 16:291–303

Prasannakumar M et al (2020) Metagenome sequencing of fingermillet-associated microbial consortia provides insights into structural and functional diversity of endophytes. 3Biotech 10:1–17

Pun B, Nongkhlaw FM, Joshi S (2022) Metaomics technologies in understanding ethnomedicinal plants and endophyte microbiome. In: Plant-microbe interactions. CRC Press, pp 129–149

Qin Y, Druzhinina IS, Pan X, Yuan Z (2016) Microbially mediated plant salt tolerance and microbiome-based solutions for saline agriculture. Biotechnol Adv 34:1245–1259. https://doi.org/10.1016/j.biotechadv.2016.08.005

Qu Z et al (2021) Exploring the symbiotic mechanism of a virus-mediated endophytic fungus in its host by dual unique molecular identifier–RNA sequencing. Msystems 6:e0081421. https://doi.org/10.1128/msystems.00814-00821

Quach NT et al (2023) Metabolic and genomic analysis deciphering biocontrol potential of endophytic Streptomyces albus RC2 against crop pathogenic fungi. Braz J Microbiol 54:2617–2626. https://doi.org/10.1007/s42770-023-01134-8

Remali J, Sarmin NIM, Ng CL, Tiong JJL, Aizat WM, Keong LK, Zin NM (2017) Genomic characterization of a new endophytic Streptomyces kebangsaanensis identifies biosynthetic pathway gene clusters for novel phenazine antibiotic production. PeerJ 5:e3738. https://doi.org/10.7717/peerj.3738

Sagita R, Quax W, Haslinger K (2021) Current state and future directions of genetics and genomics of endophytic fungi for bioprospecting efforts. Front Bioeng Biotechnol 9:649906

Santoyo G, Moreno-Hagelsieb G, del Carmen Orozco-Mosqueda M, Glick BR (2016) Plant growth-promoting bacterial endophytes. Microbiol Res 183:92–99

Satam H et al (2023) Next-generation sequencing technology: current trends and advancements. Biology 12:997

Scott K et al (2023) Endophyte genomes support greater metabolic gene cluster diversity compared with non-endophytes in Trichoderma. PloS One 18:e0289280

Sharma SK, Johri BN, Ramesh A, Joshi OP, Sai Prasad S (2011) Selection of plant growth-promoting Pseudomonas spp. that enhanced productivity of soybean-wheat cropping system in Central India. J Microbiol Biotechnol 21:1127–1142

Sheibani-Tezerji R, Rattei T, Sessitsch A, Trognitz F, Mitter B (2015) Transcriptome profiling of the endophyte Burkholderia phytofirmans PsJN indicates sensing of the plant environment and drought stress. mBio 6:e00621–e00615. https://doi.org/10.1128/mBio.00621-15

Slama HB et al (2021) Biotechnology and bioinformatics of endophytes in biocontrol, bioremediation, and plant growth promotion. In: Endophytes: mineral nutrient management, vol 3. Springer, pp 181–205

Sonali Jaiswal AO, Thakur P, Mishra SK (2023) Functional importance of endophytic microorganisms in plant growth promotion bioactive compound production for sustainable agriculture. Def Life Sci J 8:93–108. https://doi.org/10.14429/dlsj.8.17944

Trovero MF, Scavone P, Platero R, de Souza EM, Fabiano E, Rosconi F (2018) Herbaspirillum seropedicae differentially expressed genes in response to iron availability. Front Microbiol 9:1430. https://doi.org/10.3389/fmicb.2018.01430

Tsuda K, Sato M, Stoddard T, Glazebrook J, Katagiri F (2009) Network properties of robust immunity in plants. PLoS Genet 5:e1000772. https://doi.org/10.1371/journal.pgen.1000772

Udayan E, Gnanadoss Joel J (2023) Potential of endophytic fungi as therapeutics: antibiotics, antiviral and anticancer properties. Res J Biotechnol 18:6

Utturkar SM et al (2016) Enrichment of root Endophytic bacteria from Populus deltoides and single-cell-genomics analysis. Appl Environ Microbiol 82:5698–5708. https://doi.org/10.1128/aem.01285-16

Vahabi K, Sherameti I, Bakshi M, Mrozinska A, Ludwig A, Oelmüller R (2015a) Microarray analyses during early and later stages of the Arabidopsis/Piriformospora indica interaction. Genom Data 6:16–18. https://doi.org/10.1016/j.gdata.2015.07.019

Vahabi K, Sherameti I, Bakshi M, Mrozinska A, Ludwig A, Reichelt M, Oelmüller R (2015b) The interaction of Arabidopsis with Piriformospora indica shifts from initial transient stress induced by fungus-released chemical mediators to a mutualistic interaction after physical contact of the two symbionts. BMC Plant Biol 15:58. https://doi.org/10.1186/s12870-015-0419-3

van Dijk EL, Jaszczyszyn Y, Naquin D, Thermes C (2018) The third revolution in sequencing technology trends in genetics. Trends Genet 34:666–681. https://doi.org/10.1016/j.tig.2018.05.008

Vandana UK et al (2021) The endophytic microbiome as a hotspot of synergistic interactions, with prospects of plant growth promotion. Biology 10:101

Veilumuthu P, Nagarajan T, Sasikumar S, Siva R, Jose S, Christopher JG (2022) Streptomyces sp. VITGV100: an endophyte from Lycopersicon esculentum as new source of indole type compounds. Biochem Syst Ecol 105:104523. https://doi.org/10.1016/j.bse.2022.104523

Verma V, Srivastava A, Garg SK, Singh VP, Arora PK (2023) Incorporating omics-based tools into endophytic fungal research. Biotechnol Notes 5:1–7

Yuan J, Zhang W, Sun K, Tang MJ, Chen PX, Li X, Dai CC (2019a) Comparative transcriptomics and proteomics of Atractylodes lancea in response to Endophytic fungus Gilmaniella sp. AL12 reveals regulation in plant metabolism. Front Microbiol 10:1208. https://doi.org/10.3389/fmicb.2019.01208

Yuan X-L et al (2019b) Comparative mitochondrial genome analysis of the tobacco endophytic fungi Leptosphaerulina chartarum and Curvularia trifolii and the phylogenetic implications. https://doi.org/10.20944/preprints201907.0119.v1

Zhang Q, Xue X-Z, Miao S-M, Cui J-L, Qin X-M (2020) Differential relationship of fungal endophytic communities and metabolic profiling in the stems and roots of Ephedra sinica based on metagenomics and metabolomics. Symbiosis 81:115–125

Evolutionary and Ecological Factors Influencing Endophyte Interactions

7

Ayushi Malik, Mayank Suthar, Khushboo Jain, Navnidhi Sharma, Anukrati Dhabhai, Mukesh Meena, and Avinash Marwal

Abstract

Endophytes as microorganisms residing in plant tissues systemically have emerged as the most critical players in plant-microbe interactions, having significant impacts such as to the evolutionary trajectories and ecological dynamics of host plants. This chapter outlines the complex interplay between the host plants and their endophytes and unveils the evolutionary and environmental factors influencing the symbiosis. One of the factors impacting the endophyte is host specificity, another is co-evolutionary relationship with the plant, and third one is abiotic factors like temperature and nutrient availability. The molecular mechanisms which endophytes use, for example, gene expression of stress responsive genes and secondary metabolite production are very important in building of host plant tolerance to happenings of environmental stresses. High-level methods such as mass spectrometry and molecular networking determine the microbial

A. Malik · M. Suthar · A. Marwal (✉)
Department of Biotechnology, Vigyan Bhawan-Block B, New Campus, Mohanlal Sukhadia University, Udaipur, Rajasthan, India

K. Jain
Department of Biotechnology, Faculty of Science, MVSC, Janardhan Rai Nagar Rajasthan Vidyapeeth (Deemed to be University), Udaipur, Rajasthan, India

N. Sharma
School of Life Sciences, Coventry University, Coventry, UK

A. Dhabhai
National Institute for Implementation Research on Non-Communicable Diseases, Jodhpur, Rajasthan, India

M. Meena
Laboratory of Phytopathology and Microbial Biotechnology, Department of Botany, Mohanlal Sukhadia University, Udaipur, Rajasthan, India

M. Meena et al. (eds.), *Multi-omics Approach to Investigate Endophyte Diversity*, Microorganisms for Sustainability 53, https://doi.org/10.1007/978-981-96-4100-0_7

metabolites as well as their producing organisms. This makes it possible to learn more about the details of the molecular interactions that are sustaining an endophyte-host relationship. Furthermore, the review covers the impact of frontier cultivation techniques in association between symbionts and their impact on sustainable agriculture. Possibilities for the future in this field pay attention to deeper research work about mechanisms of endophyte-mediated stress tolerance and their use in environmental management and crop upgrading.

Keywords
Endophytes · Evolution · Stress factors · Plant-microbe interactions · Ecology network

7.1 Introduction

Microorganisms known as endophytes are those that colonize live plant tissue symptomless and do not immediately or overtly harm the plant. Endophytes were earlier reported as organisms that grow within plant tissues as per their Greek prefixes "endo" (within) and "phyton" (plant) by De Bary in 1866 (Giba et al. 2020). Microorganisms have an impact on plants not only through exterior interactions that occur above- or below ground, but also through internal interactions that occur when they colonize and create plant-microbe endosphere above the ground (with leaves and stems) and below the ground (with roots). There have been reports of over 200 genera from 16 phyla of bacterial species linked to endophytes; the majority of these species are found in the phyla Actinobacteria, Proteobacteria, and Firmicutes endophytes, the highly complex link between microorganisms and their plant hosts, be a very vital aspect in the study of plant-microbe interactions. The cleavages between the plant tissues which were once ignored have attracted a lot of attention because of their great impacts which have been established on plant health, growth, and ecological resilience. The tracing of interactions that affect these processes is not only important when it comes to the enlightenment of basic biological phenomena but also because they may be taken advantage of by plant breeding, biotechnology, and ecological restoration (Swapnil et al. 2023a, b).

Endophytic associations are among the key components of the evolutionary and ecological forces that regulate the interactions between host plants and their endophytes. The subtle evolutionary process, which has been designed by millions of years of co-evolutionary history, is responsible for the formation of the complex molecular conversations and the physical adaptations (Rodriguez et al. 2008). While so, ecological factors, such as the abiotic environmental conditions and competing interspecific relations, are the driving forces for the endophytic communities' establishment, diversity, and function in the plant host (Hardoim et al. 2008).

Endophytes are very successful and able to partner with plants since they have received evolutionary forces with a battery of genetic traits and metabolic abilities. The phylogenetic origin of endophytic relations can be traced back to the evolutionary ancient mutualistic alliances between plants and microbes, with signs of which such interactions played a foundational role in the colonization of land by early plants (Porras-Alfaro and Bayman 2011). Therefore, naturally, the endophytes have developed mechanisms to escape from plant defenses, get access to the nutrients, as well as confer benefit such as the fatigue tolerance and diseases resistance to their host plants (Hardoim et al. 2008).

From the ecological point of view, endophytic interactions as well as complex relations between plant ecosystem's biotic and abiotic factors are the key factors of environmental dynamics. Endophyte species colonization depends on such factors as soil type, climate, host-plant species, and microbial community in rhizosphere (Compant et al. 2010). Furthermore, the competitive interactions among the endophytic species and their host plant attract the composition and diversity of the endophytic community which have important implications on plant health and ecosystem functioning (Hardoim et al. 2008).

In this chapter, we touch upon the evolutionary and ecological factors, which predetermine the nature of endophyte-host relationships, interweaving our current knowledge that comes from different disciplines including microbiology, ecology, evolutionary biology, and plant sciences. We seek to unravel the leads behind these interactions, the adaptation strategies of endophytes, the ecological effects of endophyte-host association, and the possible uses in sustainable agriculture and environmental management.

7.2 Endophytes: The Dynamic Entity

De Bary in 1866 was the first to identify additional (non-pathogenic) creatures within plants when he found microbial cells in the tissues of the plants that had been examined under a microscope, but this finding remained unexamined until the end of the twentieth century, when the term "endophytes" was defined. The word "endophyte" was initially used by De Bary in 1866 to describe any organism that develops within plant tissues.

Plant-associated prokaryotes known as endophytes bind to their host plants by invading their interior tissues. This characteristic makes them useful as a tool in agriculture to enhance crop productivity. Numerous researches have examined the link between plants and their endophytes, particularly in the case of grasses like tall fescue. It has been demonstrated that endophytes release poisons that deter grazing animals and insects. Finding bioactive compounds such as Taxol (Paclitaxel) (Strobel et al. 1996), Camptothecin (Kusari et al. 2013), Artemisinin (Kusari and Spiteller 2012), and Cepharanthine (Zhao et al. 2011) in the endophyte has attracted a lot of attention.

Endophytes not only increase nutrient intake in plants but also boost their immunity against various types of stress, such as infections, osmotic stress, heavy metals, and pollutants. By strengthening plant defenses and employing antibiotics, hydrolytic enzymes, and nutrient restriction, endophytes can enhance plant health and target pests and diseases. Endophytes may produce growth-regulating phytohormones such auxins, cytokinins, gibberellic acid, and ethylene (ACC deaminase) or

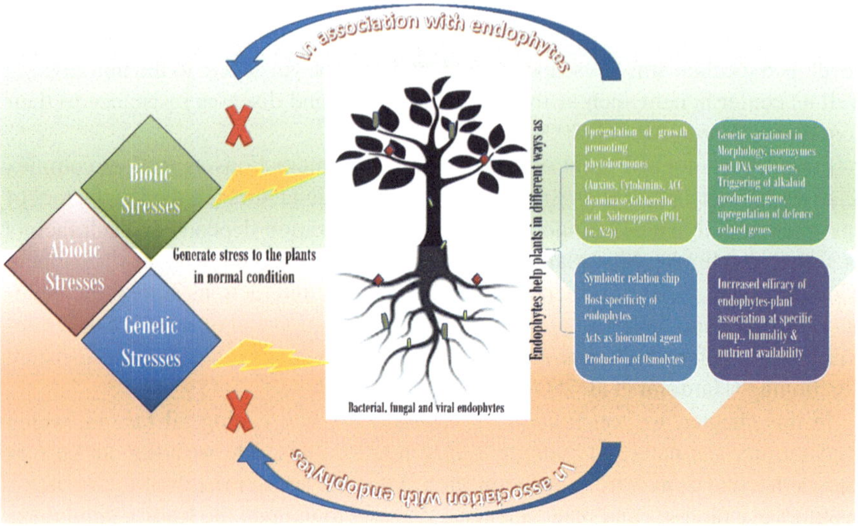

Fig. 7.1 Endophytes and their role in reduction of plant stresses

they can produce beneficial metabolites from the environment like phosphorus, iron (siderophores), and nitrogen (Sudheep et al. 2017; Aleynova et al. 2023).

These endophytes provided supplemental nitrogen from their substrates and in return created their communities around the roots of such plants (Krings et al. 2021). Due to the benefits from this mutualistic relationship, some plants even went to the extent of developing specialized structures to harbor endophytes (Tailor et al. 2023). Rhynia, an early vascular plant lacking true roots, developed rhizoids to increase endophytic substrate interface (Kerp 2017). Bryophytes possess non-photosynthetic filaments to accommodate microbial endophytes whereas these microbial colonies reside in root cells in the sporophyte stage of ferns (Chang et al. 2023) (Fig. 7.1).

7.2.1 Types of Endophytes and Their Subsequent Interactions

Various types of bacteria (such as actinomycetes or mycoplasma) or fungi belong to the endophytic communities and related to the plants in different ways. The endophytic communities have been split into subgroups, such as "obligate" or "facultative," which are connected to all kinds of plants, obligate endophytes (Mycorrhizal fungi, members of the fungal genera *Balansia, Epichloë*, and *Neotyphodium*) are defined as those that are dependent on plant metabolism for their existence and are dispersed vertically. Conversely, facultative endophytes (*Piriformospora indica*) are those that, for a portion of their life cycle, reside outside of the host and are typically linked to plants from the nearby soil and atmosphere (Gouda et al. 2016). Figure 7.2 shows the various types of endophytes and their examples.

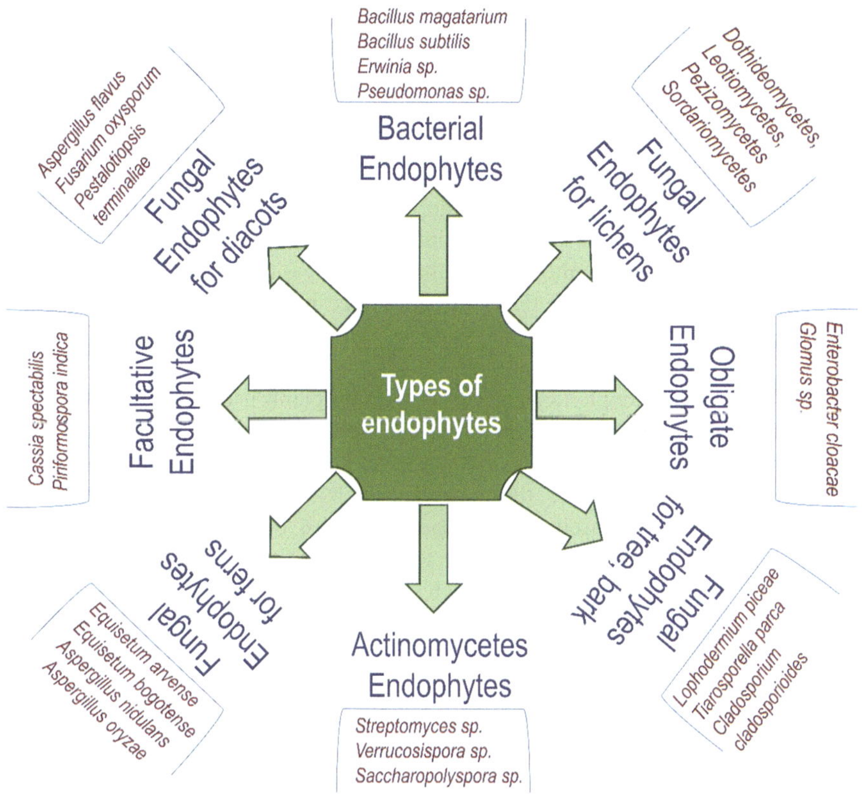

Fig. 7.2 Types of endophytes

Corresponding plant-microbe interactions are necessary for successful endophyte colonization. The plant recognizes the endophyte when it invades the host, and signal molecule cross-talk begins. Numerous investigations have demonstrated how endophytes respond chemotactically to host plant root exudates (Khare et al. 2018).

The allelopathic effects of red clover on maize, which result in decreased plant emergence and height, appear to be caused by endophytic bacteria in the host plants by inhibiting nematode multiplication; bacterial endophytes may help other crops that are grown in rotation with the host plants. The flora that flourishes in sand dunes with low nutrient levels comprises members of the genera *Pseudomonas, Stenotrophomonas*, and *Burkholderia* (Dalton et al. 2004). The presence of nitrogenase in the roots of plant cell walls of stems and rhizomes, as identified by antibodies, suggests that *Burkholderia* endophytes may be a source of N for the grasses. The host plant's genotype appears to have a significant influence on how the plant reacts to endophytes. Certain plants and wild races appear to offer sufficient in-plant circumstances that encourage and facilitate endophyte nitrogen fixing and reap its benefits.

Research on the interactions between endophytes such as *Azorhizobiumcauli nodans* and enterobacteria has been conducted using *Arabidopsis thaliana* as a model plant. *Arabidopsis thaliana*, which is a small flowering plant widely known for its usage in scientific research, has a unique advantage of the endophyte colonization study since its mutations are well characterized and specified. Such types of mutations ensure that the researchers can study plant behavior in interaction with endophytes grown in the laboratory. Endophytic bacteria find alternative means of growth and communication through the internal mediums of plants. This contrasts with the use of traditional culture media that gives a less authentic and more generalized model of bacterial genetic expression since it does not mimic the plant's native environment. The complex and multi-faceted ecosystems which have internal clocks and evolve as organized and unstable factors add to the importance of these studies. One role of endophytic bacteria is to defend plants using the "quorum sensing" mechanism. The density of the bacterial population is the determinant of this method where bacteria emit diffusible signal molecules. These molecules, again, govern the expression of genes, thus, enabling the bacteria to aggregate in response to their surroundings. Based on the examination of endophyte colonization in *Arabidopsis thaliana*, researches are able to obtain even more valuable information on the intricate networking between bacterial genetics and plant hosts, deepening the understanding of the complex nature of microbial communities within plants and their reactions to the environmental cues (Rosenblueth and Martínez-Romero 2006).

Eventually, the dependence on endophytes became so significant that their colonies were found in almost all plant organs such as roots, leaves (e.g., fungal endophytes in *Theobroma cacao*), stems (e.g., endophytes in elm plantlets), seeds (e.g., endophytes in *Helianthemum chamaecistus*), flowers (e.g., endophytic fungi of *Geodorum densiflorum*, an endangered orchid), and fruits (e.g., Endophytes from the *Vaccinium* sp. of blueberry) (Strobel 2018; Vaitiekūnaitė et al. 2022; Rahayu et al. 2021).

These types of symbiotic relationships between a plant and an endophyte have benefitted the infected host by increasing the nutrients uptake, stress tolerance, protection from pests, etc. (Clay and Schardl 2002) while in return providing photosynthates to the endophytic component for metabolism (Cheplick and Faeth 2009).

7.3 Evolutionary Origin of Endophytes

The evolution of land plants on Earth dates back to the Paleozoic era (more than half a billion years ago) where they were more susceptible to grow in harsher conditions compared to their algal predecessors (Chang et al. 2023). As the lack of nutrients and organic content was a major part of these conditions, these plants developed a mutual relationship with the fungal and bacterial communities (Wellman et al. 2003).

7.3.1 Detrimental Factors and Their Vital Role in Evolution

The establishment and dynamics of interactions between endophytes and host plants are intricately shaped by a multitude of factors, both biotic and abiotic, which play crucial roles in driving evolutionary trajectories for both parties (Kraft et al. 2015). Biotic factors encompass a spectrum of interactions, including those with host plants, competition with other microorganisms, and co-evolutionary relationships, all of which contribute to the specificity and adaptability of endophytes (Lu et al. 2021; Marwal et al. 2014). For example, studies have demonstrated the host specificity of endophytes, highlighting their ability to form mutually beneficial associations with specific plant species. In the case of *Trichoderma* fungi, known for their symbiotic bonds with plants, they act as effective biocontrol agents against soil pathogens, underscoring the importance of host specificity in endophytic relationships (Lu et al. 2021).

Similarly, abiotic factors such as temperature, humidity, and nutrient availability also exert profound influences on endophyte adaptation and survival within host plants. Recent research has shown that specific temperature ranges can significantly influence the degree of interaction between endophytes and their host plants, with certain temperatures promoting higher rates of colonization and symbiotic activity (cite specific studies) (Meena et al. 2022, 2023). For instance, findings have indicated that at particular temperatures, the establishment of endophytic interactions is more pronounced, suggesting a direct correlation between temperature and the efficacy of endophyte-plant associations.

Furthermore, exposure to abiotic stressors like drought or pathogen attacks can induce evolutionary changes in endophytes, driving the development of traits that enhance their resilience and effectiveness in symbiotic relationships with plants (Marwal et al. 2019, 2020). Notably, studies have documented the role of endophytes such as *Piriformospora indica* in enhancing host plant tolerance to abiotic stress, including drought, underscoring the adaptive potential of endophytic associations in mitigating environmental pressures (Lu et al. 2021).

Overall, the interplay between biotic and abiotic factors serves as a driving force in the evolution of endophytes, shaping their interactions with host plants and ultimately influencing the ecological dynamics of plant-microbe associations (Table 7.1).

7.3.2 Gene Pool Variation in Endophytes Co-evolution and Its Adaptation

Gene pool variation in endophytes, like any other organism, is also influenced by several other factors which can include random mutations occurring over time, competition for resources with other species, environmental factors like weather, soil pH, nutrient availabilities, horizontal gene transfer with other species living in the same environment, and most commonly, the host-endophyte interaction. It has been observed that the endophytes have evolved over time and have altered the

Table 7.1 Factors affecting endophytes

Factor	Description
Plants endophyte co-evolution relation	The endophytic community probably underwent co-evolution among themselves and with their host. Mutual relationships that are beneficial to both participants, sometimes, evolve to even more complicated forms, so that ecological factors can remain balanced (Frey-Klett et al. 2011)
Plant specificity in relation with endophytic co-existence	Endophytes exhibit varying degrees of host specificity, with some being restricted to specific host species or genera, while others show a greater degree of adaptability and they are capable of colonizing a wider range of host species (Gange et al. 2007).
Vertical transmission criteria	Endophytes have the ability to persist within plant populations as they can be passed down from parent to offspring through seeds. This ensures their continuity and presence over time (Truyens et al. 2015).
Horizontal transmission criteria	Endophytes do not just spread only upwards through parents to offspring like seeds and horizontally throughout vegetative propagation. They can be broadcast in various methods like soil, air, and various conveyors like insects and animal, thus, it is spreading as well and different plants are infected (Frank et al. 2017).
Abiotic conditions	The diversity and distribution of endophyte communities are affected by several factors, including temperature, humidity, soil type, and nutrient availability. These abiotic factors play a crucial role in shaping the composition and abundance of endophytes within a given plant community (Chen et al. 2020).
Competitive existence of various endophyte with plant	Endophytes may engage in resource competition within plant tissues, which can affect their colonization success and diversity. This competition between different endophyte species can influence their ability to establish and thrive within a plant community (Khare et al. 2018).
Symbiotic interactions	Certain endophytes exhibit mutualistic associations with their host plants, which can lead to benefits such as improved nutrient uptake or protection against pathogens. These beneficial relationships between endophytes and their host plants can contribute to the overall health and fitness of the plant community (Müller and Krauss 2005; Meena et al. 2020).
Successive ecological phenomena	As plant communities undergo succession, the endophyte communities associated with them may also undergo changes. Different endophyte species may become dominant at different stages of succession, leading to shifts in the composition and diversity of endophyte communities over time (Wani et al. 2015).
Plant defense mechanisms against competitive endophytes	Endophytes are subject to various defense mechanisms that plants have evolved over time. These mechanisms can impact the diversity and composition of endophyte communities associated with plants (Fadiji and Babalola 2020).
Endophytes supportive mechanism for plant stress tolerance	These endophytes have the ability to assist plants in tolerating different environmental stresses, including drought, salinity, or heavy metal toxicity. Through this mechanism, endophytes can support the growth and survival of plants in harsh environmental conditions (Kamran et al. 2022; Maheshwari et al. 2023).

expression of several genes within their host tissues. This genetic variation changes the probability of pathogen attack hence increasing the persistence of endophytes directly affecting their population (Schönrogge et al. 2022).

Endophytes have a diverse morphology and host scope. They have displayed considerable variations in their characteristics like host specificity, conidial morphology, genetic variations of isozymes and DNA sequences, and stroma production. All these characteristics act as adaptive markers in endophytic symbiotic relationships with the host and their reproductive strategies (Clay 2003; Marwal and Gaur 2020). Latest studies have also hinted towards alkaloid production of one or more classes in endophytes which helps the host plant in combating against various pests. It may be due to hybridization among asexual endophytes that results in the proliferation of these alkaloid-producing genes (Clay and Schardl 2002).

Similar to other relationships like host-parasite, host-predator, or host-mutualist interactions, interactions between endophytes and plants can be envisioned as unfolding on the ecological stage of a dynamic fitness landscape. This landscape features adaptive peaks and valleys representing the most and least fit combinations of fungus and plant genotypes within a population. The development and persistence of highly integrated and specialized symbioses necessitate the alignment of architectural, morphological, physiological, and life history traits between the fungus and the host plant (*Epichloë* species in grasses and *Rhizobia* species in legumes) (Thompson 2009).

Fossil evidence dating back 400 million years strongly indicates that plants formed symbiotic associations with mycorrhizal fungi. The adaptive significance of this capability has endured throughout the evolutionary history of plants (Li et al. 2023). Plants actively enlist the assistance of microbes to navigate and endure challenging environmental conditions while in return the endophytes get refuge and nutrition (Saikkonen 2004).

7.4 Endophytic Interaction and Their Relation with Ecological Network

In addition to locally occurring abiotic stress factors, it has been discovered that the host genotype affects how some endophytes change their lifestyle to a pathogenic condition (Bacon et al. 2008). Using a microbial phylogenetic molecular ecology network, the response of the rice plant microbial community to the inoculation of cadmium-resistant endophytic bacterium R5 (*Stenotrophomonas*) was examined for the first time in that study. The findings demonstrated that the introduction of R5 into rice plants altered the topological features of the microbial network. The resulting network exhibited increased complexity and interaction in the roots and above ground portions, suggesting that the introduction of R5 created an environment that was conducive to microbial interactions. Furthermore, these interactions might have something to do with how rice absorbs and transports cadmium (Zheng et al. 2022). Endophyte resources are valuable for study in a variety of domains, including ecological protection, germicide development, and multi-resistance breeding. The diversity and community composition of endophytes and rhizosphere bacteria of white radish (*Raphanus sativus*) in various compartments and cultivation conditions, such as greenhouse and open field cultivation, were analyzed at the phylum and genus levels in this study using high-throughput sequencing (Illumina-MiSeq) technology (Sun et al. 2022). The variety of bacteria and fungi found in soil around plant roots may enhance the ecological interactions between these microorganisms and plants. The underlying mechanisms and possible microbial interactions at the soil-mangrove root contact are still mostly unknown (Zhou et al. 2021).

7.4.1 Abiotic Factors Correlation with Endophytes

Bacterial endophytes are internal to plants, but their distribution and population density are affected by biotic and abiotic variables such as plant-microbe interactions, microbe–microbe interactions, and environmental pressures (Ryan et al. 2008). In addition to locally occurring abiotic stress factors, it has been discovered that the host genotype affects how some endophytes change their lifestyle to a pathogenic condition (Bacon et al. 2008). This study suggests that abiotic factors may not be the primary cause of the disease outbreak because it discovered no differences in the rhizosphere soil physicochemical parameters between healthy and afflicted plants. However, only the bulk soil physical properties of each sampled tomato were determined in this study, and no report are available to correlation between rhizosphere soil properties and each sample's *F. oxysporum* dynamics. The soil's physiochemical properties clearly displayed patterns of discrepancy with the agricultural practice used in this study (Wei et al. 2018).

7.4.2 Biotic Factors Correlation with Endophytic Interaction

Recent studies have found that there are variations in the ability of different strains of a particular type of endophyte to stimulate the production of certain chemicals in plants. Consequently, some strains of endophytes that provide plants with protection against certain biotic stresses, such as insects, do not produce compounds that are detrimental to grazing animals (Latch 1993). It was found that the arrangement of the microbiome community associated with the host is determined by biotic filters that are influenced by the host's features, which are evaluated based on the phylogenetic distance between species. However, it is still challenging to differentiate between abiotic factors such as environmental specialization or dispersal barriers and possible biotic mechanisms such as host specificity or local species interactions in microbiomes that are highly diverse and transmitted horizontally. The study aimed to investigate the influence of the distance between replicated plots in a common garden and the host's phylogenetic relatedness among eighteen native Asteraceae plant species on the organization of the population of foliar fungal endophytes (FFE) (Whitaker et al. 2021).

7.4.3 Molecular Interaction of Endophytes

Endophytes employ a range of molecular strategies to bolster stress tolerance in host plants, including the stimulation and expression of stress-responsive genes, the production of scavenger molecules like reactive oxygen species (ROS), and the synthesis of antistress metabolites (Lata et al. 2018; Swapnil et al. 2021). For instance, the upregulation of stress-responsive genes like those encoding for heat shock proteins or antioxidant enzymes facilitates the endophytic colonization of plants, enhancing their resilience against environmental pressures.

In response, plants release various signaling molecules, including phytohormones like auxins, cytokinins, and ethylene, to attract beneficial microorganisms like endophytes. These hormones play crucial roles in modulating plant growth and development, as well as in orchestrating responses to biotic and abiotic stresses. By releasing such signaling molecules, plants effectively communicate with endophytes and facilitate their establishment within their tissues.

To elucidate the intricate molecular interactions between endophytes and host plants, researchers employ various advanced techniques, including co-culture approaches coupled with mass spectrometry (MS). By combining these techniques with molecular networking and automatic database searches, scientists can identify microbial metabolites and their producing organisms (Wang et al. 2016). Tandem MS (MS/MS) is particularly instrumental in this process, as it enables the identification of known molecules, the correlation of chemical entities based on structural similarity, and the expedited discovery of novel compounds.

In summary, the molecular interactions between endophytes and host plants involve the coordinated expression and suppression of specific genes, the production of signaling molecules by both parties, and the synthesis of metabolites that confer stress tolerance. Advanced techniques such as co-culture approaches, mass spectrometry, molecular networking, and tandem MS are essential tools in unravelling these complex interactions and discovering novel bioactive compounds with potential applications in agriculture and biotechnology.

7.5 Effects of Dynamic Cultivation on the Interaction

Since Gluconobacter is one of the most prevalent endophytic genera, comparable research on endophytes and their stressed hosts might disclose the flux via many metabolic pathways, illuminating the true nature of endophyte-mediated stress tolerance at the cellular and metabolic levels. AMF-induced increases in phosphorus under dual inoculation did not change the nutritional status; nevertheless, the dynamics of nitrogen and carbon were significantly changed. In particular, adding *S. williamsii* to mycorrhizal tomato plants resulted in a considerable increase in the concentration of nitrogen in the shoots; however, this benefit came at the expense of increased carbon emissions (Hallasgo et al. 2020). The results show that the agro-environmental policies have nonlinear social-ecological effects over time, which can differ from predictions because of the intricate interactions between families and land. Additionally, the results of this study imply that the spatial agent-based model is capable of modeling human agents' adaptive decision-making and interactions in dynamic social and physical settings (Wang et al. 2021).

7.6 Endophytes Affecting Host Genetic, Physiological, and Immunological Expressions

The interaction between endophytes and their host plants is dynamic and multifaceted. Research in this field has revealed that endophytes can have profound effects on the host's genetic makeup, altering gene expression patterns and influencing the plant's response to environmental stimuli. Additionally, endophytes can modulate the physiological processes of their host, such as nutrient uptake, water regulation, and growth patterns (Mahmood et al. 2017; Khare et al. 2018).

Endophytes, such as those belonging to the genus *Clavicipitaceae*, residing within the tissues of host plants, have a range of effects on their hosts by impacting genetic, physiological, and immunological dimensions (Florea et al. 2017). Genetically, certain endophytes, like *Epichloë*, enhance the expression of defense-related genes, exemplified by their ability to confer resistance against pathogenic invaders (Caradus and Johnson 2020). Physiologically, *Piriformospora indica,* for instance, contributes to improved nutrient absorption and water retention, fostering the host's growth and resilience to environmental stresses (Varma et al. 1999). Immunologically, these microorganisms induce systemic resistance; for example, Trichoderma strains have been shown to enhance the plant's defenses against subsequent pathogen attacks (Tseng et al. 2020).

The intricate relationships between endophytes and host organisms underscore the vital role these microorganisms play in shaping the genetic, physiological, and immunological expressions of plants, thereby influencing their overall health and adaptability in diverse ecological contexts.

7.7 Future Prospects

It is expected that the future endeavor of the inquiry of endophyte-host interactions is greatly beneficial in increasing our knowledge about the plant-microbial interactions and their ecological significance. Colonization and function of endophyte in host plants is one of the subjects of the molecular mechanism investigations. It will shed new light on strategies to enhance plant stress resistance. Additionally, examining the endophyte species diversity and composition in various ecological contexts will give rise to ecological practices for sustainable agricultural techniques and ecosystem-based strategy implementation. Furthermore, the establishment of biotechnological tools and methods will speed up the discovery and also usage of new endophyte-based bioactive compounds with pharmaceutical, agricultural, and biotechnology applications (Meena et al. 2021). Specifically, the next generation of research endeavors in this field will potentially reveal the complex web of interactions between endophytes and host plants, ultimately leading to the development of revolutionary approaches in addressing global problems focusing on food security, environmental resilience, and human health.

7.8 Conclusion

The endophytes represented as a remarkable and ecologically important part of the plant-microbes' interactions, exerting various roles in regard to the host plant health, resilience as well as adaptability. Both evolutionary as well as ecological variables, including for example host specificity, co-evolutionary dynamics and environmental conditions, mediate this relationship in a very complex way. The molecular mechanisms utilized by endophytes to promote host plants' resilience to both biotic and abiotic stresses provoke the possible application of these organisms in agronomy and environmental management. Through the disentangling of the intricate network of endophyte and plant appearances, scientists can implement the microbe synergistic in the fight to food shortage, ecosystem conservation, and human development. By harnessing interdisciplinary partnerships and inventive research strategies, endophyte biology endows the field with enticing prospects of uncovering new ways to give solutions that can meet our changing world.

Conflict of Interest The authors have no conflict of interest to declare.

References

Aleynova OA, Nityagovsky NN, Ananev AA, Suprun AR, Ogneva ZV, Dneprovskaya AA et al (2023) The endophytic microbiome of wild grapevines *Vitis amurensis Rupr.* and *Vitis coignetiae Pulliat* growing in the Russian Far East. Plan Theory 12(16):2952

Bacon CW, Glenn AE, Yates IE (2008) *Fusarium verticillioides*: managing the endophytic association with maize for reduced fumonisins accumulation. Toxin Rev 27(3–4):411–446

Caradus JR, Johnson LJ (2020) Epichloë fungal endophytes-from a biological curiosity in wild grasses to an essential component of resilient high performing ryegrass and fescue pastures. J Fungi (Basel) 6(4):322

Chang X, Young B, Vaccaro N, Strickland R, Goldstein W, Struwe L, White JF (2023) Endophyte symbiosis: evolutionary development, and impacts of plant agriculture. Grass Res 3(1). https://doi.org/10.48130/GR-2023-0018

Chen J, Akutse KS, Saqib HSA, Wu X, Yang F, Xia X et al (2020) Fungal endophyte communities of crucifer crops are seasonally dynamic and structured by plant identity, plant tissue and environmental factors. Front Microbiol 11:1519

Cheplick GP, Faeth S (2009) Ecology and evolution of the grass-endophyte symbiosis. Oxford University Press

Clay K (2003) The ecology and evolution of endophytes. Agric Ecosyst Environ 44(1–4):39–64

Clay K, Schardl C (2002) Evolutionary origins and ecological consequences of endophyte symbiosis with grasses. Am Nat 160(S4):S99–S127

Compant S, Clément C, Sessitsch A (2010) Plant growth-promoting bacteria in the rhizo-and endosphere of plants: their role, colonization, mechanisms involved and prospects for utilization. Soil Biol Biochem 42(5):669–678

Dalton DA, Kramer S, Azios N, Fusaro S, Cahill E, Kennedy C (2004) Endophytic nitrogen fixation in dune grasses (*Ammophila arenaria* and *Elymus mollis*) from Oregon. FEMS Microbiol Ecol 49(3):469–479

Fadiji AE, Babalola OO (2020) Elucidating mechanisms of endophytes used in plant protection and other bioactivities with multifunctional prospects. Front Bioeng Biotechnol 8:467

Florea S, Panaccione DG, Schardl CL (2017) Ergot alkaloids of the family clavicipitaceae. Phytopathology 107(5):504–518

Frank AC, Saldierna Guzmán JP, Shay JE (2017) Transmission of bacterial endophytes. Microorganisms 5(4):70

Frey-Klett P, Burlinson P, Deveau A, Barret M, Tarkka M, Sarniguet A (2011) Bacterial-fungal interactions: hyphens between agricultural, clinical, environmental, and food microbiologists. Microbiol Mol Biol Rev 75(4):583–609

Gange AC, Dey S, Currie AF, Sutton BC (2007) Site-and species-specific differences in endophyte occurrence in two herbaceous plants. J Ecol 95(4):614–622

Giba J, Wangpan T, Tangjang S (2020) Endophytic fungi in bamboo: a review. Int J Curr Microbiol App Sci 9(11):2024–2031

Gouda S, Das G, Sen SK, Shin HS, Patra JK (2016) Endophytes: a treasure house of bioactive compounds of medicinal importance. Front Microbiol 7:219261

Hallasgo AM, Spangl B, Steinkellner S, Hage-Ahmed K (2020) The fungal endophyte Serendipita williamsii does not affect phosphorus status but carbon and nitrogen dynamics in arbuscular mycorrhizal tomato plants. J Fungi 6(4):233

Hardoim PR, van Overbeek LS, van Elsas JD (2008) Properties of bacterial endophytes and their proposed role in plant growth. Trends Microbiol 16(10):463–471

Kamran M, Imran QM, Ahmed MB, Falak N, Khatoon A, Yun BW (2022) Endophyte-mediated stress tolerance in plants: a sustainable strategy to enhance resilience and assist crop improvement. Cells 11(20):3292

Kerp H (2017) Organs and tissues of Rhynie chert plants. Philos Trans R Soc Lond B Biol Sci 373(1739):20160495

Khare E, Mishra J, Arora NK (2018) Multifaceted interactions between endophytes and plant: developments and prospects. Front Microbiol 9:2732

Kraft NJB, Adler PB, Godoy O, James EC, Fuller S, Levine JM (2015) Community assembly, coexistence and the environmental filtering metaphor. Funct Ecol 29(5):592–599

Krings M, Serbet SM, Harper CJ (2021) Rhizophyditesmatryoshkaegen. et sp. nov. (fossil Chytridiomycota) on spores of the early land plant Horneophytonlignieri from the lower Devonian Rhynie Chert. Int J Plant Sci 182(2):109–122

Kusari S, Spiteller M (2012) Are we ready for industrial production of bioactive plant secondary metabolites utilizing endophytes? Nat Prod Rep 29(7):793–804

Kusari S, Singh S, Jayabaskaran C, Spiteller M (2013) Biotechnological potential of plant-associated endophytic fungi: hope versus hype. Trends Biotechnol 31(7):297–303

Lata R, Chowdhury S, Gond SK, White JF Jr (2018) Induction of abiotic stress tolerance in plants by endophytic microbes. Lett Appl Microbiol 66(4):268–276

Latch GCM (1993) Physiological interactions of endophytic fungi and their hosts. Biotic stress tolerance imparted to grasses by endophytes. Agric Ecosyst Environ 44(1–4):143–156

Li Z, Xiong K, Wen W, Li L, Xu D (2023) Functional endophytes regulating plant secondary metabolism: current status, prospects and applications. Int J Mol Sci 24(2):1153

Lu H, Wei T, Lou H, Shu X, Chen Q (2021) A critical review on communication mechanism within plant-endophytic fungi interactions to cope with biotic and abiotic stresses. J Fungi 7(9):719

Maheshwari Y, Tailor S, Marwal A, Mishra A (2023) Plant-microbe interactions to reduce salinity stress in plants for the improvement of the agricultural system. In: Plant-microbe interaction-recent advances in molecular and biochemical approaches. Academic Press, pp 297–309. https://doi.org/10.1016/B978-0-323-91876-3.00002-6

Mahmood S, Lakra N, Marwal A, Sudheep NÁ, Anwar K (2017) Crop genetic engineering: an approach to improve fungal resistance in plant system. In: Plant-microbe interactions in agro-ecological perspectives: volume 2: microbial interactions and agro-ecological impacts, pp 581–591. https://doi.org/10.1007/978-981-10-6593-4_23

Marwal A, Gaur RK (2020) Molecular markers: tool for genetic analysis. In: Animal biotechnology. Academic Press, pp 353–372. https://doi.org/10.1016/B978-0-12-811710-1.00016-1

Marwal A, Sahu A, Gaur RK (2014) New insights in the functional genomics of plants responding to abiotic stress. In: Molecular approaches in plant abiotic stress. CRC Press, Science Publishers, Taylor & Francis Group, Boca Raton, pp 158–180. https://doi.org/10.1201/b15538

Marwal A, Kumar R, Verma RK, Mishra M, Gaur RK, Khurana SP (2019) Genomics and molecular mechanisms of plant's response to abiotic and biotic stresses. In: Plant biotechnology: progress in genomic era. Springer, pp 131–146. https://doi.org/10.1007/978-981-13-8499-8_6

Marwal A, Srivastava AK, Gaur RK (2020) Improved plant tolerance to biotic stress for agronomic management. Agrica 9:84–100. https://doi.org/10.5958/2394-448X.2020.00013.9

Meena M, Swapnil P, Divyanshu K, Kumar S, Harish TYN, Zehra A, Marwal A, Upadhyay RS (2020) PGPR-mediated induction of systemic resistance and physiochemical alterations in plants against the pathogens: current perspectives. J Basic Microbiol 60(10):828–861. https://doi.org/10.1002/jobm.202000370

Meena M, Zehra A, Swapnil P, Harish, Marwal A, Yadav G, Sonigra P (2021) Endophytic nanotechnology: an approach to study scope and potential applications. Front Chem 9:613343. https://doi.org/10.3389/fchem.2021.613343

Meena M, Yadav G, Sonigra P, Nagda A, Mehta T, Swapnil P, Marwal A (2022) Role of elicitors to initiate the induction of systemic resistance in plants to biotic stress. Plant Stress 5:100103. https://doi.org/10.1016/j.stress.2022.100103

Meena M, Yadav G, Sonigra P, Nagda A, Mehta T, Swapnil P, Marwal A, Zehra A (2023) Advantageous features of plant growth-promoting microorganisms to improve plant growth in difficult conditions. In: Plant-microbe interaction-recent advances in molecular and biochemical approaches. Academic Press, pp 279–296. https://doi.org/10.1016/B978-0-323-91876-3.00019-1

Müller CB, Krauss J (2005) Symbiosis between grasses and asexual fungal endophytes. Curr Opin Plant Biol 8(4):450–456

Porras-Alfaro A, Bayman P (2011) Hidden fungi, emergent properties: endophytes and microbiomes. Annu Rev Phytopathol 49:291–315

Rahayu ND, Sukarno N, Listiyowati S, Rafi M, Mursidawati S, Sandra E (2021) Flower endophytic fungi of *Geodorum densiflorum* endangered orchid. IOP Conf Ser Earth Environ Sci 948(1):012037

Rodriguez RJ, Henson J, Van Volkenburgh E, Hoy M, Wright L, Beckwith F, Redman RS (2008) Stress tolerance in plants via habitat-adapted symbiosis. ISME J 2(4):404–416

Rosenblueth M, Martínez-Romero E (2006) Bacterial endophytes and their interactions with hosts. Mol Plant-Microbe Interact 19(8):827–837

Ryan RP, Germaine K, Franks A, Ryan DJ, Dowling DN (2008) Bacterial endophytes: recent developments and applications. FEMS Microbiol Lett 278:1–9

Saikkonen K (2004) Evolution of endophyte–plant symbioses. Trends Plant Sci 9(6):275–280

Schönrogge K, Gibbs M, Oliver A, Cavers S, Gweon HS, Ennos RA, Cottrell J, Iason GR, Taylor J (2022) Environmental factors and host genetic variation shape the fungal endophyte communities within needles of Scots pine (Pinus sylvestris). Fungal Ecol 57–58:101162

Strobel G (2018) The emergence of endophytic microbes and their biological promise. J Fungi (Basel) 4(2):57

Strobel G, Yang X, Sears J (1996) Taxol from *Pestalotiopsis microspora*, an Endophytic fungus of *Taxus wallachiana*. Microbiology 142(2):435–440

Sudheep NÁ, Marwal A, Lakra N, Anwar K, Mahmood S (2017) Fascinating fungal endophytes role and possible beneficial applications: an overview. In: Plant-microbe interactions in agro-ecological perspectives: volume 1: fundamental mechanisms, methods and functions, pp 255–273. https://doi.org/10.1007/978-981-10-5813-4_13

Sun N, Gu Y, Jiang G, Wang Y, Wang P, Song W, Ma P, Duan Y, Jiao Z (2022) Bacterial communities in the endophyte and rhizosphere of white radish (*Raphanus sativus*) in different compartments and growth conditions. Front Microbiol 13:900779. PMID: 35847086, PMCID: PMC9277120

Swapnil P, Meena M, Singh SK, Dhuldhaj UP, Marwal A (2021) Vital roles of carotenoids in plants and humans to deteriorate stress with its structure, biosynthesis, metabolic engineering and functional aspects. Curr Plant Biol 26:100203. https://doi.org/10.1016/j.cpb.2021.100203

Swapnil P, Meena M, Marwal A, Vijayalakshmi S, Zehra A (eds) (2023a) Plant-microbe interaction-recent advances in molecular and biochemical approaches: volume 1: overview of biochemical and physiological alteration during plant-microbe interaction. https://doi.org/10.1016/C2021-0-00316-3

Swapnil P, Meena M, Marwal A, Vijayalakshmi S, Zehra A (eds) (2023b) Plant-microbe interaction-recent advances in molecular and biochemical approaches: volume 2: agricultural aspects of microbiome leading to plant defence. https://doi.org/10.1016/C2021-0-00317-5

Tailor S, Jain K, Marwal A, Meena M, Mishra A (2023) Agricultural management by improving beneficial microflora. In: Plant-microbe interaction-recent advances in molecular and biochemical approaches, pp 95–116. https://doi.org/10.1016/B978-0-323-91876-3.00004-X

Thompson JN (2009) The coevolutionary process. University of Chicago Press

Truyens S, Weyens N, Cuypers A, Vangronsveld J (2015) Bacterial seed endophytes: genera, vertical transmission and interaction with plants. Environ Microbiol Rep 7(1):40–50

Tseng Y-H, Rouina H, Groten K, Rajani P, Furch ACU, Reichelt M, Baldwin IT, Nataraja KN, Shaanker RU, Oelmüller R (2020) An endophytic trichoderma strain promotes growth of its hosts and defends against pathogen attack. Front Plant Sci 11:573670

Vaitiekūnaitė D, Bružaitė I, Snitka V (2022) Endophytes from blueberry (Vaccinium sp.) fruit: characterization of yeast and bacteria via label-free surface-enhanced Raman spectroscopy (SERS). Spectrochim Acta A Mol Biomol Spectrosc 275:121158

Varma A, Savita V, Sudha, Sahay N, Butehorn B, Franken P (1999) *Piriformospora indica*, a cultivable plant-growth-promoting root endophyte. Appl Environ Microbiol 65(6):2741–2744

Wang M, Carver JJ, Phelan VV, Sanchez LM, Garg N, Peng Y, Bandeira N (2016) Sharing and community curation of mass spectrometry data with global natural products social molecular networking. Nat Biotechnol 34(8):828–837

Wang Y, Zhang Q, Sannigrahi S, Li Q, Tao S, Bilsborrow R, Song C (2021) Understanding the effects of China's agro-environmental policies on rural households' labor and land allocation with a spatially explicit agent-based model. J Artif Soc Soc Simul 24(3):10–18564

Wani ZA, Ashraf N, Mohiuddin T, Riyaz-Ul-Hassan S (2015) Plant-endophyte symbiosis, an ecological perspective. Appl Microbiol Biotechnol 99:2955–2965

Wei W, Wang L, Bao P, Shao Y, Yue H, Yang D, Wang H (2018) Metal-free C (sp2)–H/N–H cross-dehydrogenative coupling of quinoxalinones with aliphatic amines under visible-light photoredox catalysis. Org Lett 20(22):7125–7130

Wellman CH, Osterloff PL, Mohiuddin U (2003) Fragments of the earliest land plants. Nature 425(6955):282–285

Whitaker BK, Christian N, Chai Q, Clay K (2021) Foliar fungal endophyte community structure is independent of phylogenetic relatedness in an Asteraceae common garden. Ecol Evol 10(24):13895–13912. PMID: 33391689, PMCID: PMC7771118

Zhao K, Penttinen P, Guan T, Xiao J, Chen Q, Xu J, Zhang L (2011) The diversity and antimicrobial activity of endophytic Actinomycetes isolated from medicinal plants in Panxi plateau, China. Curr Microbiol 62(1):182–190

Zheng Z, Li P, Xiong Z, Ma T, Mathivanan K, Praburaman L, Meng D, Yi Z, Ao H, Wang Q, Rang Z, Li J (2022) Integrated network analysis reveals that exogenous cadmium-tolerant endophytic bacteria inhibit cadmium uptake in rice. Chemosphere 301:134655

Zhou Z, Hu R, Ni Y, Zhuang W, Luo Z, Huang W, Yan Q, He Z, Zhong Q, Wang C (2021) Genetic Elucidation of Quorum Sensing and Cobamide Biosynthesis in Divergent Bacterial-Fungal Associations across the soil-mangrove root interface. Front Microbiol 12:698385. PMID: 34675891, PMCID: PMC8524053

Endophytes for Agricultural and Environmental Sustainability

8

Ravinder Kumar, Ajay Kamboj, Basanti Brar,
Vishnu Dayal Rajput, and Joginder Singh Duhan

Abstract

Endophytic microbes, encompassing fungi and bacteria, reside within plant tissues without causing harm, significantly benefiting host plants. These microorganisms enhance nutrient availability through nitrogen fixation, phosphorus, potassium, zinc solubilization, and the production of siderophores and phytohormones. These functions promote plant growth and protect against diverse abiotic and biotic stresses. Additionally, endophytic microbes secrete various hydrolytic enzymes and bioactive compounds, further supporting plant health and sustainability and offering a viable alternative to chemical fertilizers. Recent research has highlighted that endophytes enhance nutrient uptake, stress tolerance, and disease resistance in host plants, leading to improved yields. They exhibit capabilities to remediate heavy metals and organic pollutants and confer tolerance to stress, making them suitable for cultivation in marginal lands. Endophytes mitigate environmental risks associated with conventional agricultural practices by reducing reliance on synthetic fertilizers and pesticides. This chapter synthesizes current knowledge on the role of endophytes in agriculture, emphasizing their sustainable solutions for enhancing crop productivity and overall plant health.

R. Kumar (✉) · A. Kamboj · J. S. Duhan (✉)
Department of Biotechnology, Ch. Devi Lal University, Sirsa, India

B. Brar
Department of Microbiology, CCS Haryana Agricultural University, Hisar, India

V. D. Rajput
Academy of Biology and Biotechnology, Southern Federal University, Rostov-on-Don, Russia

M. Meena et al. (eds.), *Multi-omics Approach to Investigate Endophyte Diversity*, Microorganisms for Sustainability 53, https://doi.org/10.1007/978-981-96-4100-0_8

Keywords

Endophytes · Sustainable agriculture · Siderophore · Phytostimulation · Heavy metals

8.1 Introduction

Endophytes, predominantly bacteria and fungi, have been identified as essential components of plant microbiomes that form mutualistic connections with their hosts. Endophytes, unlike pathogens, live within plant tissues without creating visible disease symptoms. Instead, they frequently benefit their hosts, such as improved nutrient uptake and resilience to biotic and abiotic challenges (Gouda et al. 2018). As global concerns about food security, environmental degradation, and sustainable farming methods develop, there is an increasing interest in using endophytes to solve these issues (Qadri et al. 2014). Endophytic bacteria and fungi are essential for nitrogen fixation, phosphate, zinc, potassium solubilization, and other nutrient cycling processes, which promote soil fertility and plant growth (Bashan et al. 2014). Farmers can lessen their reliance on chemical fertilizers by inoculating crops with nitrogen-fixing bacteria like *Rhizobium* and *Azospirillum* or phosphate-solubilizing fungi like *Aspergillus* and *Penicillium*. This improves soil health and crop yields.

Endophytes enhance plant growth by producing chemicals such as indole-3-acetic acid (IAA), cytokinins (CK), and gibberellins (Santoyo et al. 2016). These phytohormones promote plant growth and development, including root elongation, lateral root production, and shoot proliferation. Furthermore, some endophytic bacteria have the potential to fix atmospheric nitrogen and convert it into ammonia or other nitrogenous compounds that plants may easily absorb (Bhattacharyya and Jha 2012). This nitrogen-fixing potential is especially advantageous for plants growing in nitrogen-deficient soils, where endophytic bacteria help to balance the ecosystem's nitrogen budget. The siderophore-producing microbes can also bind and transport the iron-siderophore complex by expressing specific proteins when there is an iron shortage (Dudeja et al. 1996; Duhan et al. 1998; Nudel et al. 2001; Bhupenchandra et al. 2024). These organisms are very sensitive to low-concentration metals deposited in food webs, providing a severe public health danger. Heavy metals continue accumulating in the environment due to the formation of numerous industries, posing health dangers to humans (Sharma et al. 2021). This chapter presents an overview of endophyte bioprospecting for agricultural and environmental sustainability, including their diversity, modes of action, and isolation and identification procedures.

8.1.1 Diversity of Endophytic Microorganisms

Endophytic microorganisms include many bacteria and fungi, each having distinct features and functional properties. Bacterial endophytes from genera such as *Bacillus, Pseudomonas*, and *Rhizobium* have been widely explored for their plant growth-promoting (PGP) properties, which include nitrogen fixation, phosphate solubilization, and phytohormone synthesis (Kandel et al. 2017; Chandran et al. 2021; Kumar et al. 2022). Similarly, fungal endophytes, primarily from the phyla Ascomycota and Basidiomycota, are extremely diverse and play critical roles in plant health and productivity (Hardoim et al. 2015). Fungal endophytes include *Claviceps, Fusarium,* and *Colletotrichum* species, which have been demonstrated to promote plant development, provide stress tolerance, and aid in disease suppression (Saikkonen et al. 2010).

The dominant genera associated with endophytic functions include *Achromobacter, Burkholderia, Bacillus, Enterobacter, Herbaspirillum, Pseudomonas, Pantoea, Rhizobium,* and *Streptomyces* across various host plants. This diversity underscores their agricultural potential to enhance crop productivity and sustainably maintain soil health (Negi et al. 2024). Various factors influence endophytic microbe diversity, including plant species, tissue type, geographic location, and environmental circumstances. Certain plant species support specific endophytic communities, while others have more diversified endophyte assemblages (Finkel et al. 2017). For example, tropical regions have higher endophytic diversity than temperate zones, indicating that climatic factors influence microbial colonization patterns (Vandenkoornhuyse et al. 2015).

8.2 Applications in Sustainable Agriculture

Endophytes have emerged as essential resources for sustainable agriculture, providing alternatives for increasing crop yield, lowering synthetic input dependency, and mitigating traditional agricultural practices' environmental implications. Endophytes improve plant health and performance through various methods, making them ideal candidates for biofertilization, biocontrol, and phytoremediation applications.

8.2.1 Biofertilization

Biofertilization is one of the most important applications of endophytes in sustainable agriculture, which involves using microorganisms to improve nutrient availability and plant uptake.

According to Ludueña et al. (2019), *Enterobacter* sp. J49 is an endophytic strain that stimulates plant development in maize and peanut crops. This strain encourages plant growth using various strategies, the most significant of which is the provision of soluble phosphorus. In addition to raising the plant's phosphorus content, *Enterobacter* sp. J49 also improves soil fertility, benefiting other plants typically grown in rotation with these commodities. Furthermore, it was noted that this strain is resilient to several abiotic stimuli, including salt, high temperatures, severe pHs, and a wide range of pesticides typically used on maize and peanut crops in Argentina (Anzuay et al. 2017).

8.2.2 Biocontrol/Pest and Disease Resistance

Endophytes promote plant development and nutrient intake and help biologically control plant diseases and pests, providing environmentally benign alternatives to synthetic pesticides (Berendsen et al. 2012). Many endophytic bacteria and fungi produce antimicrobial chemicals, enzymes, and volatile organic compounds (VOCs) that limit plant pathogens' development and activity, including fungi, bacteria, and nematodes (Pang et al. 2009; Seth et al. 2023).

The two most widely used mechanisms are the generation of antibiotics and induced systemic resistance. Bacterial species that are endophytes, including *Bacillus amyloliquifasciens, B. pumilus, B. subtilis, Pseudomonas fluorescens, P. syringae,* and *Serratia marcescens*, are capable of inducing systemic resistance (Kloepper and Ryu 2006). Furthermore, Marian et al. (2020) revealed a new endophytic *Streptomyces* sp. MBFA-172 from strawberries exhibited biocontrol potential against *Glomerella cingulata.* Similar to this, the olive endophyte *Bacillus* sp. has demonstrated antagonistic activity against *Xylella fastidiosa*, the cause of olive rapid decline syndrome (OQDS), and the ginseng endophyte *Burkholderia stabilis* has demonstrated activity against ginseng root rot disease (Kim et al. 2020a; Zicca et al. 2020).

Fungal endophytes may possess biocontrol properties because they synthesize inhibitory chemicals mycoparasitism, release of chitinolytic enzymes, and competition for nutrients and space (Biocontrol of phytopathogens like *Rhizoctonia solani*, a powerful phytopathogen, *Fusarium oxysporum*, which causes Fusarium wilt of lentil, and *Gaeumannomyces graminis*, which causes take-all disease of wheat, is aided by *Piriformospora indica, Sebacina vermifera*, and Trichoderma species (Melo and Faull 2000; Boonphong et al. 2001; Kari Dolatabadi et al. 2012.) According to Kemp et al. (2020), *Sarocladium zeae*, a systemic wheat endophyte, has been linked to biocontrol action against Fusarium head blight disease. Several methods for screening novel biocontrol drugs that exhibit direct or indirect biocontrol activity have recently been developed (Raymaekers et al. 2020). Farmers can efficiently manage diseases and pests in agricultural systems by leveraging endophyte antagonistic qualities while reducing the dangers associated with chemical pesticides.

8.2.3 Plant Growth Promotion and Enhanced Yield

Endophytes use a variety of ways to increase plant growth, improve nutrient uptake, and provide stress resistance to their host plants. These methods frequently entail extensive interactions between endophytes, plants, and the surrounding environment, resulting in complex physiological and biochemical reactions.

8.2.3.1 Mechanisms of Action

Endophytes not only promote plant development, but they also help their host plants absorb nutrients more efficiently. Certain endophytic fungi are known to solubilize phosphate and mobilize other critical minerals like iron and zinc, boosting their availability to plants (Nadeem et al. 2014). Endophytes improve plant nutrition and productivity by improving nutrient absorption and use efficiency, especially in nutrient-limited circumstances.

Toppo et al. (2024) isolated endophytic fungi from the leaves of *Anisomeles indica,* and five were identified as *Colletotrichum yulongense* Ai1, *Colletotrichum cobbittiense* Ai2, *Colletotrichum alienum* Ai2.1, *Colletotrichum cobbittiense* Ai3, and *Fusarium equiseti* using sequencing based on their ability for plant growth promotion, secondary metabolite production, and extracellular enzyme production. Five isolates tested positive for their plant growth promotion potential, while isolates Ai4. Ai1, Ai2, and Ai2.1 showed significant production of secondary metabolites, viz. alkaloids, phenolics, flavonoids, saponins etc. Isolate Ai2 showed maximum total phenolic concentration (25.98 mg/g), while isolate Ai4 showed maximum total flavonoid concentration (20.10 mg/g). Significant results were observed for producing extracellular enzymes such as cellulases, amylases, laccases, and lipases. The isolates significantly influenced the seed germination percentage of tomato seedlings and augmented their growth and development under in vitro assay. Following are the mechanisms through which endophytic fungi promote plant growth and enhance yield.

Nitrogen Fixation

The usage of agrochemical fertilizers has increased in the worldwide food production process to increase human consumption. Urea, or nitrogen fertilizer, is primarily used to supply plants with nitrogen nutrition to increase output. However, according to Bhattacharjee et al. (2008), overuse of chemical fertilizers has an adverse effect on soil fertility, aquatic creature growth, and greenhouse gas output. The microorganisms are crucial in transforming atmospheric nitrogen into ammonia, which the plant can use. One potential substitute for N-fertilizers is a bacterium that fixes nitrogen, which could result in more productive and sustainable agriculture. According to Gupta et al. (2012), various endophytic and associative bacteria are known to fix atmospheric N_2 and provide it to the host plant. Liaqat and Eltem (2016) described the nitrogenase activity, phosphate solubilization, and IAA generation of *Pseudoxanthomonas mexicana* and *Stenotrophomonas rhizophilia* isolated from pears. According to a study, endophytic bacteria of rice, specifically *Flavobacterium* sp., *Microbacterium* sp., and *Xanthomonas* sp., demonstrated encouraging plant growth promoting (PGP) activities such as fixing nitrogen and producing siderophores and phosphate solubilization (Walitang et al. 2017). *Bradyrhizobium* sp., an endophyte, was shown to have nitrogenase activity in another publication (Schneijderberg et al. 2018). According to Ke et al. (2019), inoculating maize with *Pseudomonas stutzeri*, an endophytic bacterium with nitrogenase activity, increased plant growth. These strains have been injected into maize, which has improved crop growth and physiological characteristics.

Another crucial feature of symbiotic and free-living microorganisms that promotes plant growth is biological N_2 fixation, which involves transforming atmospheric N_2 into ammonia that plants absorb (Zhang et al. 2020). There have been reports of several bacterial species living inside the plant that fix atmospheric nitrogen. According to a study, endophytic bacteria *Serratia* sp. that were isolated from rice were found to fix nitrogen, and when injected into plants, they increased biomass and biochemical components (Sandhiya et al. 2005). According to Lindström and Mousavi (2020), N_2 fixing was mostly seen in rhizobial and actinorhizal plant symbionts such as *Rhizobia, Frankia,* and *Azospirillum* species. *Gluconacetobacter diazotrophicus*, an endophyte, is a remarkable bacterium that efficiently fixes nitrogen (Dong et al. 1994; Bertalan et al. 2009).

Apart from bacteria, endophytic fungi are also recognized for effectively fixing atmospheric N_2 (Genre et al. 2020). For example, it has been observed that endophytic *Phomopsis liquidambri* increases nodulation and nitrogen assimilation via H_2NO signalling, hence promoting the growth of peanut plants (Xie et al. 2017). Dobereiner et al. (1993) reported that *Gluconacetobacter diazotrophicus* is essential to sugarcane's nitrogen fixation process, which can fix around 150 kg N/ha/year. The obligatory endophyte *Azoarcus* is another possible nitrogen-fixing organism. *Burkholderia vietnamiensis*, an endophyte of sugarcane, was reported by Govindarajan et al. (2006) to be effective at fixing nitrogen. Compared to the untreated control, the inoculation of sugar plantlets with this specific strain increased plant biomass. This natural ability of endophytes to fix N_2 diminishes the use of chemical fertilizers and thus helps in sustainable agriculture production.

Nutrient Solubilization

To meet the world's ever-increasing demand for food, conventional fertilizers containing excess phosphorus or other essential elements/nutrients are used to increase agricultural productivity. However, this practice can pollute surface and ground waters, eutrophicate waterways, deplete soil fertility, and accumulate harmful elements like high concentrations of arsenic (As) and selenium (Se) in the soil. Many soil microorganisms can liberate soluble phosphate (P) and make it available to plants by solubilizing or mineralizing insoluble soil phosphate. Numerous crops benefit from the increased development and yield caused by these bacteria. Nutrient-solubilizing microorganisms can increase global food production without endangering the environment through soil, crops, or seed inoculation (Alori et al. 2017).

Endophytic bacteria are vital for providing the necessary nutrients for bean plant growth, particularly those that take part in nitrogen fixation. Even while bacterial microbiomes in legume roots and nodules have been the subject of countless studies, little is known about the embryonic bacteria that help plants use nutrients following seed germination. Certain nitrogen-fixing bacteria isolates are also capable of solubilizing potassium or phosphorus through the use of inorganic or organic phosphorus. Odd levels of indole-3-acetic acid (IAA) secretion (0.41–7.10 μg/ml) were observed in all 26 nitrogen-fixing bacteria (Hao et al. 2024). Similarly, another researcher (Rana et al. 2020) isolated 159 endophytic bacteria from surface-sterilized roots and stems of different genotypes of wheat growing in the Divine Valley of Baru Sahib, Himachal Pradesh, and found that hydrogen cyanide production (19%) was higher among activities that promote plant growth than siderophores (16%), P-solubilizers (16%), ammonia (14%), K-solubilizers (14%), IAA producers (12%), Zn-solubilizers (5%), N_2-fixers (2%), and biocontrol (2%). Based on 16S rRNA gene sequencing and BLAST analysis, one of the isolates, EU-B2RT.R1, revealed a notable degree of nitrogenase activity, P-solubilization, and IAA generation. This isolate was identified as *Acinetobacter guillouiae* EU-B2RT.R1.

Teotia et al. (2016) stated that some rhizosphere microbes, such as bacteria and fungi, are responsible for potassium solubilization; these include *Bacillus* spp. (*B. megaterium, B. mucilaginosus, B. edaphicus, B. circulans, Acidithiobacillus ferrooxidans, Pseudomonas putida, Arthrobacter* sp., and *Paenibacillus* sp.), Aspergillus spp., and *Aspergillus terreus*, respectively. *Pseudomonas, Bacillus*, and *Aspergillus* are possible rhizosphere K-solubilizing microorganisms that release organic acids that solubilize inaccessible potassium and make it available to plant roots. Most research has focused on bacteria that fix nitrogen and dissolve phosphate; however, greater attention needs to be paid on a commercial scale to the readily accessible biofertilizer that contains solubilized potassium. Goteti et al. (2013) in their pot culture experiment with maize revealed that seed bacterization with *Pseudomonas* P29 @ 10 $g \cdot kg^{-1}$ significantly enhanced total dry mass (12.96 g) and uptake of N (2.268%), K (2.0%), Mn (60 ppm), and Zn (278.8 ppm).

Siderophore Production

According to Sharma and Johri (2003), bacterial siderophores are low-molecular-weight substances with a strong Fe^{3+} chelating affinity that solubilize and transfer iron elements into bacterial cells. Plants benefit from bacteria-producing siderophores because they can stop plant diseases from growing (Duhan 2013; Duhan and Dudeja 1999; Sharma and Johri 2003).

The term "iron bearer," from the Greek "siderophore," refers to molecules with a particularly high affinity for metal. Creating high-affinity siderophore-mediated iron transport systems is one of the main mechanisms developed for iron acquisition to counteract the iron-stressed state. This high-affinity transport mechanism consists of two primary parts: the creation of membrane receptor molecules, which mediates the internalization of iron bound to siderophores, and the generation of siderophores (Duhan and Dudeja 1998a, 1998b; Desai and Archana 2011). Moreover, metals other than iron can be bound by bacterial siderophores, increasing the metals' bioavailability in the plant's rhizosphere. Overall, the efficiency of phytoremediation procedures will be further increased by increased plant growth and metal uptake (Rajkumar et al. 2010).

The synthesis of siderophores gives bacteria a competitive edge over other microbes in the same ecological niche, allowing them to invade plant tissues. According to Loaces et al. (2011), pathogenic fungi *A. brasilense* and *H. seropedicae* were inhibited by the community of endophytic siderophore-producing bacteria

(SPB) linked to Oryza sativa grown in Uruguay and SPB isolate of the genus *Pseudomonas*. Throughout the plant's growth, the density of these bacteria, which were found in grains, roots, and leaves, varied between $\log_{10}$ 3.44 and $\log_{10}$ 5.52 cfu/g fresh weight.

Phytohormones Synthesis

Plants create molecules called phytohormones that control various aspects of their life, including growth, development, reproduction, lifespan, and even death. Accordingly, a class of tiny chemicals known as "Phytohormones" was created, including gibberellins/gibberellic acid (GA), cytokinins (CK), indole-3-acetic acid (IAA; the most well-known auxin), and abscisic acid (ABA) (Kim et al. 2020). Plant growth, development, and responses to the environment are all influenced by phytohormones, which include auxins, abscisic acid, brassinosteroids, cytokinins, ethylene, and gibberellins (Piotrowska and Bajguz 2011).

Tetracyclic diterpenoid acids, known as gibberellins or gibberellic acid (GA), are involved in several physiological and developmental processes (Bottini et al. 2004): fruit and flower growth, stem and floral induction, seed germination, seedling emergence, and leaf growth (King and Evans 2003). According to a study, rice plants treated with endophytic bacterial strain application showed significantly higher growth parameters than plants treated with exogenous GA3, and *Bacillus amyloliquefaciens*, which was isolated from rice seeds, can produce gibberellin (Shahzad et al. 2016). In a different study, four bacterial endophytes were isolated from wild pistachio trees; under in vivo conditions, *Pseudomonas protegens* produced more GA and caused carrot slices to root (Etminani and Harighi, 2018). Sandhya et al. (2017) reported endophytic bacteria isolated from maize identified as *Pseudomonas aeruginosa*, *P. monteilii*, and *P. putida* have produced gibberellin and promote plant growth.

Fusarium avenaceum, *Sarocladium terricola*, and *Xylariaceae* sp. were the three endophytes that enhanced the germination of *Lolium multiflorum* seeds by over 70%. Compared to controls, the application of fungal filtrates and extracts favored shoot and root length, plant dry weight, and the number of roots. The mechanisms of *L. multiflorum* plant growth stimulation following the application of fungal filtrates and/or extracts may be partially explained by the tentative HPLC-MS identification of phytohormone-like compounds, such as gibberellin A2 and zeatin or the antioxidant acetyl eugenol (García-Latorre et al. 2023). According to Bilal et al. (2018), inoculating leaves with a rhizobacterial isolate called *Pseudocitrobacter anthropi* C18 not only restored soybean growth but also enhanced the production of enzymatic antioxidants like catalases, ascorbic acid oxidase, peroxidase activity, and 1,1-diphenyl-2-picrylhydrazyl, as well as nonenzymatic antioxidants like flavonoids, phenolic, and proline contents. It reduced damage and subsequent secondary oxidative stress as a result. *Pseudocitrobacter anthropi* is an excellent phytostimulant, biofertilizer, and heavy metal bio-remediating agent because it can withstand metals up to 1200 ppm and is effective in detoxifying the effects of arsenic and chromium by regulating phytohormones (IAA 59.02 μg/ml and GA 101.88 nM/ml) and solubilizing inorganic phosphates.

Plant hormones known as cytokinins play a significant role in several processes, including apical dominance, chloroplast maturation, cell division and proliferation, seed germination, senescence prevention, and plant-pathogen signaling pathways. Several cytokinins (cis-zeatin cytokinin, riboside type zeatin, isopentyladenine, and isopentenyladenosine) larger than 30 pmol/ml were formed by bacterial endophytes *Pseudomonas, Sphingomonas, Stenotrophomonas*, and *Arthrobacter* sp. isolated from humic-treated cucumber plants (De Hita et al. 2020).

Because they encourage morphogenesis while postponing plant differentiation and/or senescence, cytokinins (CKs) are frequently referred to be "juvenile" factors (Wybouw and De Rybel 2019). UmaMaheswari et al. (2013) identified the endophytic bacteria isolated from legume crops, specifically red gram, black gram, green gram, cowpea, and chickpea. These bacteria were found to be *Bacillus* sp., *Micrococcus* sp., *Flavobacterium* sp., *Pseudomonas* sp., and *Serratia* sp. The cytokinin synthesis of these bacteria ranged from 0.52 to 2.96 g ml^{-1}. Endophytic bacteria that produce beneficial phytohormones may be employed as bio-stimulants to promote plant development and provide protection against phytopathogenic organisms.

Using the IAA biosynthesis route, the root endophyte *Piriformospora indica* generated auxin (Xu et al. 2018). Endophytes play a significant role in controlling plant development by producing IAA. According to Khan et al. (2014), indole acetic acid (11.23 μmml^{-1}), an endophyte called *Sphingomonas* sp. isolated from the foliar region of *Tephrosia apollinea,* enhanced growth activity in tomato plants. In a different investigation, GC-MS analysis revealed that *Micrococcus luteus* RWL-3, *Micrococcus yunnanensis* RWL-2, *Pantoea dispersa* RWL-3, and *Staphylococcus epidermidis* RWL-7 produced IAA (11.50–38.80 μgml^{-1}). The primary growth-promoting characteristics of rice plants, such as dry biomass, shoot and root length, chlorophyll, and protein contents, were markedly boosted upon inoculation (Shahzad et al. 2017). Sun et al. (2020) found that endophytic fungus, specifically *Falciphora oryzae*, aided in forming lateral roots while decreasing the height of the parent root. Defez et al. (2016) revealed that endophyte IAA activity also showed transcriptional alterations in nitrogen-fixing root nodules, increasing nitrogenase activity in rice. Sun et al. (2020) found that endophytic fungus, specifically *Falciphora oryzae*, aided in forming lateral roots while decreasing the height of the parent root.

In addition, fungi may create cytokinins, auxins, and gibberellins, which are crucial for chemical communication. By encouraging seed germination and raising chlorophyll content, the endophytic fungus *Porostereum spadiceum* generates gibberellins, which help soybean growth in normal and salt-affected conditions (Hamayun et al. 2017). Numerous endophytic fungi, such as *Rhizopus stolonifer, Pochonia chlamydosporia, A. flavus, Paecilomyces formosus, P. glomerata,* and *Penicillium corylophilum* via gibberellins and indole acetic acid production can improve plant growth under abiotic stress (Khan et al. 2012). While the majority of fungal endophytes that create gibberellic acid are members of the Ascomycetes group, *P. spadiceum,* a member of the Basidiomycota, is the first endophyte that produces gibberellic acid and is engaged in phytostimulation (Waqas et al. 2012).

Antimicrobial Compounds Synthesis

There are numerous examples of secondary metabolites from endophytic fungi that are either being utilized or have the potential to be employed against horticultural and agricultural plant infections and pests. This comprises fungal endophytes that generate Ambuic acid is an antifungal and anti-oomycete substance derived from the endophytic fungus *Pestalotiopsis microspora*, is efficacious against many *Fusarium* species as well as *Pythium ultimum*. Colletonoic acid is derived from a *Colletotrichum* sp. isolated from *Artemisia annua* and has good antibacterial, antifungal, and anti-algal properties (Hussain et al. 2014). Colletonoic acid is antimicrobial against bacteria and the fungus *Helminthosporium sativum* (Zou et al. 2000). Cordycepsidone A, derived from the endophytic fungus *Cordyceps dipterigena*, exhibits potent antifungal action against the plant pathogenic fungus *Gibberella fujikuroi* (Varughese et al. 2012).

The majority of the research on endophytic improvement in response to biotic stresses has focused on Epichloe (Ascomycete fungi forming a symbiosis with grass) endophytes, a family of fungi with strong interactions with grasses that have both positive and harmful impacts (Kou et al. 2021; Zhu et al. 2022). Epichloe endophytes promote biotic resistance through a variety of methods, primarily by upregulating the jasmonic (JA) and salicylic acid (SA) production pathways (Kou et al. 2021; Zhu et al. 2022; Li et al. 2020). Xia et al. (2016) evaluated the influence of Epichloe endophytes on plants infected with powdery mildew (*Blumeria graminis*), revealing that endophytic fungi can decrease the chlorophyll loss of activity and damage (a measure of disease severity) compared to non-endophytic fungi-containing plants. Along with the antibacterial qualities of endophytes, improvements in metabolic and photosynthetic rates in EAP mitigate the consequences of pathogen presence. EAP has been proven to have higher proliferation rates under pathogen infection than non-EAP (Zhu et al. 2022). It might be argued that larger and healthier plants have less pathogen impact and more ability to resist infection (Chen et al. 2022). This is demonstrated by endophytic increases in calcium (Ca) acquisition, which is implicated in systematic acquired resistance signalling, whereby higher Ca levels result in a better ability to respond to stimuli (Thor 2019; Dodds and Rathjen 2010; Vadassery et al. 2009). Table 8.1 illustrates the role of endophytes in agricultural sustainability and plant growth promotion.

8.3 Applications in Sustainable Environment

Endophytes are vital in phytoremediation, a green method that uses plants and microbes to clean up contaminated soils, sediments, and water bodies (Ma et al. 2016). Endophytic bacteria and fungi can degrade or detoxify contaminants, including heavy metals, organic pollutants, and agrochemicals, via many metabolic pathways (Khan et al. 2014). Endophytes colonize plant roots and tissues, facilitating pollutant uptake, transformation, and sequestration, lowering bioavailability and environmental impact. Endophytic microorganism-based phytoremediation technologies have been used successfully in various environments, including industrial

Table 8.1 Endophytes in agricultural sustainability and plant growth promotion

Function	Plants	Endophytic organism	Bioactive compounds/ phytocompounds	References
Phytostimulation	*Phragmites communis, Potamogeton crispus, Nymphaea tetragona, Najas marina*	*Aeromonas caviae, A. salmonicida, Bacillus niacin, B. aryabhattai, B. sphaericus, B. simplex, Delttia tsuruhatensis, Paenibacillus lautus, Enterobacter hormaechei, Flavobacterium oceanosedimentum, Klebsiella terrigena, Lactococcus lactis, Microbacterium flavescens, Paenibacillus barcinonensis, Pantoea agglomerans, Pseudomonas jessenii, P. taiwanensis, P. xanthomarina, Staphylococcus epidermidis*	Aromatic phytocompounds, Biodegradation (Chlorpyrifos, Fenpropathrin, Naphthalene, Bifenthrin), P-solubilization	Chen et al. (2012), Baldan et al. (2015), Bresson et al. (2013), and Liu et al. (2013)
Growth promotion	*Glycine max, Phaseolus vulgaris L., Simmondsia chinensis, Triticum aestivum, Solanum lycopersicum, Saccharum officinarum, Mulberry, Glycine max*	*Pseudomonas putida, Gibberellin, Trichoderma polysporum, T. atroviridae, T. harzianum, Streptomyces nobilis, S. kunmingensis, S. mutabilis, S. enissocaesilis, S. djakartensis, Streptomyces* sp., *Micromonospora* sp., *Rhizobium* sp., *Pseudomonas* sp., *Xanthomonas* sp., *Salinococcus* sp., *Marinomonas* sp., *Microbacterium arborescens, Enterobacter* sp., *Pantoea stewartia, Bacillus subtilis 7PJ-16, Streptomyces* sp. *NEAU-S7GS2*	Growth hormones, Proteolytic enzymes, Phosphate solubilization factors, Enhanced, GA, IAA and ACC deaminase, Siderophore production, Biosolubilizing activities, Enhanced seed germination, Agricultural biofertilizer, biocontrol agent, antifungal and plant growth promotion, Controls sclerotinia stem rot of soybean.	Anwar et al. (2016), Salam et al. (2017), Zahoor et al. (2017), You et al. (2012), Gangwar et al. (2014), Xu et al. (2019) and Zhang et al. (2020)

(continued)

Table 8.1 (continued)

Function	Plants	Endophytic organism	Bioactive compounds/ phytocompounds	References
Indole-3-acetic acid (IAA)	*Lavandula dentate, Suaeda japonica*	*Bacillus thuringiensis, Penicillium* sp.	Indole-3-acetic acid (IAA), bioactive gibberellins (GAs), other inactive GAs, triterpenoid saponins, ginsenosides, enhanced seed germination	You et al. (2012) and Wu et al. (2013)
Phosphate solubilization	*Arabidopsis thaliana, Platycladus orientalis, Brassica campestris, Phaseolus vulgaris, Tinospora cordifolia, Calotropis procera, Albizia lebbeck, Leptochloa fusca, Z. mays L., Microbispora, Streptomyces, Streptosporangium, Aloe vera, Mentha arvensis, Ocimum sanctum, Trichoderma atroviridae, T. polysporum, T. harzianum, Nocardia* sp., *Rhodococcus* sp., *Kocuria* sp., *Nocardioides* sp., *Pseudonocardia* sp., *Micromonospora* sp., *Saccharopolyspora* sp.	*Phyllobacterium brassicacearum, B. subtilis, Mucor* sp., *Actinomycetes, Piriformospora indica*	Abscisic acid (ABA), Growth hormones, Enhanced IAA production, ACC deaminase, HCN production, Phosphate solubilization, Enhanced seed germination, Plant growth promotion	Bresson et al. (2013), Zahoor et al. (2017), de Araujo et al. (2000), Gangwar et al. (2014)
Phyto-immobilization	*Alstonia boonei, Enantia chlorantha, Kigelia africana, Brassica campestris, Triticum aestivum, Solanum lycopersicum, Saccharum officinarum*	*Aspergilus niger, Macrophomina* sp., *Trichoderma* sp., *Penicillium citrinum, P. nigricans, Burkholderia cepacia, Enterobacter, Saccharum* sp.	Secondary metabolites, Enhanced IAA production, Siderophore production, Phosphate solubilization factors, Active volatile and non-volatile metabolites	Manikandan et al. (2015), Anwar et al. (2016) and Sinma et al. (2015)

Phytotransformation	*Achillea millefolium, Dactylis glomerata, Solidago canadensis, Trifolium aureum, Lolium multiflorum, Zea mays, Poplar*	*Microbacterium foliorum, Plantibacter flavus, Psudomonas, Burkholderia cepacia, Trichoderma atroviridae, T. polysporum, T. harzianum, S. pisciphila, Enterobacter*	Petroleum tolerance, Alkanes, Diesel degradation, Phenol, Toluene, Co-contamination of nickel and trichloroethylene tolerance and degradation, Pyrenes tolerance and degradation	Lumactud et al. (2016), Andria et al. (2009), Wang et al. (2010), Weyens et al. (2015) and Sheng et al. (2008)

ABA Abscisic acid, *GA* Gibberellic acid, *IAA* Indole-3-acetic acid, *ACC* 1-aminocyclopropane-1-carboxylic acid sites, mining areas, and agricultural fields, proving their potential for long-term environmental cleanup.

8.3.1 Heavy Metal Remediation/Phytoremediation

Unfortunately, the number of contaminants in our environment rapidly increases due to increased heavy metals and organic matter (Bharagava and Chowdhary 2018). Due to their carcinogenic, teratogenic, and mutagenic qualities, heavy metals in pesticides and fertilizers are toxic to living creatures. A huge amount of trash generated by diverse anthropogenic activities is kept and disposed of in the environment, posing public health risks (Saxena et al. 2010). Since the last century, rapid industrialization has resulted in a massive increase in numerous pollutants in the environment, including pesticides, heavy metals, halogenated hydrocarbons, and polycyclic aromatic hydrocarbons (PAH), posing substantial health risks (Ma et al. 2011).

Environmental contamination by heavy metals has become hazardous to living species in an ecosystem, posing a major threat to all forms of life due to the harmful effects of long-term environmental pollution. Metals are not recyclable, and various human activities cause heavy metals to accumulate in the soil. Various metals deposited in soil include Al^{3+}, Au^{+}, Hg^{2+}, Pb^{2+}, and Cd^{2+}, which play minor roles in metabolic functions but are hazardous to humans and other living organisms (Siddiquee et al. 2015). Among the microbes, endophytes are now considered a promising means of the remediation of different pollutants from their natural environments (Digra and Nonzom 2024).

The word "bioremediation" is defined as the use of living organisms (plants or microbes such as bacteria or fungi) to remediate various contaminants found in the environment to improve human health (Bharagava and Chowdhary 2018). Using bacteria in bioremediation of toxic substances is a more straightforward, cost-effective, and environmentally acceptable strategy. Many microbial communities employ these chemical contaminants as growth boosters, aiding their breakdown. Many studies have shown that microorganisms such as bacteria and fungi play important roles in decomposing biodegradable compounds by producing extracellular enzymes for nutrition acquisition under stress situations. Microorganisms use several processes in the presence of inorganic metal pollution. They use biotransformation, extrusion, exopolysaccharide creation surrounding the cell wall, enzyme production, and metallothionein production to counteract metal toxicity (Sharma et al. 2021). Endophytes, among the documented bacteria, have enormous potential to degrade dangerous chemicals in the atmosphere.

Interestingly, they are frequently used to remove environmental pollutants due to their ability to survive inside the host under extreme habitat conditions and to assist the plant in battling various contaminants found in the soil habitat (Young et al. 2006). Furthermore, because of their enzyme-producing capabilities, endophytes can transform complex polymers into monomeric forms once the host plant dies (Szink et al. 2016). Strong bioresources for biological clean-up have also been

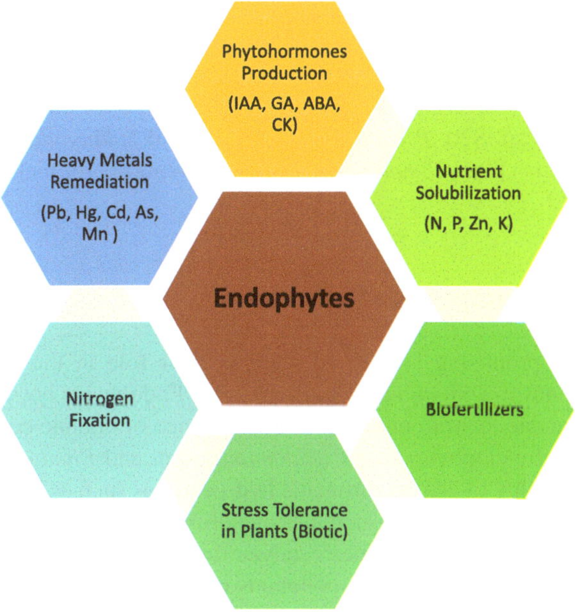

Fig. 8.1 Endophytes and their role in sustainable agriculture and environment

found in a variety of bacterial species, including *Arthrobacter, Bacillus, Burkholderia, Pseudomonas*, and *Stenotrophomonas*; a few fungi, including *Mucor, Talaromyces, Trichoderma*, and *Archaea*, including *Natrialba* and *Haloferax*. Saharan et al. 2023 also highlighted the function of plant growth-promoting bacteria (PGPB) in facilitating the low-cost and ecologically benign bioremediation of heavy and hazardous metals. Figure 8.1 represents a graphical representation of endophytes' roles in protecting agriculture and the environment.

Similarly, in vitro and in vivo studies on various endophytic fungi to investigate their heavy metal resistance yielded remarkable results (Pietro-Souza et al. 2020). For example, four endophytes, *Curvularia geniculata* P1, *Aspergillus* sp. 31, *Westerdykella* sp. P71 and *Lindgomycetaceae* P87, removed approximately 100% of the mercury in the culture medium. Following their metal-removal activity, they were chosen for bioaugmentation in *Zea mays* and *Aeschynomene fluminensis*, where they displayed phytoremediation by boosting mercury bioaccumulation within host root tissues and, therefore, increasing host biomass (Pietro-Souza et al. 2020). Endophytes of the same taxon isolated from different plant species or portions of the plant demonstrated varied metal tolerance, which could be attributed to their long-term connection with plants exposed to varying metal concentrations (Li et al. 2012).

According to Zhang et al. (2011), endophytic bacteria can also create indole-3-acetic acid (IAA) and other plant growth-promoting chemicals, which aid in cleaning heavy metals found in varied ecosystems. They can also lessen induced heavy metal stress in plants and indirectly benefit them by promoting their growth. For example, an endophytic bacterium *Paenibacillus* sp. RM isolated from the plant *Tridax procumbens* demonstrated heavy metal resistance against Cu (59.4%) and Zn (51.4%) in vitro (Govarthanan et al. 2016). Recently, three endophytic bacteria, *Stenotrophomonas* sp. S20, *Pseudomonas* sp. P21, and *Sphingobium* sp. S42, isolated from the marsh plant *Tamarix chinensis*, were reported for nickel remediation (Chen et al. 2020).

8.3.2 Organic Compounds Remediation

One particularly promising area of research is their role in the remediation of organic compounds in contaminated environments. Endophytes possess enzymatic pathways that enable them to metabolize a wide range of organic compounds. For instance, bacterial endophytes like *Pseudomonas* spp. and *Rhodococcus* spp. are known for their ability to degrade aromatic hydrocarbons such as benzene, toluene, ethylbenzene, and xylene through enzymes like dioxygenases and monooxygenases (Kahlon et al. 2016). These enzymes catalyze the initial steps in the degradation process, converting complex organic pollutants into simpler, less toxic metabolites. Moreover, fungal endophytes such as *Phoma* spp. and *Fusarium* spp. contribute to organic compound remediation through mechanisms such as ligninolytic enzymes and peroxidases, which break down recalcitrant compounds like polycyclic aromatic hydrocarbons (PAHs) and pesticides (Kumar et al. 2021). The collaborative efforts of these endophytes with their host plants further enhance their remediation capabilities by facilitating nutrient exchange and metabolic support. Table 8.2 shows the biodegrading potential of endophytes against heavy metals.

8.3.3 Stress Tolerance

Furthermore, endophytes play an important role in providing stress tolerance to their host plants, allowing them to tolerate a variety of biotic and abiotic challenges such as pathogen infection, drought, salt, and heavy metal toxicity (Marasco et al. 2018). Endophytes use a variety of processes to reduce stress-induced damage and increase plant resilience, including the synthesis of extracellular enzymes, antioxidants, and osmolytes. Certain endophytic bacteria, for example, produce enzymes like catalase, peroxidase, and superoxide dismutase that scavenge reactive oxygen species (ROS) produced during stress, protecting plant cells from oxidative damage (Glick 2014; Yadav and Meena 2022; Abhijith Shankar et al. 2024). Similarly, some endophytic fungi produce suitable solutes, such as trehalose and proline, which operate as osmoprotectants, preserving cellular turgor and integrity during osmotic stress (Verma et al. 2016).

One well-known method of plant stress alleviation induced by bacterial endophytes is the development of the enzyme 1-aminocyclopropane-1- carboxylase (ACC) deaminase, which decreases the ethylene levels in plants (Glick et al. 1998).

Table 8.2 Role of endophytes in biodegradation of toxic or heavy metals

S. No.	Endophytes	Host plant	Resistance against metals	Reference
1.	*Exiguobacterium* sp.	*Vigna radiata*	As	Pandey and Bhatt (2016)
2.	*Rahnella* sp. *JN27*	*Zea mays*	Cd	Yuan et al. (2014)
3.	*Endophytes belonged to Firmicutes, Proteobacteria and Actinobacteria*	*Pteris vittata* and *P. multifida*	As	Zhu et al. (2014)
4.	*Microbacterium* sp., *and Pseudomonas* sp.	*Rumex acetosa*	Numerous	Gaiero et al. (2013)
5.	*Galactomyces geotrichum WLL1*	*Trapa japonica*	Zn	Waqas et al. (2014)
6.	*B. pumilus E2S2, Bacillus* sp. *E1S2, Bacillus* sp. *E4S1, Achromobacter* sp. *E4L5,* and *Stenotrophomonas* sp. *E1L*	*Sedum plumbizincicola*	Cd, Zn, and Pb	Ma et al. (2015)
7.	*Alternaria* sp. *CBSF68, Penicillium* sp. *CBRF65, Fusarium* sp. *CBRF44*	*Brassica napus*	Cd and Pb	Shi et al. (2017)
8.	*Pseudomonas* sp. Lk9	Solanum nigrum	Cd, Zn, Cu, and Cr	Chen et al. (2014)
9.	*Serendipita vermifera* sp. *P04*	*Populus clone INRA 717-1B4*	Cu, Cd, Zn, Pb, Cu	Lacercat-Didier et al. (2016)
10.	*Piriformospora indica*	*Helianthus annuus cv. Zaria*	Cd	Shahabivand et al. (2017)
11.	*Penicillium roqueforti Thom.*	*Triticum aestivum var. Bhakkar-2000 (KJ672075)*	Zn, Pb, Ni, Cd, Cu	Ikram et al. (2018)
12.	*Serendipita indica*	*Ocimum basilicum*	Cu, Pb	Sabra et al. (2018)
13.	*Pseudomonas koreensis AGB-1*	*Miscanthus sinensis*	Cd, Pb, Cu, Zn and As	Babu et al. (2015)
14.	*Lindgomycetaceae P87, Aspergillus* sp. *A31*	*Aeschynomene fluminensis*	Hg	Pietro-Souza et al. (2020)
15.	*Aspergillus flavus*	*Tomato*	Cd and Cr	Aziz et al. (2021)

As Arsenic, *Cd* Cadmium, *Cr* Chromium, *Cu* Copper, *Hg* Mercury, *Pb* Lead, *Ni* Nickle, *Zn* Zinc

For instance, it is known that *Burkholderia phytofirmans* strain PsJN improves grapevine's ability to withstand low temperatures and reduces the effects of drought in wheat and maize (Barka et al. 2006; Naveed et al. 2014a, b). Salt resistance in *Oryza sativa* L. cv. KDML105 was demonstrated to be conferred by Streptomyces sp. GMKU 336 generates 1-aminocyclopropane-1-carboxylate (ACC) deaminase (Jaemsaeng et al. 2018).

While it has been demonstrated that fungi endophytes like *Neotyphodium* spp. and *Trichoderma* spp. can enhance drought tolerance in grass (Bacon and Hill 1996) and cocoa plants (Bae et al. 2009), *Piriformospora indica* can also cause drought tolerance in Chinese cabbage plants (Sun et al. 2010) and salt tolerance in barley (Baltruschat et al. 2008). It has been demonstrated that the endophytic *Aspergillus fumigatus* SG-17 from *Myricaria laxiflora* increases rice's resistance to drought by generating (Z)-N-(4-hydroxyethyl) formamide (NFA), an analogue of coumarin (Qin et al. 2019). Furthermore, it has been proposed that endophytes that occupy the phyllosphere can provide resistance against a range of abiotic stressors, including frost damage, severe temperatures, salt, UV radiation, and drought (Daisy et al. 2002).

This approach mitigates pollution and promotes soil health and biodiversity by restoring contaminated ecosystems to their natural state (Li et al. 2019). Such ecological restoration efforts are crucial for achieving sustainable development goals and ensuring the long-term health of terrestrial and aquatic environments. Research efforts should focus on developing practical guidelines and protocols for integrating endophytes into remediation strategies, considering factors such as host plant selection, microbial inoculation methods, and monitoring techniques (Kusari et al. 2022). Collaborative interdisciplinary research combining microbiology, plant sciences, environmental engineering, and biotechnology will be essential to overcoming these challenges and advancing the field.

8.4 Challenges and Future Directions

Despite their enormous potential, endophyte bioprospecting for agricultural and environmental sustainability presents obstacles and constraints. Several difficulties, including technical constraints, regulatory barriers, and socioeconomic considerations, impede endophyte-based technology's mainstream acceptance and commercialization.

8.4.1 Technical Challenges

When endophytic microbes are isolated from their natural habitats, they may experience morphological or physiological alterations that impact their functional properties and ecological responsibilities. To overcome these technical obstacles, new isolation techniques, culture media, and high-throughput screening assays must be developed to discover and characterize viable endophyte candidates for agricultural and environmental applications.

Streamlining the regulatory clearance process and establishing clear criteria for registering and commercializing endophyte-based products will help them integrate into mainstream agricultural and environmental practices.

8.4.2 Socioeconomic Considerations

Policymakers and funding organizations may prioritize investments in traditional technologies or research fields that provide more immediate returns on investment, pushing endophyte research to the sidelines. To address these socioeconomic constraints, concerted outreach and education activities will be required to promote awareness about endophyte benefits and prospective agricultural and environmental management applications.

8.4.3 Future Directions

Future studies should improve our understanding of endophyte diversity, function, and interactions within plant microbiomes. Integrating multi-omics methodologies, advanced imaging techniques, and computational modelling will allow researchers to understand the intricate dynamics of endophyte-plant interactions better and maximize their performance in real-world contexts.

8.5 Conclusion

Finally, bioprospecting of endophytes provides significant potential for enhancing agricultural and environmental sustainability by N_2 fixation, nutrient solubilization, phytohormones synthesis, stress tolerance, and phytoremediation of heavy metals and organic compounds in the face of global concerns. Researchers can develop creative methods for increasing crop yield, lowering environmental consequences, and fostering ecosystem resilience by utilizing endophytic microbes' various processes and roles. However, to realize endophytes' full potential, the scientific community, governments, industrial partners, and civil society must work together to overcome technical, regulatory, and socioeconomic constraints. Working together to achieve a common objective, we can harness the potential of endophytes to build a more sustainable and resilient future for agriculture and the environment.

References

Abhijith Shankar PS, Parida P, Bhardwaj R, Yadav A, Swapnil P, Seth CS, Meena M (2024) Deciphering molecular regulation of reactive oxygen species (ROS) and reactive nitrogen species (RNS) signalling networks in *Oryza* genus amid environmental stress. Plant Cell Rep 43(7):185. https://doi.org/10.1007/s00299-024-03264-1

Alori ET, Glick BR, Babalola OO (2017) Microbial phosphorus solubilization and its potential for use in sustainable agriculture. Front Microbiol 8:971. https://doi.org/10.3389/fmicb.2017.00971

Andria V, Reichenauer TG, Sessitsch A (2009) Expression of alkane monooxygenase (alkB) genes by plant-associated bacteria in the rhizosphere and endosphere of Italian ryegrass (*Lolium multiflorum* L.) grown in diesel contaminated soil. Environ Pollut 157(12):3347–3350. https://doi.org/10.1016/j.envpol.2009.08.023

Anwar S, Ali B, Sajid I (2016) Screening of rhizospheric actinomycetes for various in-vitro and in-vivo plant growth promoting (PGP) traits and for agroactive compounds. Front Microbiol 7(1334):1334. https://doi.org/10.3389/fmicb.2016.01334

Anzuay MS, Ciancio MGR, Ludueña LM, Angelini JG, Barros G, Pastor N, Taurian T (2017) Growth promotion of peanut (*Arachis hypogaea* L.) and maize (*Zea mays* L.) plants by single and mixed cultures of efficient phosphate solubilizing bacteria that are tolerant to abiotic stress and pesticides. Microbiol Res 199:98–109. https://doi.org/10.1016/j.micres.2017.03.006

Aziz L, Hamayun M, Rauf M, Iqbal A, Husssin A, Khan SA et al (2021) *Aspergillus Flavus* reprogrammed morphological and chemical attributes of *Solanum lycopersicum* through SlGSH1 and SlPCS1 genes modulation under heavy metal stress. J Plant Interact 16(1):104–115. https://doi.org/10.1080/17429145.2021.1903105

Babu AG, Shea PJ, Sudhakar D, Jung IB, Oh BT (2015) Potential use of *Pseudomonas koreensis* AGB-1 in associated with *Miscanthus sinensis* to remediate heavy metal (loid)-contaminated mining site soil. Aust J Environ Manag 151:160–166

Bacon C, Hill N (1996) Symptomless grass endophytes: products of coevolutionary symbioses and their role in the ecological adaptations of grasses. In: Redlin SC, Carris LM (eds) Endophytic fungi in grasses and woody plants. APS Publishing, St. Paul, pp 155–178

Bae H, Sicher RC, Kim MS, Kim S-H, Strem MD, Melnick RL, Bailey BA (2009) The beneficial endophyte *Trichoderma hamatum* isolate DIS 219b promotes growth and delays the onset of the drought response in *Theobroma cacao*. J Exp Bot 60(11):3279–3295. https://doi.org/10.1093/jxb/erp165

Baldan E, Nigris S, Romualdi C, D'Alessandro S, Clocchiatti A, Zottini M et al (2015) Beneficial bacteria isolated from grapevine inner tissues shape *Arabidopsis thaliana* roots. PLoS One 10(10):e0140252. https://doi.org/10.1371/journal.pone.0140252

Baltruschat H, Fodor J, Harrach BD, Niemczyk E, Barna B, Gullner G et al (2008) Salt tolerance of barley induced by the root endophyte *Piriformospora indica* is associated with a strong increase in antioxidants. New Phytol 180(2):501–510. https://doi.org/10.1111/j.1469-8137.2008.02583.x

Barka EA, Nowak J, Clément C (2006) Enhancement of chilling resistance of inoculated grapevine plantlets with a plant growth-promoting rhizobacterium, *Burkholderia phytofirmans* strain PsJN. Appl Environ Microbiol 72(11):7246–7252. https://doi.org/10.1128/AEM.01047-06

Bashan Y, Kamnev AA, de-Bashan LE. (2014) Tricalcium phosphate is inappropriate as a universal selection factor for isolating and testing phosphate-solubilizing bacteria that enhance plant growth: a proposal for an alternative procedure. Biol Fertil Soils 50(5):459–471

Berendsen RL, Pieterse CMJ, Bakker PAHM (2012) The rhizosphere microbiome and plant health. Trends Plant Sci 17(8):478–486. https://doi.org/10.1016/j.tplants.2012.04.001. Pubmed:22564542

Bertalan M, Albano R, de Pádua V, Rouws L, Rojas C, Hemerly A et al (2009) Complete genome sequence of the sugarcane nitrogen-fixing endophyte *Gluconacetobacter diazotrophicus* Pal5. BMC Genomics 10(1):V. https://doi.org/10.1186/1471-2164-10-450

Bharagava R, Chowdhary P (2018) Emerging and eco-friendly approaches for waste management. Springer, Singapore, pp 189–217

Bhattacharjee RB, Singh A, Mukhopadhyay SN (2008) Use of nitrogen-fixing bacteria as biofertiliser for nonlegumes: prospects and challenges. Appl Microbiol Biotechnol 80(2):199–209. https://doi.org/10.1007/s00253-008-1567-2

Bhattacharyya PN, Jha DK (2012) Plant growth-promoting rhizobacteria (PGPR): emergence in agriculture. World J Microbiol Biotechnol 28(4):1327–1350. https://doi.org/10.1007/s11274-011-0979-9

Bhupenchandra I, Chongtham SK, Devi AG, Dutta P, Sahoo MR, Mohanty S, Kumar S, Choudhary AK, Devi EL, Sinyorita S, Devi SH, Mahanta M, Kumari A, Devi HL, Josmee RK, Pusparani A, Pathaw N, Gupta S, Meena M, Janmeda P, Seth CS, Arora A, Sahoo A, Swapnil P (2024) Unlocking the potential of arbuscular mycorrhizal fungi: exploring role in plant growth promotion, nutrient uptake mechanisms, biotic stress alleviation, and sustaining agricultural production systems. J Plant Growth Regul 43(9):1–39. https://doi.org/10.1007/s00344-024-11467-9

Bilal L, Asaf S, Hamayun M, Gul H, Iqbal A, Ullah I et al (2018) Plant growth-promoting endophytic fungi *Asprgillus fumigatus* TS1 and *Fusarium proliferatum* BRL1 produce gibberellins and regulates plant endogenous hormones. Symbiosis 76(2):117–127. https://doi.org/10.1007/s13199-018-0545-4

Boonphong S, Kittakoop P, Isaka M, Pittayakhajonwut D, Tanticharoen M, Thebtaranonth Y (2001) Multiplolides a and B, new antifungal 10-membered lactones from *Xylaria multiplex*. J Nat Prod 64(7):965–967

Bottini R, Cassán F, Piccol P (2004) Gibberellin production by bacteria and its involvement in plant growth promotion and yield increase. Appl Microbiol Biotechnol 65:497–503. https://doi.org/10.1007/s00253-004-1696-1

Bresson J, Varoquaux F, Bontpart T et al (2013) The P.G.P.R., strain *Phyllobacterium brassicacearum* STM196 induces a reproductive delay and physiological changes that result in improved drought tolerance in Arabidopsis. New Phytol 200:558–569

Chandran H, Meena M, Swapnil P (2021) Plant growth-promoting rhizobacteria as a green alternative for sustainable agriculture. Sustain For 13(19):10986. https://doi.org/10.3390/su131910986

Chen WM, Tang YQ, Mori K, Xiao-Lei W (2012) Distribution of culturable endophytic bacteria in aquatic plants and their potential for bioremediation in polluted waters. Aquat Biol 15:99–110

Chen B, Zhang Y, Rafiq MT, Khan KY, Pan F, Yang X, Feng Y (2014) Improvement of cadmium uptake and accumulation in *Sedum alfredii* by endophytic bacteria *Sphingomonas* SaMR12: effects on plant growth and root exudates. Chemosphere 117:367–373

Chen J, Li N, Han S, Sun Y, Wang L, Qu Z et al (2020) Characterization and bioremediation potential of nickel-resistant endophytic bacteria isolated from the wetland plant *Tamarix chinensis*. FEMS Microbiol Lett *367*(12):fnaa098

Chen Q, Yu JJ, He J, Feng T, Liu JK (2022) Isobenzofuranones and isocoumarins from kiwi endophytic fungus *Paraphaeosphaeria sporulosa* and their antibacterial activity against *Pseudomonas syringae* pv. actinidiae. Phytochemistry 195:113050

Daisy BH, Strobel GA, Castillo U, Ezra D, Sears J, Weaver DK, Runyon JB (2002) Naphthalene, an insect repellent, is produced by *Muscodor vitigenus*, a novel endophytic fungus. Microbiology 148(11):3737–3741

De Araujo JM, Da Silva AC, Azevedo JL (2000) Isolation of endophytic actinomycetes from roots and leaves of maize (*Zea mays* L.). Braz Arch Biol Technol 43:4. https://doi.org/10.1590/S1516-89132000000400016

De Hita D, Fuentes M, Zamarreño AM, Ruiz Y, Garcia-Mina JM (2020) Culturable bacterial endophytes from sedimentary humic acid-treated plants. Front Plant Sci 11:837. https://doi.org/10.3389/fpls.2020.00837

Defez R, Esposito R, Angelini C, Bianco C (2016) Overproduction of indole-3-acetic acid in free-living rhizobia induces transcriptional changes resembling those occurring in nodule bacteroids. Mol Plant-Microbe Interact 29(6):484–495

Desai A, Archana G (2011) Role of siderophores in crop improvement. In: Maheshwari D (ed) Bacteria in agrobiology: plant nutrient management. Springer, Berlin. https://doi.org/10.1007/978-3-642-21061-7_6

Digra S, Nonzom S (2024) Bioremediation potential of endophytes: a promising tool. App Biochem Microbiol 60:694–714

Dobereiner J, Reis V, Paula M, Olivares F (1993) Endophytic diazotrophs in sugar cane, cereals and tuber plants. In: Palacios R, Mora J, Newton WE (eds) New horizons in nitrogen fixation. Current plant science and biotechnology in agriculture, vol 17. Springer, Dordrecht. https://doi.org/10.1007/978-94-017-2416-6_55

Dodds PN, Rathjen JP (2010) Plant immunity: towards an integrated view of plant-pathogen interactions. Nat Rev Genet 11:539–548

Dong Z, Canny MJ, McCully ME, Roboredo MR, Cabadilla CF, Ortega E, Rodes R (1994) A nitrogen-fixing endophyte of sugarcane stems (a new role for the apoplast). Plant Physiol 105(4):1139

Dudeja SS, Duhan JS, Khurana AL (1996) Siderophore-mediated iron acquisition in rhizobia under free-living and symbiotic conditions. In: Behl RK, Gupta AP, Khurana AL, Singh A (eds) Resource management in the fragile environment. CCS HAU, Hisar and Maxmullar Bhavan, New Delhi, pp 212–229

Duhan JS (2013) Tn5 mutants of *Rhizobium* sp. (*Cajanus*) and its role in nitrogen fixation and iron uptake in pigeonpea. Afr J Microbiol Res 79(16):1459–1464. https://doi.org/10.5897/AJMR12.223

Duhan JS, Dudeja SS (1998a) Effect of elevated temperature on growth, siderophore production and survival of pigeonpea rhizobia. Annals Agric Bio-Res 3(2):235–238

Duhan JS, Dudeja SS (1998b) Effect of exogenous iron, synthetic chelator and rhizobial siderophore on iron acquisition by pigeonpea host in pigeonpea- *Rhizobium* symbiosis. Microbial Res 153:37–45

Duhan JS, Dudeja SS (1999) Competitiveness does not correlate with siderophore production in *rhizobium-Cajanus cajan* symbiosis. Symbiosis 26:79–87

Duhan JS, Dudeja SS, Khurana AL (1998) Siderophore production in relation to nitrogen fixation and iron uptake in pigeonpea-*Rhizobium* symbiosis. Folia Microbiol 43(4):421–426

Etminani F, Harighi B (2018) Isolation and identification of endophytic bacteria with plant growth promoting activity and biocontrol potential from wild pistachio trees. Plant Pathol J 34(3):208–217

Esposito R, Angelini A, Bianco C (2016) Overproduction of indole-3-acetic acid in free-living rhizobia induces transcriptional changes resembling those occurring inside nodule bacteroids. Mol Plant-Microbe Interact 29:484–495. https://doi.org/10.1094/MPMI-01-16-0010-R

Finkel OM, Castrillo G, Herrera PS, Salas GI, Dangl JL, Brotman Y (2017) Understanding and exploiting plant beneficial microbes. Curr Opin Plant Biol 38:155–163

Gaiero JR, McCall CA, Thompson KA, Day NJ, Best AS (2013) and Dunfield KE. Am J Bot 100:1738–1750. https://doi.org/10.3732/ajb.1200572

Gangwar M, Dogra S, Gupta UP, Kharwar RN (2014) Diversity and biopotential of endophytic actinomycetes from three medicinal plants in India. Afr J Microbiol Res 8(2):184–191

García-Latorre C, Rodrigo S, Marin-Felix Y et al (2023) Plant-growth promoting activity of three fungal endophytes isolated from plants living in dehesas and their effect on *Lolium multiflorum*. Sci Rep 13:7354. https://doi.org/10.1038/s41598-023-34036-8

Genre A, Lanfranco L, Perotto S, Bonfante P (2020) Unique and common traits in mycorrhizal symbioses. Nat Rev Microbiol 18(11):649–660

Glick BR (2014) Bacteria with ACC deaminase can promote plant growth and help to feed the world. Microbiol Res 169(1):30–39

Glick BR, Penrose DM, Li J (1998) A model for the lowering of plant ethylene concentrations by plant growth-promoting bacteria. J Theor Biol 190(1):63–68

Goteti PK, Emmanuel LDA, Desai SS, Mir HA (2013) Prospective zinc solubilising bacteria for enhanced nutrient uptake and growth promotion in maize (*Zea mays* L.). Int J Microbiol 2013:869697

Gouda S, Kerry RG, Das G, Paramithiotis S, Shin HS, Patra JK (2018) Revitalization of plant growth promoting rhizobacteria for sustainable development in agriculture. Microbiol Res 206:131–140

Govarthanan M, Mythili R, Selvankumar T, Kamala KS, Rajasekar A, Chang YC (2016) Bioremediation of heavy metals using an endophytic bacterium *Paenibacillus* sp RM isolated from the roots of *Tridax procumbens*. 3 Biotech 6:1–7

Govindarajan M, Balandreau J, Muthukumarasamy R, Revathi G, Lakshminarasimhan C (2006) Improved yield of micropropagated sugarcane following inoculation by endophytic *Burkholderia vietnamiensis*. Plant Soil 280:239–252. https://doi.org/10.1007/s11104-005-3223-2

Gupta G, Panwar J, Akhtar MS, Jha PN (2012) Endophytic nitrogen-fixing bacteria as biofertilizer. In: Lichtfouse E (ed) Sustainable agriculture reviews. Sustainable agriculture reviews, vol 11. Springer, Dordrecht. https://doi.org/10.1007/978-94-007-5449-2_8

Hamayun H, Khan K, Khan W et al (2017) Gibberellins producing endophytic fungus *Porostereum spadiceum* AGH786 rescues growth of salt-affected soybean. Front Microbiol 8:686. https://doi.org/10.3389/fmicb.2017.00686

Hao J, Zhang X, Qi S, Song F, Lyu X, Ma Y, Peng H (2024) Species diversity, nitrogen fixation, and nutrient solubilization activities of endophytic bacteria in pea embryos. App Sci 14:788. https://doi.org/10.3390/app14020788

Hardoim PR, van Overbeek LS, Berg G, Pirttilä AM, Compant S, Campisano A et al (2015) The hidden world within plants: ecological and evolutionary considerations for defining functioning of microbial endophytes. Microbiol Mol Biol Rev 79(3):293–320

Heyland DK, Johnson AP, Reynolds SC, Muscedere J (2011) Procalcitonin for reduced antibiotic exposure in the critical care setting: a systematic review and an economic evaluation. Crit Care Med 39(7):1792–1799. https://doi.org/10.1097/CCM.0b013e31821201a5

Hussain H, Root N, Jabeen F, Al-Harrasi A, Al-Rawahi A, Ahmad M et al (2014) Seimatoric acid and colletonoic acid: two new compounds from the endophytic fungi, *Seimatosporium* sp. and *Colletotrichum* sp. Chin Chem Lett 25(12):1577–1589. https://doi.org/10.1016/j.cclet.2014.06.006

Ikram M, Ali N, Jan G, Jan FG, Rahman IU, Iqbal A, Hamayun M (2018) IAA producing fungal endophyte *Penicillium roqueforti* Thom., enhances stress tolerance and nutrients uptake in wheat plants grown on heavy metal contaminated soils. PLoS One 13(11):e0208150. https://doi.org/10.1371/journal.pone.0208150

Jaemsaeng R, Jantasuriyarat C, Thamchaipenet A (2018) Molecular interaction of 1- aminocyclopropane-1-carboxylate deaminase (ACCD)-producing endophytic *Streptomyces* sp. GMKU 336 towards salt-stress resistance of *Oryza sativa* L. cv. KDML105. Sci Rep 8(1):1950. https://doi.org/10.1038/s41598-018-19799-9

Kahlon RS (2016) Biodegradation and bioremediation of organic chemical pollutants by Pseudomonas. In Pseudomonas: Molecular and applied biology (pp. 343–417). Springer, Cham

Kandel SL, Joubert PM, Doty SL (2017) Bacterial endophyte colonization and distribution within plants. Microorganisms 5(4):77. https://doi.org/10.3390/microorganisms5040077

Kari Dolatabadi H, Mohammadi Goltapeh E, Mohammadi N, Rabiey M, Rohani N, Varma A (2012) Biocontrol potential of root endophytic fungi and Trichoderma species against fusarium wilt of lentil under in vitro and greenhouse conditions. J Agric Sci Technol 14(2):407–420

Ke X, Feng S, Wang J, Lu W, Zhang W, Chen M, Lin M (2019) Effect of inoculation with nitrogen fixing bacterium *Pseudomonas stutzeri* A1501 on maize plant growth and the microbiome indigenous to the rhizosphere. Syst Appl Microbiol 42(2):248–260. https://doi.org/10.1016/j.syapm.2018.10.010

Kemp ND, Vaughan MM, McCormick SP, Brown JA, Bakker MG (2020) *Sarocladium zeae* is a systemic endophyte of wheat and an effective biocontrol agent against Fusarium head blight. Biol Control 149:104329. https://doi.org/10.1016/j.biocontrol.2020.104329

Khan AL, Hamayun M, Kang SM, Kim YH, Jung HY, Lee JH, Lee IJ (2012) Endophytic fungal association via gibberellins and indole acetic acid can improve plant growth under abiotic stress: an example of *Paecilomyces formosus* LHL10. BMC Microbiol 12:3. https://doi.org/10.1186/1471-2180-12-3

Khan AL, Waqas M, Kang SM, Al-Harrasi A, Hussain J, Al-Rawahi A et al (2014) Bacterial endophyte *Sphingomonas* sp. LK11 produces gibberellins and IAA and promotes tomato plant growth. J Microbiol 52(8):689–695. https://doi.org/10.1007/s12275-014-4002-7

Kim JA, Jeon J, Kim KT, Choi G, Park SY, Lee HJ et al (2017) Draft genome sequence of an endophytic fungus, *Gaeumannomyces* sp. strain JS-464, isolated from a reed plant, Phragmites communis. Genome Announc 5(31):e00734–e00717. https://doi.org/10.1128/genomeA.00734-17

Kim SW, Goossens A, Libert C, Van Immerseel F, Staal J, Beyaert R (2020) Phytohormones: multifunctional nutraceuticals against metabolic syndrome and comorbid diseases. Biochem Pharmacol 175:113866. https://doi.org/10.1016/j.bcp.2020.113866

King RW, Evans LT (2003) Gibberellins and flowering of grasses and cereals: prizing open the lid of the "florigen" black box. Annu Rev Plant Biol 54:307–328. https://doi.org/10.1146/annurev.arplant.54.031902.135029

Kloepper JW, Ryu CM (2006) Bacterial endophytes as elicitors of induced systemic resistance. In: Schulz BJE, Boyle CJC, Sieber TN (eds) Microbial root endophytes. Springer, Berlin, pp 33–52. https://doi.org/10.1007/3-540-33526-9_3

Kou M-Z, Bastías DA, Christensen MJ, Zhong R, Nan ZB, Zhang XX (2021) The plant salicylic acid signalling pathway regulates the infection of a biotrophic pathogen in grasses associated with an Epichloe endophyte. J Fungi (Basel) 7(8):633. https://doi.org/10.3390/jof7080633. Pubmed:34436172

Kumar A et al (2021) Fungal endophytes: role in sustainable agriculture and environmental remediation. Front Microbiol 12:642468

Kumar R, Swapnil P, Meena M, Selpair S, Yadav BG (2022) Plant growth-promoting rhizobacteria (PGPR): approaches to alleviate abiotic stresses for enhancement of growth and development of medicinal plants. Sustain For 14(23):15514. https://doi.org/10.3390/su142315514

Kusari P et al (2022) Harnessing endophytes for bioremediation of organic pollutants: challenges and perspectives. J Hazard Mater 424(A):127424

Lacercat-Didier L, Berthelot C, Foulon J, Errard A, Martino E, Chalot M, Blaudez D (2016) New mutualistic fungal endophytes isolated from poplar roots display high metal tolerance. Mycorrhiza 26(7):657–671. https://doi.org/10.1007/s00572-016-0699-y

Lee JC, Son Y-O, Pratheeshkumar P, Shi X (2012) Oxidative stress and metal carcinogenesis. Free Radic Biol Med 53(4):742–757. https://doi.org/10.1016/j.freeradbiomed.2012.06.002

Li HY, Li DW, He CM, Zhou ZP, Mei T, Xu HM (2012) Diversity and heavy metal tolerance of endophytic fungi from six dominant plant species in a Pb-Zn mine wasteland in China. Fungal Ecol 5(3):309–315. https://doi.org/10.1016/j.funeco.2011.06.002

Li X et al (2019) Plant-endophyte symbiosis in phytoremediation. Environ Pollut 249:153–160

Li XZ, Simpson WR, Song ML, Bao GS, Niu XL, Zhang ZH et al (2020) Effects of seed moisture content and Epichloe endophyte on germination and physiology of *Achnatherum inebrians*. S Afr J Bot 134:407–414. https://doi.org/10.1016/j.sajb.2020.03.022

Liaqat F, Eltem R (2016) Identification and characterization of endophytic bacteria isolated from in vitro cultures of peach and pear rootstocks. 3 Biotech 6(2):1–8. https://doi.org/10.1007/s13205-016-0442-6

Lindström K, Mousavi SA (2020) Effectiveness of nitrogen fixation in rhizobia. Microb Biotechnol 13(5):1314–1335. https://doi.org/10.1111/1751-7915.13517

Liu F, Xing S, Ma H, Du Z, Ma B (2013) Cytokinin producing, plant growth promoting rhizobacteria that confer resistance to drought stress in *Platycladus orientalis* container seedlings. Appl Microbiol Biotechnol 97(20):9155–9164. https://doi.org/10.1007/s00253-013-5193-2

Loaces I, Ferrando L, Scavino AF (2011) Dynamics, diversity and function of endophytic siderophore-producing bacteria in rice. Microb Ecol 61(3):606–618. https://doi.org/10.1007/s00248-010-9780-9

Ludueña LM, Anzuay MS, Angelini JG, McIntosh M, Becker A, Rupp O et al (2019) Genome sequence of the endophytic strain *Enterobacter* sp. J49, a potential biofertilizer for peanut and maize. Genomics 111(4):913–920. https://doi.org/10.1016/j.ygeno.2018.05.021. Pubmed:29857118

Lumactud R, Shen SY, Lau M, Fulthorpe (2016) Bacterial endophytes isolated from plants in natural oil seep soils with chronic hydrocarbon contamination. Front Microbiol 7(755):755. https://doi.org/10.3389/fmicb.2016.00755

Ma Y, Rajkumar M, Luo Y, Freitas H, Hazard J (2011) Inoculation of endophytic bacteria on host and non-host plants–effects on plant growth and Ni uptake. J Hazard Mater 195:230–237. https://doi.org/10.1016/j.jhazmat.2011.08.034

Ma Y, Oliveira RS, Nai FJ, Rajkumar M, Luo YM, Rocha I, Freitas H (2015) The hyperaccumulator Sedum *Plumbizincicola harbors* metal-resistant endophytic bacteria that improve its phytoextraction capacity in multi-metal contaminated soil. J Environ Manag 156:62–69. https://doi.org/10.1016/j.jenvman.2015.03.024

Ma Y, Rajkumar M, Zhang C, Freitas H (2016) Beneficial role of bacterial endophytes in heavy metal phytoremediation. J Environ Manag 174:14–25. https://doi.org/10.1016/j.jenvman.2016.02.047

Makaras T, Montvydienė D, Kazlauskienė N, Stankevičiūtė M, Raudonytė-Svirbutavičienė E (2020) Juvenile fish responses to sublethal leachate concentrations: comparison of sensitivity of different behavioral endpoints. Environ Sci Pollut Res 27(5):4876–4890. https://doi.org/10.1007/s11356-019-07211-6

Manikandan M, Kannan V, Mendoza OH et al (2015) The contribution of endophytic bacteria to *Albizia lebbeck*-mediated phytoremediation of tannery effluent contaminated soil. Int J Phytoremediation 18(1):77–86

Marasco R, Rolli E, Ettoumi B et al (2018) A drought resistance-promoting microbiome is selected by root system under desert farming. PLoS One 13(5):e48479

Marian M, Ohno T, Suzuki H et al (2020) A novel strain of endophytic Streptomyces for the biocontrol of strawberry anthracnose caused by *Glomerella cingulata*. Microbiol Res 234:126428

Melo IS, Faull JL (2000) Parasitism of *Rhizoctonia solani* by strains of *Trichoderma* spp. Sci Agric 57(1):55–59

Mohd S, Shukla J, Kushwaha AS et al (2017) Front Microbiol 8:754. https://doi.org/10.3389/fmicb.2017.00754

Nadeem SM, Zahir ZA, Naveed M, Arshad M (2014) Rhizobacteria containing ACC-deaminase confer salt tolerance in maize grown on salt-affected fields. Can J Microbiol 60(2):71–78

Naveed M, Hussain MB, Zahir ZA, Mitter B, Sessitsch A (2014a) Drought stress amelioration in wheat through inoculation with *Burkholderia phytofirmans* strain PsJN. Plant Growth Regul 73(2):121–131

Naveed M, Mitter B, Reichenauer TG, Wieczorek K, Sessitsch A (2014b) Increased drought stress resilience of maize through endophytic colonization by *Burkholderia phytofirmans* PsJN and Enterobacter sp. FD17. Environ Exp Bot 97:30–39

Negi R, Sharma B, Kumar S, Chaubey KK, Kaur T, Devi R et al (2024) Plant endophytes: unveiling hidden applications toward agro-environment sustainability. Folia Microbiol (Praha) 69(1):181–206

Nudel C, Gonzalez R, Castaneda N, Mahler G, Actis LA (2001) Influence of iron on growth, production of siderophore compounds, membrane proteins, and lipase activity in *Acinetobacter calcoaceticus* BD 413. Microbiol Res 155:263–269

Pandey N, Bhatt N (2016) Role of soil associated *Exiguobacterium* in reducing arsenic toxicity and promoting plant growth in *Vigna radiate*. Eur J Soil Biol 75:142–150

Pang QY, Zhang AQ, Zang LL, Liu C, Wang JS (2009) Role of endophytic fungi in biocontrol of tomato bacterial wilt disease caused by *Ralstonia solanacearum*. Biotechnology 8(4):380–386

Pietro-Souza W, de Campos PF, Mello IS et al (2020) Mercury resistance and bioremediation mediated by endophytic fungi. Chemosphere 240:124874

Piotrowska A, Bajguz A (2011) Conjugates of abscisic acid, brassinosteroids, ethylene, gibberellins, and jasmonates. Phytochemistry 72(17):2097–2112

Qadri M, Rajput L, Hussain A et al (2014) Enhancing crop yield and productivity by balancing plant metabolism and enhancing soil health. In: Microbial metabolism in sustainable agroecosystem, pp 153–182

Qin W, Liu C, Jiang W, Xue Y, Wang G, Liu S (2019) A coumarin analogue NFA from endophytic *Aspergillus fumigatus* improves drought resistance in rice as an antioxidant. BMC Microbiol 19(1):1–11

Rajkumar M, Ae N, Prasad MNV, Freitas H (2010) Potential of siderophore-producing bacteria for improving heavy metal phytoextraction. Trends Biotechnol 28(3):142–149

Rana KL, Kour D, Kaur T et al (2020) Endophytic microbes from diverse wheat genotypes and their potential biotechnological applications in plant growth promotion and nutrient uptake. Proc Natl Acad Sci India Sect B Biol Sci 90:969–979. https://doi.org/10.1007/s40011-020-01168-0

Raymaekers K, Ponet L, Holtappels D, Berckmans B, Cammue BPA (2020) Screening for novel biocontrol agents applicable in plant disease management–a review. Biol Control 144:104240

Sabra M, Aboulnasr A, Franken P, Perreca E, Wright LP, Camehl I et al (2018) Front Plant Sci 9:1726. https://doi.org/10.3389/fpls.2018.01726

Saharan BS, Chaudhary T, Mandal BS, Kumar D, Kumar R, Sadh PK, Duhan JS (2023) Microbe-plant interactions targeting metal stress: new dimensions for bioremediation applications. J Xenobiot 13:252–269. https://doi.org/10.3390/jox13020019

Saikkonen K, Wäli P, Helander M, Faeth SH (2010) Evolution of endophyte-plant symbioses. Trends Plant Sci 15(4):275–228

Salam N, Khieu TN, Liu MJ (2017) Endophytic actinobacteria associated with *Dracaena cochinchinensis* Lour.: isolation, diversity, and their cytotoxic activities. Biomed Res Int 2017:1308563

Sandhiya G, Sugitha T, Balachandar D, Kumar K (2005) Endophytic colonization and in planta nitrogen fixation by a diazotrophic Serratia sp. in rice. Indian J Exp Biol 43:802–807

Sandhya V, Shrivastava M, Ali SZ, Prasad VSSK (2017) Endophytes from maize with plant growth promotion and biocontrol activity under drought stress. Russ Agric Sci 43:22–34. https://doi.org/10.3103/S1068367417010116

Santoyo G, Moreno-Hagelsieb G, Orozco-Mosqueda MDC, Glick BR (2016) Plant growth-promoting bacterial endophytes. Microbiol Res 183:92–99

Saxena S, Srivastava RK, Samaddar AB (2010) IUP. J Soil Water Sci 3:72–90

Schneijderberg M, Schmitz L, Cheng X, Polman S, Franken C, Geurts R et al (2018) A genetically and functionally diverse group of non-diazotrophic *Bradyrhizobium* spp. colonizes the root endophytic compartment of *Arabidopsis thaliana*. BMC Plant Biol 18:1–9. https://doi.org/10.1186/s12870-018-1272-y

Seth K, Vyas P, Deora S, Gupta AK, Meena M, Swapnil P, Harish (2023) Understanding plant-plant growth-promoting rhizobacteria (PGPR) interactions for inducing plant defense. In: Swapnil P, Meena M, Harish, Marwal A, Vijayalakshmi S, Zehra A (eds), Plant-microbe interaction—recent advances in molecular and biochemical approaches, volume 2. Academic Press, Elsevier, 201–226. https://doi.org/10.1016/B978-0-323-91876-3.00010-

Shahabivand S, Parvaneh A, Aliloo AA (2017) Ecotoxicol Environ Saf 145:496–502. https://doi.org/10.1016/j.ecoenv.2017.07.064

Shahzad R, Waqas M, Khan AL, Asaf S, Khan MA, Kang SM et al (2016) Seed-borne endophytic *Bacillus amyloliquefaciens* RWL-1 produces gibberellins and regulates endogenous phytohormones of Oryza sativa. Plant Physiol Biochem 106:236–243. https://doi.org/10.1016/j.plaphy.2016.05.006

Shahzad R, Waqas M, Khan AL et al (2017) Indoleacetic acid production and plant growth promoting the potential of bacterial endophytes isolated from rice (*Oryza sativa* L.) seeds. Acta Biol Hung 68:175–186. https://doi.org/10.1556/018.68.2017.2.5

Sharma A, Johri BN (2003) Growth promoting influence of siderophore-producing *Pseudomonas* strains GRP3A and PRS_9 in maize (*Zea mays* L.) under iron limiting conditions. Microbiol Res 158(3):243–248. https://doi.org/10.1078/0944-5013-00197. Pubmed:14521234

Sharma P, Kumar S (2021) Bioremediation of heavy metals from industrial effluents by endophytes and their metabolic activity: recent advances. Bioresour Technol 339:125589. https://doi.org/10.1016/j.biortech.2021.125589

Sharma P, Tripathi S, Vadakedath N, Chandra R (2021) In-situ toxicity assessment of pulp and paper industry wastewater on *Trigonella foenum-graecum* L: potential source of cytotoxicity and chromosomal damage. Environ Technol Innov 21:101251. https://doi.org/10.1016/j.eti.2020.101251

Shelly S, Pritam C, Ieshita P et al (2009) Microbial transformation of xenobiotics for environmental bioremediation. Afr J Biotechnol 8(22):6016–6027. https://doi.org/10.5897/AJB09.740

Sheng X, Chen X, He L (2008) Characteristics of an endophytic pyrene-degrading bacterium of Enterobacter sp. 12J1 from *Allium Macrostemon* Bunge. Int Biodeterior Biodegrad 62(2):88–95. https://doi.org/10.1016/j.ibiod.2007.12.003

Shi Y, Xie H, Cao L, Zhang R, Xu Z, Wang Z, Deng Z (2017) Effects of Cd- and Pb-resistant endophytic fungi on growth and phytoextraction of *Brassica napus* in metal-contaminated soils. Environ Sci Pollut Res Int 24(1):417–426. https://doi.org/10.1007/s11356-016-7693-y

Siddiquee S, Rovina K, Azad SA (2015) Heavy metal contaminants removal from wastewater using the potential filamentous fungi biomass: a review. J Microb Biochem Technol 07(6). https://doi.org/10.4172/1948-5948.1000243

Sinma K, Nurak T, Khucharoenphaisan K (2015) Potentiality of endophytic actinomycetes isolated from sugar cane. KMITL Sci Tech J15:88–97

Sun C, Johnson JM, Cai D, Sherameti I, Oelmüller R, Lou B (2010) *Piriformospora indica* confers drought tolerance in Chinese cabbage leaves by stimulating antioxidant enzymes, the expression of drought-related genes and the plastid-localized CAS protein. J Plant Physiol 167(12):1009–1017. https://doi.org/10.1016/j.jplph.2010.02.013

Sun X, Wang N, Li P, Jiang Z, Liu X, Wang M et al (2020) Endophytic fungus *Falciphora oryzae* promotes lateral root growth by producing indole derivatives after sensing plant signals. Plant Cell Environ 43(2):358–373. https://doi.org/10.1111/pce.13667

Szink I, Davis EL, Ricks KD, Koide RT (2016) New evidence for broad trophic status of leaf endophytic fungi of *Quercus gambelii*. Fungal Ecol 22:2–9. https://doi.org/10.1016/j.funeco.2016.04.003

Teotia P, Kumar V, Kumar M, Shrivastava N, Varma A (2016) Rhizosphere microbes: potassium solubilization and crop productivity- present and future aspects. In: Meena V, Maurya B, Verma J, Meena R (eds) Potassium solubilizing microorganisms for sustainable agriculture. Springer, Berlin. https://doi.org/10.1007/978-81-322-2776-2_22

Thor K (2019) Calcium-Nutrient and Messenger. Front. Plant Sci 10:440. https://doi.org/10.3389/fpls.2019.00440

Toppo P, Jangir P, Mehra N, Kapoor R, Mathur P (2024) Bioprospecting of endophytic fungi from medicinal plant *Anisomeles indica* L. for their diverse role in agricultural and industrial sectors. Sci Rep 14(1):588. https://doi.org/10.1038/s41598-023-51057-5

UmaMaheswari T, Anbukkarasi K, Hemalatha T, Chendrayan K (2013) Studies on phytohormone producing ability of indigenous endophytic bacteria isolated from tropical legume crops. Int J Curr Microbiol Appl Sci 2:127–136

Vadassery J, Ranf S, Drzewiecki C, Mithöfer A, Mazars C, Scheel D et al (2009) A cell wall extract from the endophytic fungus *Piriformospora indica* promotes growth of *Arabidopsis* seedlings and induces intracellular calcium elevation in roots. Plant J 59(2):193–206. https://doi.org/10.1111/j.1365-313X.2009.03867.x

Vandenkoornhuyse P, Quaiser A, Duhamel M, Le Van A, Dufresne A (2015) The importance of the microbiome of the plant holobiont. New Phytol 206(4):1196–1206. https://doi.org/10.1111/nph.13312

Varughese T, Riosa N, Higginbotham S, Arnold AE, Coley PD, Kursar TA et al (2012) Antifungal depsidone metabolites from *Cordyceps dipterigena*, an endophytic fungus antagonistic to the phytopathogen *Gibberella fujikuroi*. Tetrahedron Lett 53(13):1624–1626. https://doi.org/10.1016/j.tetlet.2012.01.076

Verma J, Bhatt A, Agrawal PK (2016) In-vitro study on bioaccumulation and tolerance of heavy metals by endophytic fungi Alternaria alternata isolated from Cupressus torulosa D. DON. Octa J Environ Res 4:146–154

Verma SK, Kingsley KL, Bergen MS, Kowalski KP, White JF (2018) Fungal disease prevention in seedlings of rice (*Oryza sativa*) and other grasses by growth-promoting seed-associated endophytic bacteria from invasive *Phragmites australis*. Microorganisms 6(1):21. https://doi.org/10.3390/microorganisms6010021

Verma SK, Wei X, Song T, Wang J, Li Y (2020) Potential applications of endophytic fungi for sustainable agriculture and biotechnology. Fungal Biol Rev 34(1):26–40

Walitang DI, Kim K, Madhaiyan M, Kim YK, Kang Y, Sa T (2017) Characterizing endophytic competence and plant growth promotion of bacterial endophytes inhabiting the seed endosphere of rice. BMC Microbiol 17(1):209. https://doi.org/10.1186/s12866-017-1117-0

Wang Y, Li H, Zhao W, He X, Chen J, Geng X, Xiao M (2010) Induction of toluene degradation and growth promotion in corn and wheat by horizontal gene transfer within endophytic bacteria. Soil Biol Biochem 42(7):1051–1057. https://doi.org/10.1016/j.soilbio.2010.03.002

Waqas M, Khan AL, Kamran M, Hamayun M, Kang SM, Kim YH, Lee IJ (2012) Endophytic fungi produce gibberellins and indoleacetic acid and promote host-plant growth during stress. Molecules 17(9):10754–10773. https://doi.org/10.3390/molecules170910754

Waqas M, Khan AL, Kang SM, Kim YH, Lee IJ (2014) Phytohormone-producing fungal endophytes and hardwood-derived biochar interact to ameliorate heavy metal stress in soybeans. Biol Fertil Soils 50(7):1155–1167. https://doi.org/10.1007/s00374-014-0937-4

Weyens N, Beckers B, Schellingen K, Ceulemans R, van der Lelie D, Newman L et al (2015) The potential of the Ni-resistant TCE-degrading *Pseudomonas putida* W619-TCE to reduce phytotoxicity and improve phytoremediation efficiency of poplar cuttings on a Ni-TCE Co-contamination. Int J Phytoremediation 17(1–6):40–48. https://doi.org/10.1080/15226514.2013.828016. Pubmed:25174423

Wu H, Yang HY, You XL, Li YH (2013) Diversity of endophytic fungi from roots of panax ginseng and their saponin yield capacities. Springerplus 2(1):107. https://doi.org/10.1186/2193-1801-2-107

Wybouw B, De Rybel B (2019) Cytokinin—a developing story. Trends Plant Sci 24(2):177–185. https://doi.org/10.1016/j.tplants.2018.10.012

Xia C, Li N, Zhang X, Feng Y, Christensen MJ, Nan Z (2016) An Epichloe endophyte improves photosynthetic ability and dry matter production of its host *Achnatherum inebrians* infected by *Blumeria graminis* under various soil water conditions. Fungal Ecol 22:26–34. https://doi.org/10.1016/j.funeco.2016.04.002

Xie XG, Fu WQ, Zhang FM, Shi XM, Zeng YT, Li H et al (2017) The endophytic fungus *Phomopsis liquidambari* increases nodulation and N_2 fixation in *Arachis hypogaea* by enhancing hydrogen peroxide and nitric oxide signalling. Microb Ecol 74(2):427–440. https://doi.org/10.1007/s00248-017-0944-8

Xu L, Wu C, Oelmüller R, Zhang W (2018) Role of phytohormones in *Piriformospora indica*-induced growth promotion and stress tolerance in plants: more questions than answers. Front Microbiol 9:1646. https://doi.org/10.3389/fmicb.2018.01646

Xu WF, Ren HS, Ou T, Lei T, Wei JH, Huang CS et al (2019) Genomic and functional characterization of the endophytic *Bacillus subtilis* 7PJ-16 strain, a potential biocontrol agent of mulberry fruit sclerotiniose. Microb Ecol 77(3):651–663. https://doi.org/10.1007/s00248-018-1247-4

Yadav G, Meena M (2022) Biological control of plant diseases by endophytes. In: Shah MP, Deka D (eds) Endophytic association: what, why and how (volume: developments in applied microbiology and biotechnology). Academic Press, Elsevier Inc., pp 119–135. https://doi.org/10.1016/B978-0-323-91245-7.00007-9

You YH, Yoon H, Kang SM, Shin JH, Choo YS, Lee I-J et al (2012) Fungal diversity and plant growth promotion of endophytic fungi from six halophytes in Suncheon bay. J Microbiol Biotechnol 22(11):1549–1556. https://doi.org/10.4014/jmb.1205.05010

Young CA, Felitti S, Shields K, Spangenberg G, Johnson RD, Bryan GT et al (2006) A complex gene cluster for indole-diterpene biosynthesis in the grass endophyte *Neotyphodium lolii*. Fungal Genet Biol 43(10):679–693. https://doi.org/10.1016/j.fgb.2006.04.004

Yuan M, He H, Xiao L, Zhong T, Liu H, Li S et al (2014) Enhancement of Cd phytoextraction by two *Amaranthus* species with endophytic *Rahnella* sp. JN27. Chemosphere 103:99–104. https://doi.org/10.1016/j.chemosphere.2013.11.040

Zahoor M, Irshad M, Rahman H, Qasim M, Afridi SG, Qadir M, Hussain A (2017) Alleviation of heavy metal toxicity and phytostimulation of *Brassica campestris* L. by endophytic Mucor sp. MHR-7. Ecotoxicol Environ Saf 142:139–149. https://doi.org/10.1016/j.ecoenv.2017.04.005

Zhang YF, He LY, Chen ZJ, Zhang WH, Wang QY, Qian M, Sheng XF (2011) Characterization of lead-resistant and ACC deaminase-producing endophytic bacteria and their potential in promoting lead accumulation of rape. J Hazard Mater 186(2–3):1720–1725. https://doi.org/10.1016/j.jhazmat.2010.12.069

Zhang W, Li XG, Sun K, Tang MJ, Xu FJ, Zhang M, Dai CC (2020) Mycelial network-mediated rhizobial dispersal enhances legume nodulation. ISME J 14(4):1015–1029. https://doi.org/10.1038/s41396-020-0587-5

Zhu LJ, Guan DX, Luo J, Rathinasabapathi B, Ma LQ (2014) Characterization of arsenic-resistant endophytic bacteria from hyperaccumulators *Pteris vittata* and *Pteris multifida*. Chemosphere 113:9–16. https://doi.org/10.1016/j.chemosphere.2014.03.081

Zhu Y, Zhu S, Zhang F, Zhao Z, Christensen MJ, Nan Z, Zhang X (2022) Transcriptomic analyses reveal molecular regulation of photosynthesis by epichloe endophyte in *Achnatherum inebrians* under *Blumeria graminis* infection. J Fungi (Basel) 8(11):1201. https://doi.org/10.3390/jof8111201

Zicca S, De Bellis P, Masiello M, Saponari M, Saldarelli P, Boscia D, Sisto A (2020) Antagonistic activity of olive endophytic bacteria and of *Bacillus* spp. Strains against *Xylella fastidiosa*. Microbiol Res 236:126467. https://doi.org/10.1016/j.micres.2020.126467

Zou WX, Meng JC, Lu H, Chen GX, Shi GX, Zhang TY, Tan RX (2000) Metabolites of *Colletotrichum gloeosporioides*, an endophytic fungus in *Artemisia mongolica*. J Nat Prod 63(11):1529–1530. https://doi.org/10.1021/np000204t

Host-Induced Gene Expression in Endophytes

9

Sana Sheikh, Sareen Sheikh, V. S. Swaroop Kumar, and Rosana Babu

Abstract

Endophytic fungi are asymptomatic organisms residing in the plant host tissues. Endophytic fungi play a vital role in the modulation of gene expression in host plants by enhancing plant growth, activation of defense mechanisms, resistance against biotic and abiotic stress, and secondary metabolite production. This chapter portrays the role of versatile endophytic fungi in inducing gene expression that involves mechanisms in the host plant growth and improvement.

Keywords

Endophytes · Fungi · Plant · Signal · Gene expression

9.1 Introduction

Every known species of plant contains microbial endophytes. Endophytes are special because they can enter and flourish in plant tissues, displaying complex interactions with their host (Khare et al. 2018). Fungal endophytes reside in the plant host

S. Sheikh (✉)
Department of Botany, St Aloysius (Deemed to be University), Mangaluru, India

S. Sheikh
Department of Chemistry, Yenepoya Institute of Arts, Science and Management, Yenepoya (Deemed to be University), Mangaluru, India

V. S. Swaroop Kumar
Department of Plant Science, Central University of Kerala, Kasaragod, India

R. Babu
Computational Biology, Oxford University, Oxford, UK

M. Meena et al. (eds.), *Multi-omics Approach to Investigate Endophyte Diversity*, Microorganisms for Sustainability 53, https://doi.org/10.1007/978-981-96-4100-0_9

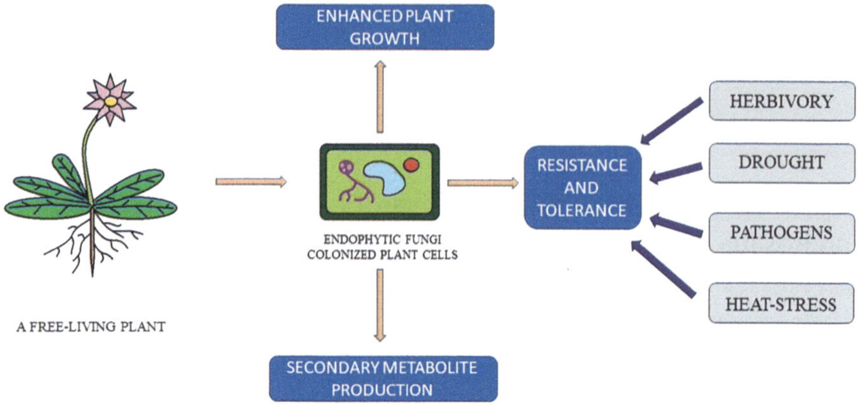

Fig. 9.1 Role of endophytic fungi

causing apparently no harm. These are fascinating groups entrapped as hidden treasures within host tissues. They tend to mimic the host metabolites and therefore possess potential applications (Fig. 9.1) in terms of bioprospecting. Fungal endophytes are incredibly diverse and occur in all kinds of plants providing wide array of potential sources of bioactive compounds for pharmaceutical, industrial, and agricultural sector, lead compounds for drug development in terms of antimicrobial, antidiabetic, anti-inflammatory, as well as anticancer agents. Isolation, culture, and extraction of secondary metabolites from fungal endophytes are among the challenging tasks in terms of research.

The term "endophyte" is derived from the Greek (endon = inside and phyte = plant) referring to life forms that colonize internal plant tissues (De Bary 1866). A wide range of microorganisms, actinomycetes, mycoplasma, and fungi exist as endophytes of plants (Khiralla et al. 2016; Seth et al. 2023; Bhupenchandra et al. 2024). Endophytes are additionally grouped as obligate and facultative based on their need for a suitable host for their survival (Gouda et al. 2016). Contrasting with the mycorrhizal organisms that occupy the root tissues and the rhizosphere, the endophytes occupy the internal plant tissues of root, stem, and leaves seldom arising out except if it is for sporulation during the passing of the host tissues (Rodriguez et al. 2009). Practically, all major groups of plants including liverworts, mosses, pteridophytes, gymnosperms, and angiosperms show harmonious relationships with endophytes (Arnold and Lutzoni 2007). These contagious endophytes are a polyphyletic group of ascomycetous fungi and a few basidiomycetes (Rodriguez and Redman 2008) showing variety in its occurrence and functions (Arnold 2007).

Plant genes that are expressed greater in presence of endophytes can reveal information about how they affect the host plant (Berendsen et al. 2015). Fungal endophytes and host plants interact to form symbionts, which have undergone significant evolution in recognition, selection, and development of generally stable and reciprocal connections between both partners (White et al. 2019). In the course of long-term evolution, it was discovered that endophytic fungi and their hosts form a "holobiont" to react to both abiotic and biotic stresses together (Vandenkoornhuyse et al. 2015). Because of this, endophytes are sometimes seen as the host plant's accessory organs (Christian et al. 2016), and because of their apparent necessity in natural environments, their genes are regarded as a part of an expanded genome. Thus, harnessing species interactions to tackle contemporary problems pertaining to plant health and yield requires an understanding of the formation process and mechanism of the plant-endophyte "holobiont" (Yan et al. 2018). This chapter intrigues the symbiotic relationship between plant host and endophytic fungi.

9.2 Plant Endophyte Interactions

9.2.1 Recognition of Fungi

When the fungus invades a plant, plant receptors precisely recognize the secreted elicitors in addition to the fungal hyphae and cell walls, which are the exterior structures that the plant recognizes specifically. Fungal elicitors that promote defense mechanism in plants may include unsaturated fatty acids, polysaccharides, polypeptides and proteins, glycoproteins and glycolipids, and oligosaccharides (chitin, chitosan, and cyclodextrin). Plant receptors can be divided into two groups based on how they recognize elicitors: intracellular nucleotide-binding leucine-rich repeat proteins (NLRs) and cell-surface pattern recognition receptors (PRRs) (Saijo et al. 2018). Three domains make up PRRs: an extracellular domain, a single-pass transmembrane domain, and an intracellular domain. These domains are found in receptor-like kinases (RLKs) and receptor-like proteins (RLPs). Plant pattern-triggered immunity (PTI) is induced by the particular domain's perception and binding of pathogen/microbe-associated molecular patterns (PAMPs/MAMPs) (Khare et al. 2018). Plants' NLRs either directly or indirectly identify some fungal effectors produced into plant cells to inhibit PTI, and this generally instigates the stronger effector-triggered immunity (ETI) in plants.

Endophytes have to move through the plant immune system's first line of defense in order to penetrate the host plant. This entails plants recognize conserved molecules that are common to multiple microbes and are often referred to as pathogen- or microbe-associated molecular patterns (MAMPs or PAMPs) (Newman et al. 2013). Pattern recognition receptors (PRRs) on the surface of plant cells identify these MAMPs. When plants are exposed to fungal endophytes, they develop chitin-specific receptors (PR-3) that identify chitin oligomers on the fungal cell wall. This recognition sets off additional defensive responses in the plant (Sanchez-Vallet et al. 2015).

9.2.2 Colonization of Endophyte

Compatible plant-microbe interactions are crucial for successful endophyte colonization. A typical endophyte can colonize a plant through two methods, either by horizontal transmission or via the vertical transmission (Hodgson et al. 2014). Horizontal transmission being the most common occurs mainly across through the soil or air with help of spores. Another method is the vertical transmission or transfer of endophytes from the parent to the offspring through seeds which is quite a restricted phenomenon. Most of the grass species demonstrate both ways of transmission (Aharwal et al. 2021). Successful colonization by endophytes is affected by several factors including the plant tissue type, the plant genotype, biotic and abiotic environmental conditions, plant age and species, and the fungal species involved in colonization (Mengistu 2020). It was observed in case of *Theobroma cacao*, endophyte colonization was also affected by the canopy cover, leaf phytochemistry, and water availability in tropical areas (Arnold and Herre 2003). Colonization patterns in case of the root restricted non-claviciptaceous endophytes were dependent on the host plant and the cultivation system used (Andrade-Linares and Franken 2013). In Barley, expression of the high-affinity ammonium transporter *PiAMT1* gene and VPE gene (induced during cell death) caused a shift in the biotrophic nature of *P. indica* to a saprophytic lifestyle during colonization. In case of *Arabidopsis,* both genes had no effect on the lifestyle of the symbiont (Lahrmann et al. 2013) and hence strengthen the proof that the species of host plays a major role in colonization.

Reception of endophytes by the host is mainly by their specific molecular patterns which are known as microbe-associated molecular patterns (MAMPs). The plants are able to perceive these through the presence of pattern reception receptors (PRRs) which constitute a major part of the plant's innate immunity (Doehlemann et al. 2014; Plett and Martin 2018). The PRRs once activated affect by inducing phosphorylation of substrates by receptor kinases leading to an intracellular signal transduction (Macho and Zipfel 2014). Some of the fungal MAMPs may be chitin, cerebrosides, ergosterol, elicitins, NEP1-like proteins, EIX protein, cell wall glucans, and cell wall glycoprotein CBEL with CBD motifs (Tsuda and Katagiri 2010). These receptors recognize whether the patterns are positioned towards a pathogen or a symbiont. Overall, the response is similar in both cases but in case of the mutualistic microbes, there is dampening of these responses (Khan et al. 2010; Mattoo and Nonzom 2021). During the infection, there is also increase in the ROS production in the infected cells, additionally there is also an influx of calcium ions (called "calcium spiking") which act as secondary messengers inducing defense responses. Plant hormones such as Jasmonic acid, Salicylic acid, and ethylene act as signaling molecules during the endophytic colonization (Plett and Martin 2018, Mengistu 2020, Mattoo and Nonzom 2021; Yadav and Meena 2021, 2022).

During *Trichoderma* colonization of *T. cacao*, it was observed that a major part of genes expressed in addition to the house-keeping genes were related to carbohydrate hydrolysis, protein digestion, and lipid metabolism which were helpful for nutrition and colonization of the symbiont. Expression of several genes such as alcohol oxidase suggested that during colonization, many of the genes expressed were also expressed by fungi during the plant disease process and starvation conditions (Bailey et al. 2006). RNAi and DNA methylation by endophytes could also affect expression of genes in the host (Dalakouras et al. 2023).

In case of barley, *Piriformospora indica* showed colonization of dead cells in the root tissues by manipulating expression of HvBI-1, a cell death regulator by downregulation (Deshmukh et al. 2006); it was also inferred in further studies that there is suppression of callose deposition and MAMP-triggered plant immunity responses in the colonized living root tissues of the differentiation zone. In several fungal endophytes, the mycelia found in living roots showed low abundance, patchy colonization, and variable values of growth direction indices and produced knot, coil, and vesicle like structures whereas the same mycelia in dead roots showed a highly regular growth patterns parallel to the root axis (Andrade-Linares and Franken 2013).

With rising omics technologies, the role of non-coding RNAs in regulation of cellular process are analyzed especially of miRNAs which are basically small 18–24 nucleotide long non-coding RNAs that are processed from almost 100 bp long precursors by AGO pathway. These miRNAs bind to mRNA by complementary fashion, which results in either their degradation via the AGO pathway or the prevention of further translation (Plett and Martin 2018).

The role of miRNAs in endophytic colonization associations is also well studied; in *Arabidopsis* and *Phialophora* association, it was observed that miRNAs undergo downregulation causing enhanced expression of ROS scavenging enzymes, and few miRNAs were mildly upregulated also which were helpful in eliciting defense and growth responses (Mishra et al. 2024). Similarly in Tomato, during colonization by *Pochonia chlamydosporia,* four out of the six miRNAs (miR482) targeting PHAS loci of NBS-LRR type resistance gene were downregulated at an early colonization stage (Pentimone et al. 2018). Most of the studies on mutualistic interactions have shown that a majority of miRNAs produced during reception of endophytes by hosts targets hormone-response pathways, protein methylation, and innate immune function.

9.2.3 Induced Defense Mechanisms

Fungal invasion activates plant defense networks, which results in the synthesis of defense substances, the production or increase of related signal molecules, the opening or closing of certain ion channels, the start of signaling responses, and the mediation of a number of signal pathways. Chemical signals released by plants and fungal endophytes mediate their recognition of one another. However, Endophytes have been shown to influence an array of fundamental host plant processes (Khare et al. 2018; Chandran et al. 2021). They have the ability to alleviate abiotic stressors, stimulate defense responses against pathogen invasion, and promote plant development.

Due to the greater versatility, microorganisms may have genes coding for unique characteristics advantageous to the host plant (Ali et al. 2014). One of the primary ways of which fungal endophytes benefit their host plants is by inducing defense mechanisms. When the plant recognizes the presence of the endophyte, it can activate specific genes involved in defense responses, such as the production of antimicrobial compounds or strengthening of cell walls. According to Das and Varma

(2009), endophytes create or induce metabolites in the host plant to facilitate development and improve environmental adaptation, hence maintaining stable symbiosis. According to Cui et al. (2015), the endophytic community appears to be influenced by factors within the host plants, which in turn affect the available nutrients for the fungal endophytes. Plant growth, development, and stress tolerance can be greatly impacted by the ability of endophytic fungi to control the physiology and metabolic activities of their host plants (Fontana et al. 2021).

When the endophyte penetrates the host plant, the plant recognizes it and begins to communicate via signal molecules. Chemotactic responses are often observed in endophytes especially towards root exudates. Additionally nutrient- and water-rich, root exudates draw a wide variety of microorganisms. One such metabolite, flavonoids, is classified as a chemoattractant and is released by a variety of plants. It plays a role for establishing the association between the endophyte and the root hair. The Nod factors, or lipo-chitooligosaccharides (LCO), are well-known signal molecules that activate the common symbiotic route (CSP) in arbuscular mycorrhizal associations and rhizobia-legume relationships (Gough and Cullimore 2011). Rozpadek et al. (2018) discovered that strigolactone (SL), which is released by *Arabidopsis thaliana* roots, functions as a signal molecule for endophytic *Mucor* sp. colonization. When *Gastrodia elata* first starts to form a relationship with endophytic fungi, such as *Mycena osmundicola* and *Armillaria mellea*, it releases strigolactones (SLs) to draw in the fungi and encourage their hyphal branching and development. This increases the probability that the fungi will come into contact with the host plants (Kim et al. 2006).

According to reports, the majority of miRNAs that are induced in the host during the endophyte establishment also target pathways involved in hormone response. The miRNA E4D3Z3Y01BW0TQ becomes more prevalent during Arbuscular Mycorrhizal Fungi infection and interferes with the gibberellic acid (GA) signaling pathway, which is known to have a repressive effect on mutualistic interactions (Formey et al. 2014). Most of the pathways targeted by miRNAs for the plant defense system are shut off during the development of symbiosis, which would have otherwise prevented endophyte proliferation (Plett and Martin 2018). According to Hardoim et al. (2015), anabolic pathway-related genes are more varied and plentiful in endophytes, whereas catabolism-related genes—particularly those connected to host invasion—are more prevalent in phytopathogens. Research on mutualistic symbioses has advanced to the point where plants can detect beneficial or parasitic invading microbes by monitoring their nutrition levels (Plett and Martin 2018). According to Bacon et al. (2008), changes in the host's genotype as well as abiotic stressors that are specific to the area can influence an endophyte's transition from a living state towards a pathogenic one. For instance, *Fusarium verticillioides* can exist as an endophyte or a pathogen in maize (Oren et al. 2003). Similar to this, *Ramularia collo-cygni* lives as an asymptomatic endophyte during crop growth but has the ability to change to a necrotrophic pathogen later in the growing season (Walters et al. 2008). Comparison studies are necessary to determine the gene expressions and conditions (in both plants and endophytes) under which the same microorganism acts as a pathogen or a mutualist in order to have a better understanding of the dynamics of endophytism.

Endophytes, however, also devise defense strategies to keep themselves safe from plant defenses. For instance, Cord-Landwehr et al. (2016) found that fungal endophytes generate chitin deacetylases that deacetylate chitosan oligomers, which are invisible to plant receptors and therefore help the endophytes avoid detection. Zeidler et al. (2004) reported that endophytes produce enzymes such as superoxide dismutases (SOD), catalases (CatA), peroxidases (POD), alkyl hydroperoxide reductases (AhpC), and glutathione-S-transferases (GSTs) to protect themselves against oxidative bursts or the generation of reactive oxygen species (ROS) as part of a plant defense system.

9.2.4 Tolerance to Stress

It has also been revealed that foliar endophytes control the host's genetic expression, which impacts the defense and physiological processes of plants (Estrada et al. 2013). Thus, endophytes can strengthen host tissues' defenses against infections by inducing the host response or by generating their own antagonistic chemicals. Thus, the host plant's gene pool and the endophyte gene pool cooperate to shield the plant against parasites (Khare et al. 2018). Induction and expression of stress-responsive genes, production of scavenger molecules like ROS, and synthesis of antistress metabolites are among the molecular strategies used by endophytes to increase stress tolerance in plants (Lata et al. 2018).

In order for the beneficial fungus *Piriformospora indica* and *A. thaliana* roots to create a mutualistic symbiosis, Peskan-Berghofer et al. (2015) showed that plant hormone abscisic acid was necessary. The study conducted by Pandey et al. (2016) reports that the inoculation of endophytic *Trichoderma harzianum* in rice results in the upregulation of genes related to aquaporin, dehydrin, and malondehyde, which helps to mitigate abiotic stressors.

9.2.5 Signaling Molecules and Secondary Metabolites

Various signals can be used by plants to trigger the fungal signaling pathways, which control the expression of virulence genes and the production of virulence agents, such as proteins, enzymes, and poisonous secondary metabolites, to mediate a range of fungal responses. Upon induction of plant immune responses, only those fungi capable of avoiding or mitigating the damage and attack caused by the protective secondary metabolites generated by the plants would be able to survive and grow into endophytes. In order to counteract the defensive secondary metabolites derived from plants, fungi generate a greater quantity of toxic metabolites than fungal elicitors. This causes the defensive secondary metabolites to become inactive. Certain fungi have the ability to neutralize plants' protective secondary metabolites. They release the corresponding lytic enzymes, such as cellulase and protease, which are detoxifying enzymes that break down the secondary metabolites from the plants before they can enter the host plant's defense mechanism. Certain endophytic fungi, similar to infectious pathogens, can alter the RNA splicing process of their hosts to inhibit their immune responses.

The cohabitation and evolution of endophytes with their host plants over an extended period of time have developed a unique interaction that has a substantial impact on the production of bioactive metabolites in plants (Jia et al. 2016; Kumar et al. 2022). Endophytic communities' interactions with the host plant have a major impact on physiological processes involving the plant, including the activation of quiet gene clusters that result in the manufacture of new secondary metabolites. The Yew plant and its fungal endophyte *Taxomyces andreanae* synthesize taxol are among the most researched bioactive chemicals with anticancer properties (Zhou et al. 2010). According to Heinig et al. (2013), the genome of endophytes did not exhibit any substantial sequence homology with the genes of *Taxus* spp. that are involved in taxol production, suggesting that the fungus may have independently evolved the mechanism. Additionally, research shows that the fungal endophytes *Curvularia* sp. and *Choanephora infundibulifera*, which harbor *Catharanthus roseus*, promote the expression of genes involved in the terpenoid indole alkaloid pathway (Sreekanth et al. 2017).

9.3 Conclusion

Endophytic fungi are thought to have a profound impact on plant physiology, growth, and metabolism, and as such, they are becoming a crucial area of study for researchers studying plants. Further research is required to understand the regulatory signals and particular signal pathways that control the expression of virulence mechanisms and endophytic fungal genes. Fungal endophytes can also improve the host plant's ability to take up nutrients, particularly phosphorus and nitrogen. This is often achieved through the secretion of enzymes that can solubilize and release bound nutrients from the soil, making them more available to the plant.

Some fungal endophytes help their host plants tolerate various environmental stresses, such as drought, salinity, or extreme temperatures. This tolerance is often mediated by changes in gene expression that allow the plant to adapt and survive under adverse conditions. Apart from defense and stress tolerance, fungal endophytes can also promote plant growth by influencing the expression of genes involved in plant development and hormone signaling pathways. The interaction between fungal endophytes and their host plants involves complex signaling networks. Both the fungus and the plant exchange signals that modulate gene expression in both organisms, leading to a symbiotic relationship that benefits both parties.

The presence of fungal endophytes can induce significant changes in the gene expression of their host plants, leading to enhanced defense mechanisms, improved nutrient uptake, stress tolerance, and overall growth promotion. Understanding these interactions at the molecular level can provide insights into the development of sustainable agricultural practices and the utilization of fungal endophytes for crop improvement.

References

Aharwal RP, Kumar S, Sandhu SS (2021) Endophytic mycoflora: antibacterial secondary metabolites and their therapeutic potential. Curr Pharmacol Rep 7(4):150–170

Ali S, Duan J, Charles TC, Glick BR (2014) A bioinformatics approach to the determination of genes involved in endophytic behavior in Burkholderia spp. J Theor Biol 343:193–198. https://doi.org/10.1016/j.jtbi.2013.10.007

Andrade-Linares DR, Franken P (2013) Fungal endophytes in plant roots: taxonomy, colonization patterns, and functions. In: Symbiotic endophytes. Springer, Berlin, Heidelberg, pp 311–334

Arnold AE (2007) Understanding the diversity of foliar endophytic fungi: progress, challenges, and frontiers. Fungal Biol Rev 21(2–3):51–66

Arnold AE, Herre EA (2003) Canopy cover and leaf age affect colonization by tropical fungal endophytes: ecological pattern and process in Theobroma cacao (Malvaceae). Mycologia 95(3):388–398

Arnold AE, Lutzoni F (2007) Diversity and host range of foliar fungal endophytes: are tropical leaves biodiversity hotspots? Ecology 88(3):541–549

Bacon CW, Glenn AE, Yates IE (2008) Fusarium verticillioides: managing the endophytic association with maize for reduced fumonisins accumulation. Toxin Rev 27:411–446. https://doi.org/10.1080/15569540802497889

Bailey BA, Bae H, Strem MD, Roberts DP, Thomas SE, Crozier J et al (2006) Fungal and plant gene expression during the colonization of cacao seedlings by endophytic isolates of four Trichoderma species. Planta 224:1449–1464

Berendsen RL, van Verk MC, Stringlis IA, Zamioudis C, Tommassen J, Pieterse CM et al (2015) Unearthing the genomes of plant-beneficial *Pseudomonas* model strains WCS358, WCS374 and WCS417. BMC Genomics 16:539. https://doi.org/10.1186/s12864-015-1632-z

Bhupenchandra I, Chongtham SK, Devi AG, Dutta P, Sahoo MR, Mohanty S, Kumar S, Choudhary AK, Devi EL, Sinyorita S, Devi SH, Mahanta M, Kumari A, Devi HL, Josmee RK, Pusparani A, Pathaw N, Gupta S, Meena M, Janmeda P, Seth CS, Arora A, Sahoo A, Swapnil P (2024) Unlocking the potential of arbuscular mycorrhizal fungi: exploring role in plant growth promotion, nutrient uptake mechanisms, biotic stress alleviation, and sustaining agricultural production systems. J Plant Growth Regul 43(9):1–39. https://doi.org/10.1007/s00344-024-11467-9

Chandran H, Meena M, Swapnil P (2021) Plant growth-promoting rhizobacteria as a green alternative for sustainable agriculture. Sustain For 13(19):10986. https://doi.org/10.3390/su131910986

Christian N, Sullivan C, Visser ND et al (2016) Plant host and geographic location drive endophyte community composition in the face of perturbation. Microb Ecol 72:621–632. https://doi.org/10.1007/s00248-016-0804-y

Cord-Landwehr S, Melcher RLJ, Kolkenbrock S, Moerschbacher BM (2016) A chitin deacetylase from the endophytic fungus *Pestalotiopsis sp.* efficiently inactivates the elicitor activity of chitin oligomers in rice cells. Sci Rep 6:38018. https://doi.org/10.1038/srep38018

Cui J-L, Guo T-T, Ren Z-X, Zhang N-S, Wang M-L (2015) Diversity and antioxidant activity of culturable endophytic fungi from alpine plants of *Rhodiola crenulata*, *R. angusta*, and *R. sachalinensis*. PLoS One 10(3):e0118204. https://doi.org/10.1371/journal.pone.0118204

Dalakouras A, Katsaouni A, Avramidou M, Dadami E, Tsiouri O, Vasileiadis S et al (2023) A beneficial fungal root endophyte triggers systemic RNA silencing and DNA methylation of a host reporter gene. RNA Biol 20(1):20–30

Das A, Varma A (2009) Symbiosis: the art of living. In: Varma A, Kharkwal AC (eds) Symbiotic fungi principles and practice. Springer, Berlin, pp 1–28. https://doi.org/10.1007/978-3-540-95894-9

De Bary A (1866) Morphologie und physiologie der pilze, flechten und myxomyceten, vol 1. Engelmann

Deshmukh S, Hückelhoven R, Schäfer P, Imani J, Sharma M, Weiss M et al (2006) The root endophytic fungus Piriformospora indica requires host cell death for proliferation during mutualistic symbiosis with barley. Proc Natl Acad Sci 103(49):18450–18457

Doehlemann G, Requena N, Schaefer P, Brunner F, O'Connell R, Parker JE (2014) Reprogramming of plant cells by filamentous plant-colonizing microbes. New Phytol 204(4):803–814

Estrada C, Wcislo WT, Van Bael SA (2013) Symbiotic fungi alter plant chemistry that discourages leaf-cutting ants. New Phytol 198:241–251. https://doi.org/10.1111/nph.12140

Fontana DC, de Paula S, Torres AG, Moura VH, de Souza S, Pascholati F, Schmidt D, Neto DD (2021) Endophytic fungi: biological control and induced resistance to phytopathogens and abiotic stresses. Pathogens 10(5):570. https://doi.org/10.3390/pathogens10050570

Formey D, Sallet E, Lelandais-Briere C, Ben C, Bustos-Sanmamed P, Niebel A et al (2014) The small RNA diversity from Medicago truncatula roots under biotic interactions evidences the environmental plasticity of the miRNAome. Genome Biol 15:457. https://doi.org/10.1186/s13059-014-0457-4

Gouda S, Das G, Sen SK, Shin HS, Patra JK (2016) Endophytes: a treasure house of bioactive compounds of medicinal importance. Front Microbiol 7:1538

Gough C, Cullimore J (2011) Lipo-chitooligosaccharide signaling in endosymbiotic plant-microbe interactions. Mol Plant-Microbe Interact 24:867–878. https://doi.org/10.1094/mpmi-01-11-0019

Hardoim PR, van Overbeek LS, Berg G, Pirttilä AM, Compant S, Campisano A et al (2015) The hidden world within plants: ecological and evolutionary considerations for defining functioning of microbial endophytes. Microbiol Mol Biol Rev 79:293–320

Heinig U, Scholz S, Jennewein S (2013) Getting to the bottom of taxol biosynthesis by fungi. Fungal Divers 60:161–170. https://doi.org/10.1007/s13225-013-0228-7

Hodgson S, de Cates C, Hodgson J, Morley NJ, Sutton BC, Gange AC (2014) Vertical transmission of fungal endophytes is widespread in forbs. Ecol Evol 4(8):1199–1208

Jia M, Chen L, Xin HL, Zheng CJ, Rahman K, Han T et al (2016) A friendly relationship between endophytic fungi and medicinal plants: a systematic review. Front Microbiol 7:906. https://doi.org/10.3389/fmicb.2016.00906

Khan A, Bassett S, Voisey C, Gaborit C, Johnson L, Christensen M et al (2010) Gene expression profiling of the endophytic fungus Neotyphodium lolii in association with its host plant perennial ryegrass. Australas Plant Pathol 39:467–476

Khare E, Mishra J, Arora NK (2018) Multifaceted interactions between endophytes and plant: developments and prospects. Front Microbiol 9:2732. https://doi.org/10.3389/fmicb.2018.02732. PMID: 30498482; PMCID: PMC6249440

Khiralla A, Spina R, Yagi S, Mohamed I, Laurain-Mattar D (2016) Endophytic fungi: occurrence, classification, function and natural products. In: Endophytic fungi: diversity, characterization and biocontrol. Nova Science Publisher's, Inc, pp 1–19

Kim YI, Chang KJ, Ka KH, Hur H, Hong IP, Shim JO et al (2006) Seed germination of Gastrodia elata using symbiotic fungi, Mycena osmundicola. Mycobiology 34(2):79–82

Kumar R, Swapnil P, Meena M, Selpair S, Yadav BG (2022) Plant growth-promoting rhizobacteria (PGPR): approaches to alleviate abiotic stresses for enhancement of growth and development of medicinal plants. Sustain For 14(23):15514. https://doi.org/10.3390/su142315514

Lahrmann U, Ding Y, Banhara A, Rath M, Hajirezaei MR, Döhlemann S et al (2013) Host-related metabolic cues affect colonization strategies of a root endophyte. Proc Natl Acad Sci 110(34):13965–13970

Lata R, Chowdhury S, Gond SK, White JF Jr (2018) Induction of abiotic stress tolerance in plants by endophytic microbes. Lett Appl Microbiol 66:268–276. https://doi.org/10.1111/lam.12855

Macho AP, Zipfel C (2014) Plant PRRs and the activation of innate immune signaling. Mol Cell 54(2):263–272

Mattoo AJ, Nonzom S (2021) Endophytic fungi: understanding complex cross-talks. Symbiosis 83(3):237–264

Mengistu AA (2020) Endophytes: colonization, behaviour, and their role in defense mechanism. Int J Microbiol 2020:6927219

Mishra A, Yadav P, Singh K (2024) Host response of Arabidopsis thaliana interaction with fungal endophytes involves microRNAs. Mol Biotechnol 67:294–303

Newman MA, Sundelin T, Nielsen JT, Erbs G (2013) MAMP (microbe-associated molecular pattern) triggered immunity in plants. Front Plant Sci 4:139. https://doi.org/10.3389/fpls.2013.00139

Oren L, Ezrati S, Cohen D, Sharon A (2003) Early events in the Fusarium verticillioides-maize interaction characterized by using a green fluorescent protein-expressing transgenic isolate. Appl Environ Microbiol 69:1695–1701. https://doi.org/10.1128/AEM.69.3.1695-1701.2003

Pandey V, Ansari MW, Tula S, Yadav S, Sahoo RK, Shukla N et al (2016) Dose-dependent response of *Trichoderma harzianum* in improving drought tolerance in rice genotypes. Planta 243:1251–1264. https://doi.org/10.1007/s00425-016-2482-x

Pentimone I, Lebrón R, Hackenberg M, Rosso LC, Colagiero M, Nigro F, Ciancio A (2018) Identification of tomato miRNAs responsive to root colonization by endophytic Pochonia chlamydosporia. Appl Microbiol Biotechnol 102:907–919

Peskan-Berghofer T, Vilches-Barro A, Muller TM, Glawischnig E, Reichelt M, Gershenzon J et al (2015) Sustained exposure to abscisic acid enhances the colonization potential of the mutualist fungus *Piriformospora indica* on *Arabidopsis thaliana* roots. New Phytol 208:873–886. https://doi.org/10.1111/nph.13504

Plett JM, Martin FM (2018) Know your enemy, embrace your friend: using omics to understand how plants respond differently to pathogenic and mutualistic microorganisms. Plant J 93:729–746. https://doi.org/10.1111/tpj.13802

Rodriguez R, Redman R (2008) More than 400 million years of evolution and some plants still can't make it on their own: plant stress tolerance via fungal symbiosis. J Exp Bot 59(5):1109–1114

Rodriguez RJ, White JF Jr, Arnold AE, Redman RS (2009) Fungal endophytes: diversity and functional roles. New Phytol 182(2):314–330

Rozpądek P, Domka AM, Nosek M, Ważny R, Jędrzejczyk RJ, Wiciarz M et al (2018) The role of strigolactone in the cross-talk between *Arabidopsis thaliana* and the endophytic fungus Mucor sp. Front Microbiol 9:441. https://doi.org/10.3389/fmicb.2018.00441

Saijo Y, Loo EP-I, Yasuda S (2018) Pattern recognition receptors and signaling in plant–microbe interactions. Plant J 93:592–613. https://doi.org/10.1111/tpj.13808

Sanchez-Vallet A, Mesters JR, Thomma BP (2015) The battle for chitin recognition in plant-microbe interactions. FEMS Microbiol Rev 39:171–183. https://doi.org/10.1093/femsre/fuu003

Seth K, Vyas P, Deora S, Gupta AK, Meena M, Swapnil P, Harish (2023) Understanding plant-plant growth-promoting rhizobacteria (PGPR) interactions for inducing plant defense. In: Swapnil P, Meena M, Harish, Marwal A, Vijayalakshmi S, Zehra A (eds) Plant-microbe interaction—recent advances in molecular and biochemical approaches, vol 2. Academic Press, Elsevier, pp 201–226. https://doi.org/10.1016/B978-0-323-91876-3.00010-5

Sreekanth D, Kristin IM, Brett AN (2017) Endophytic fungi from *Cathranthus roseus*: a potential resource for the discovery of antimicrobial polyketides. Nat Prod Chem Res 5:256. https://doi.org/10.4172/2329-6836.1000256

Tsuda K, Katagiri F (2010) Comparing signaling mechanisms engaged in pattern-triggered and effector-triggered immunity. Curr Opin Plant Biol 13(4):459–465

Vandenkoornhuyse P, Quaiser A, Duhamel M, Le Van A, Dufresne A (2015) The importance of the microbiome of the plant holobiont. New Phytol 206:1196–1206. https://doi.org/10.1111/nph.13312

Walters DR, Havis ND, Oxley SJP (2008) Ramularia collo-cygni: the biology of an emerging pathogen of barley. FEMS Microbiol Lett 279:1–7. https://doi.org/10.1111/j.1574-6968.2007.00986.x

White JF, Kingsley KL, Zhang Q, Verma R, Obi N, Dvinskikh S, Elmore MT, Verma SK, Gond SK, Kowalski KP (2019) Review: endophytic microbes and their potential applications in crop management. Pest Manag Sci 75:2558–2565. https://doi.org/10.1002/ps.5527

Yadav G, Meena M (2021) Bioprospecting of endophytes in medicinal plants of Thar Desert: an attractive resource for biopharmaceuticals. Biotechnol Rep 30:e00629. https://doi.org/10.1016/j.btre.2021.e00629

Yadav G, Meena M (2022) Biological control of plant diseases by endophytes. In: Shah MP, Deka D (eds) Endophytic association: what, why and how (volume: developments in applied microbiology and biotechnology). Academic Press, Elsevier Inc., pp 119–135. https://doi.org/10.1016/B978-0-323-91245-7.00007-9

Yan L, Zhao H, Zhao X et al (2018) Production of bioproducts by endophytic fungi: chemical ecology, biotechnological applications, bottlenecks, and solutions. Appl Microbiol Biotechnol 102:6279–6298. https://doi.org/10.1007/s00253-018-9101-7

Zeidler D, Zahringer U, Gerber I, Dubery I, Hartung T, Bors W et al (2004) Innate immunity in *Arabidopsis thaliana*: lipopolysaccharides activate nitric oxide synthase (NOS) and induce defense genes. Proc Natl Acad Sci USA 101:15811–15816. https://doi.org/10.1073/pnas.0404536101

Zhou X, Zhu H, Liu L, Lin J, Tang K (2010) A review: recent advances and future prospects of taxol-producing endophytic fungi. Appl Microbiol Biotechnol 86:1707–1717. https://doi.org/10.1007/s00253-010-2546-y

Host Recognition Mechanisms: Unraveling the Signaling Pathways in Endophyte Communication

10

Priyanka Choudhary, Sanjay Kumar Jaiswal, and Priti Upadhyay

Abstract

Endophytes within plant tissues significantly impact plant health and productivity. The intricate signaling of molecular communication between endophytes and their host plants remains a captivating enigma in the realm of plant-microbe interactions. This chapter is focused on identifying key molecular components governing the initiation and modulation of endophyte communication within host plants, which in turn utilizes advanced molecular biology techniques, gene expression profiling, functional assays, etc. It meticulously elucidates the specific signaling cascades activated during host recognition events, emphasizing the significance of signals from both plants and endophytes for successful symbiotic interactions. Additionally, this chapter probes the role of secondary metabolites and small signaling molecules as potential mediators in host recognition. Through integrating transcriptomic and metabolomic approaches, this chapter aims to unveil the dynamic interplay between host and endophyte, shedding light on intricate regulatory networks. This chapter will provide valuable insights into the fundamental processes governing endophyte-host interactions. This knowledge has the potential to inform strategies for optimizing plant-microbe associations in agriculture, promoting sustainable practices, and harnessing the full

P. Choudhary
Department of Plant Pathology, Institute of Agricultural Sciences, Banaras Hindu University, Varanasi, Uttar Pradesh, India

S. K. Jaiswal
Department of Chemistry, Tshwane University of Technology, Pretoria, South Africa

P. Upadhyay (✉)
Crop Advisor, Borlaug Institute for South Asia, Ludhiana, Punjab, India

M. Meena et al. (eds.), *Multi-omics Approach to Investigate Endophyte Diversity*, Microorganisms for Sustainability 53, https://doi.org/10.1007/978-981-96-4100-0_10

potential of endophytes for enhancing plant growth and resilience across diverse environmental conditions.

Keywords

Endophyte · Host recognition · Secondary metabolites · Effector proteins · Gene mutations · Signaling pathways

10.1 Introduction

Endophytism characterizes the symbiotic bond where a microbe resides in plant tissues without causing apparent harm. Endophytes possess unique abilities to penetrate, colonize, and flourish within plants through complex interactions with the host. They also navigate through existing endospheric microbiota for successful establishment. The relationship between endophyte with host plants can be mutualistic, parasitic, or nonsymbiotic. In essence, endophytes, residing within plant tissues, establish sophisticated communication networks with their hosts, influencing plant growth, development, and stress responses (Wilson 1995; Schulz and Boyle 2006; Yadav and Meena 2021, 2022). Figure 10.1 represents the few common endophytes that interacts with plants.

Host recognition by endophytes involves a complex interplay of molecular signals. Initial recognition often occurs through chemical cues emitted by both the plant and the endophyte. Plant-produced compounds such as flavonoids, phenolic acids, and strigolactones act as signaling molecules, attracting compatible endophytes (Mandal et al. 2010; Rochange 2010; Boyno et al. 2023; Bhupenchandra et al. 2024; Seth et al. 2023). Conversely, endophytes release diffusible molecules, including lipochitooligosaccharides (LCOs) and exopolysaccharides (EPS), which elicit specific responses in host plants (Limpens et al. 2015). Upon recognition, signal transduction pathways are activated within the plant. Receptor kinases and G-protein-coupled receptors (GPCRs) play crucial roles in perceiving endophyte-derived signals (Chiu and Paszkowski 2020; El-Defrawy and Hesham 2020). Perception triggers a cascade of events, including calcium influx, activation of mitogen-activated protein kinase (MAPK) cascades, and modulation of gene expression. These responses prime the plant for symbiotic interactions, facilitating nutrient exchange and stress tolerance. Understanding the mechanisms behind host recognition is crucial for unraveling the symbiotic relationships between endophytes and plants (Clúa et al. 2018; Wang et al. 2023; Sinha et al. 2024). Increasing evidence indicates that Reactive oxygen species (ROS) are crucial in programmed cell death, stress responses, plant defense against pathogens, and systemic stress signaling, often working in tandem with antioxidants (Choudhary and Upadhyay 2023). Hamilton et al. (2012) suggest that ROS and antioxidants, being both universal and evolutionarily conserved, likely play significant roles in symbiotic interactions.

The elucidation of signaling pathways involved in endophyte recognition has advanced significantly in recent years. Studies have identified key players,

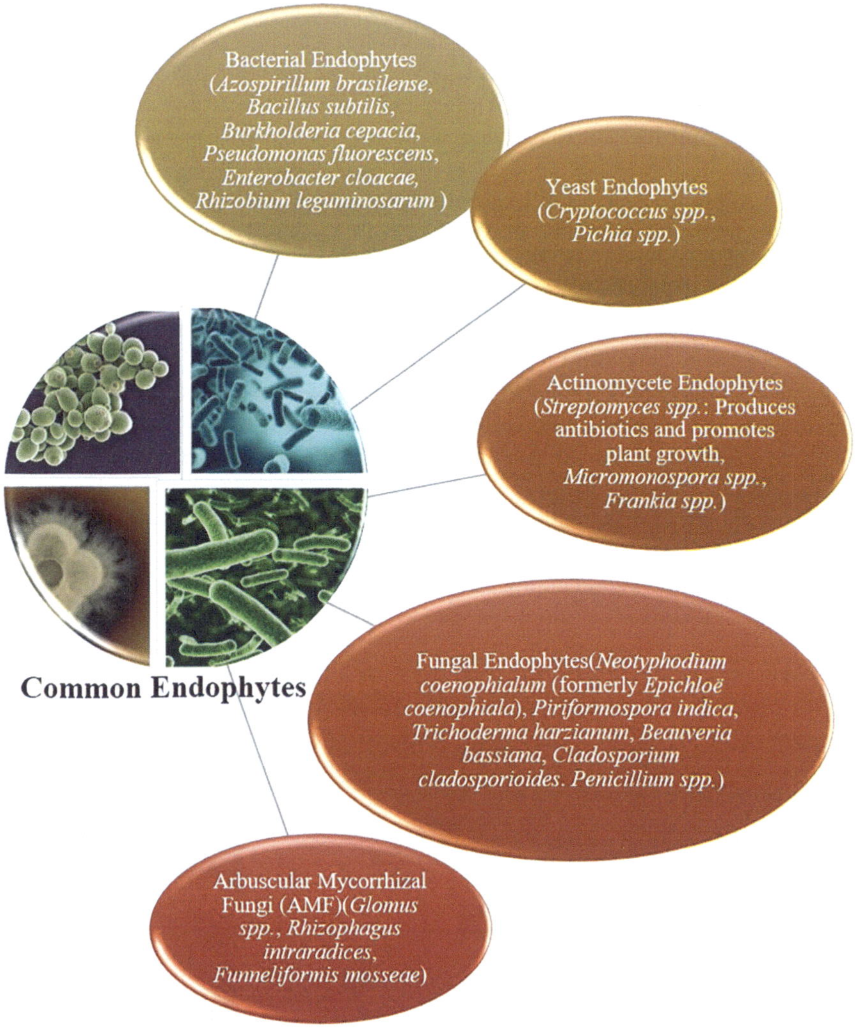

Fig. 10.1 Few common endophytes

including symbiosis receptor-like kinases (SYMRK) and Nod factor perception (NFP) proteins, which mediate recognition of endophyte-derived signals. Upon ligand binding, these receptors form complexes with other proteins, initiating downstream signaling cascades (Holsters 2008; Clúa et al. 2018). One such cascade involves the activation of symbiosis-related transcription factors, such as Nuclear Factor Y (NF-Y) and Ethylene Response Factor (ERF), which regulate the expression of symbiosis-related genes (Camehl et al. 2010; Yan et al. 2023). Additionally, calcium oscillations triggered by endophyte signals modulate the activity of calcium-dependent protein kinases (CDPKs), further regulating downstream responses (Peiter et al. 2007; Chabaud et al. 2011).

Furthermore, recent studies have highlighted the role of phytohormones, such as auxin, cytokinins, and jasmonic acid, in mediating endophyte recognition and establishment of symbiosis. Crosstalk between hormonal signaling pathways and endophyte-induced responses fine-tunes the plant's physiological and developmental processes, optimizing its interaction with the endophytic partner (Vadassery et al. 2008).

However, our understanding of the intricate communication among plants, their endophytic microbiomes, and endophytic microbes themselves is still in its infancy. Despite endophytes evolving various strategies for covert cohabitation within hosts, the molecular mechanisms orchestrating this cross-kingdom relationship remain largely unexplored. Research on plant-endophyte interactions often relies on controlled, optimized conditions for growing host plants and rarely considers the variable, field-realistic conditions. The effects of endophytes observed in healthy plants might change when these plants face less favorable or stressful conditions. A schematic representation of some benefits of plant endophyte interactions is illustrated in Fig. 10.2.

Therefore, our understanding of endophytes is based on a narrow set of experimental conditions, highlighting the need for more varied experimental settings to gain deeper insights into endophyte functioning. Additionally, the focus on easily

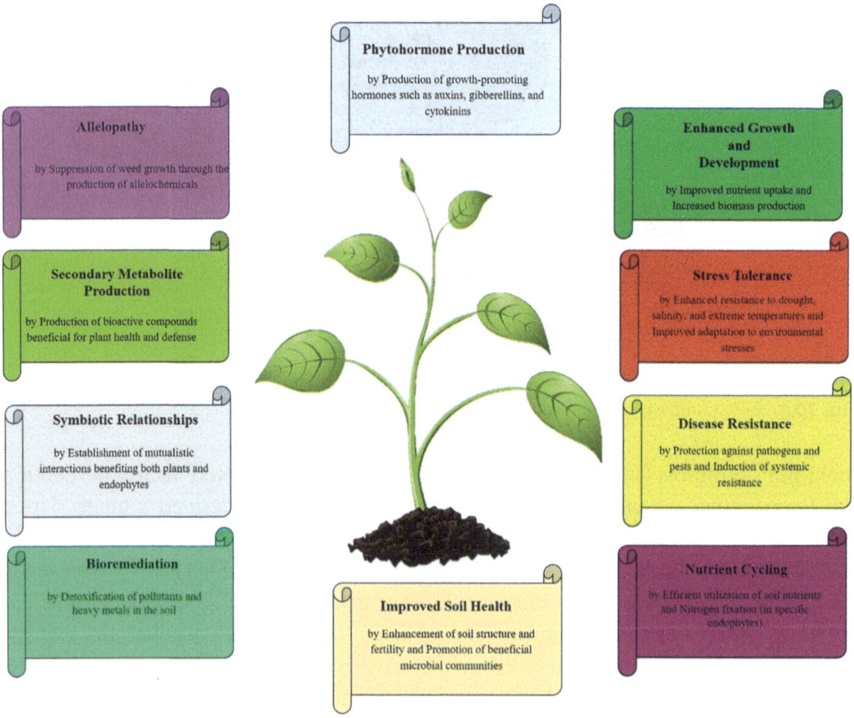

Fig. 10.2 Benefits of plant-endophyte interaction

cultivable microbial species has resulted in a biased understanding of the ecology and interactions of endophytes in plants. Although scientists are interested in endophytes, there is still much to learn about their diverse survival tactics. This review aims to consolidate current knowledge about the mechanisms governing endophytic recognition by hosts, offering insights into this complex phenomenon. This article dives into the signaling pathways involved in endophyte recognition by their host plants, illuminating the molecular interactions that govern this intriguing communication process.

Although interactions between plants and endophytes are typically advantageous, they can potentially have negative impacts on the host plant. Some endophytes live in plant tissues and can act as latent pathogens, which means they only become harmful when exposed to certain environmental factors or under stress from the plant. These dangerous endophytes can release poisons and enzymes that hurt plant tissue, impede the absorption of nutrients, and interfere with metabolic functions. Furthermore, endophytes may fight with the plant for carbon and other vital resources, weakening it and increasing its vulnerability to additional stresses. In addition, they have the potential to cause hormonal abnormalities and oxidative stress, which might further harm plants. Furthermore, endophytes can obstruct the uptake of nutrients and the general development of plants by interfering with advantageous symbiotic connections, such as those with mycorrhizal fungus or nitrogen-fixing bacteria. Lastly, endophytes can facilitate the transmission and breakout of disease by serving as reservoirs or vectors for plant diseases. For plants to remain healthy and productive, it is therefore essential to comprehend and control the factors that cause these negative consequences. The potential harms of plant-endophyte interaction have been illustrated in Fig. 10.3:

Deciphering the signaling pathways involved in endophyte recognition provides insights into the intricate molecular dialogue between plants and their microbial partners. Understanding these mechanisms holds immense potential for enhancing agricultural sustainability, as harnessing beneficial endophytes can improve crop productivity and resilience to environmental stresses. Continued research in this field will deepen our understanding of plant-microbe interactions and pave the way for innovative strategies in agriculture and biotechnology.

10.2 Fungal Endophytes: Interactions and Adaptations Within Host Plants

Endophytic fungi and their host plants form a "holobiont" that cooperatively responds to biotic and abiotic stresses, a relationship that has evolved over time (Vandenkoornhuyse et al. 2015). These endophytes are often seen as essential extensions of the host plant, with their genes being part of an extended genome (Kaul et al. 2016; Chandran et al. 2021). Endophytic fungi primarily come from the plant's environment. They invade seeds and roots via enzymes, spreading as the plant grows. Fungi can also enter through natural openings like stomata or wounds, aided by biotic or abiotic factors. This is termed "horizontal transmission." Alternatively,

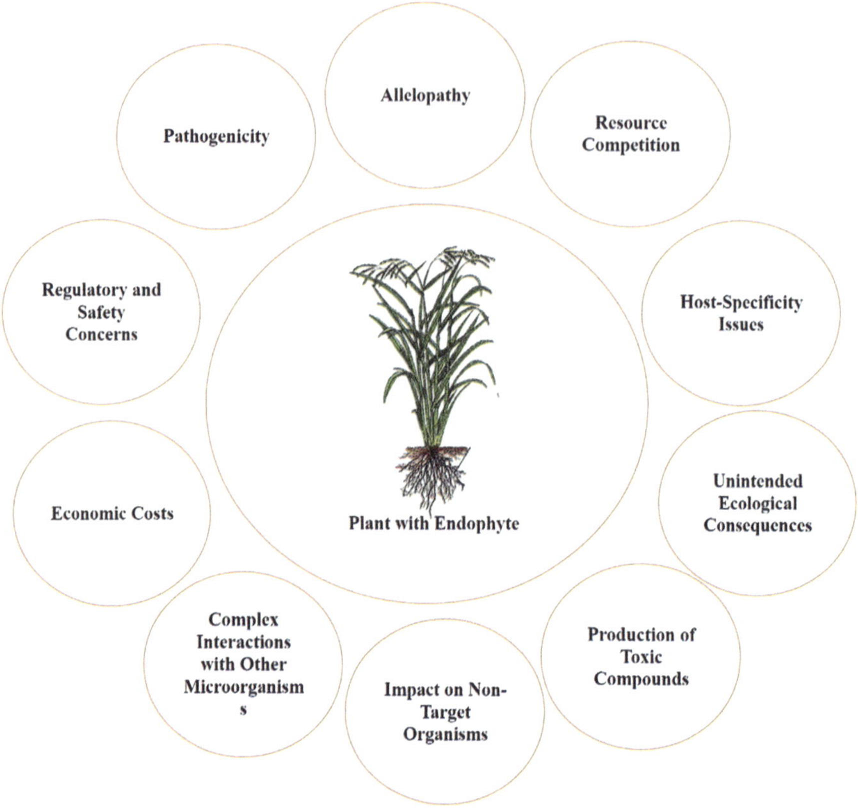

Fig. 10.3 Potential harms of plant- endophyte interaction

fungi can be passed to offspring through seeds from parent plants, known as "vertical transmission" (Saikkonen et al. 2004). Establishing the symbiotic relationship between plants and endophytic fungi is a complex process, guided by intricate signaling mechanisms. Plants and fungi recognize each other through secreted chemical signals. Root systems play a crucial role, with plant-produced substances exuded into the rhizosphere, attracting specific fungi. Fungi grow toward these root exudates, branching and invading upon contact (Yuan et al. 2010). For instance, *Gastrodia elata* secretes strigolactones to attract and stimulate *Mycena osmundicola* and *Armillaria mellea*, facilitating symbiosis (Kim et al. 2006).

The symbiotic effects between endophytes and hosts can be synergistic or antagonistic.

Synergistic Effects

1. Endophytes share coping strategies, responding together to host defenses (Lu et al. 2021).

2. Elicitors from various endophytes can work together during invasions (Gao et al. 2010).
3. Endophytes may lower host antifungal and antibacterial activities, enhancing microbial biodiversity (Mousa and Raizada 2013).

Antagonistic Effects

1. Endophytes stimulate hosts to produce antimicrobial metabolites, strengthening defenses (Kadam et al. 2020).
2. Endophytes produce antimicrobial metabolites to counter other endophytes (Zaher et al. 2015; Ancheeva et al. 2020).
3. Endophytes compete for colonization sites and nutrients (Wani et al. 2015).

Endophytic fungi and plants regulate plant-specific symbiotic signal pathways through various mechanisms. They can activate, upregulate, or inhibit signal transductions (Limpens et al. 2015) and interact with transcription factors (TFs) to indirectly regulate gene expression. They also mutate defense-related genes to downregulate plant defenses (Yan et al. 2023).

1. *Fungal Effector Proteins*: Fungi secrete small effector-like proteins (MiSSPs) that directly downregulate defense signal pathways or interact with plant TFs to regulate defense gene expression. For example, *Glomus intraradices* secretes MiSSP7, interacting with ERF19 in the ETH signaling pathway to enhance mycorrhization in *Medicago truncatula* (Lo Presti et al. 2015). Endophytic fungi, like *Epichloë festucae* in *Lolium perenne*, control virulence by expressing the stress-activated MAP kinase gene (Becker et al. 2015).
2. *Plant Endogenous Mechanisms*: Plants use "braking" mechanisms to suppress immune signaling and promote symbiosis by upregulating specific microRNAs that regulate defense gene expression. For instance, miR171e-5p in tomatoes downregulates glucan endo-1,3-beta-glucosidase to protect *Rhizophagus irregularis* (Wu et al. 2016).
3. *Gene Mutations*: Plants control defense responses through gene mutations. *Gastrodia elata* reduces strong defense responses by losing genes associated with disease resistance, such as the nucleotide-binding site and pathogenesis-related protein families (Yuan et al. 2018). *Rhizophagus irregularis* adapts to endosymbiosis by losing genes that degrade the host's cell walls (Rich et al. 2014).

Upon recognizing recruitment signals from a plant, the fungus initiates a dynamic process of communication and interaction to invade and colonize the host through one of the two methods:

- After attaching to the plant surface, the fungus may use specialized hyphal structures, such as appressoria, penetration pegs, and intracellular haustoria, to infiltrate the host (Schulz and Boyle 2005, 2006).

- Alternatively, the fungus can directly penetrate the plant cell wall by secreting degradative enzymes and utilizing turgor pressure. It releases proteases, cellulases, and other extracellular enzymes and lyases to degrade the cell walls of the plant's epidermal layers or penetrate plant cuticles (Schulz and Boyle 2005, 2006).

Environmental factors influence fungal-plant symbiosis through plant defense, fungal virulence, development and spread.

1. *Internal Environment*: Chemical substances from hosts and endophytes affect symbiosis (Schulz and Boyle 2005).
2. *External Environment*: Pathogens and environmental factors impact growth, defense, and symbiosis establishment, with endophytes often enhancing host defense (Vandenkoornhuyse et al. 2015).

Stable mutualistic symbiosis requires multiple interactions and balancing antagonisms, adapting to environmental influences.

10.3 Bacterial Endophytes: Insights into Symbiotic Communication and Adaptation

Endophytic bacteria often constitute a specialized subgroup within rhizospheric bacteria. They possess the capacity to colonize their plant hosts, distinguishing them from other rhizobacteria (Reinhold-Hurek et al. 2015; Kumar et al. 2022; Meena et al. 2022). Endophytes exhibit traits associated with promoting plant growth similar to rhizobacteria. Rhizobacterial strains have adaptive traits for root colonization, such as biofilm formation, chemotaxis, and nutrient utilization from root exudates. Nitrogen fixation, microbial metabolites, soil enzyme activities, vesicular arbuscular mycorrhiza (VAM), host genotype variability, the competitive prowess, and persistence of inoculants are key factors shaping the dynamics of rhizospheric bacteria-plant interactions and field responses (Choudhary 2023). These interactions drive shifts in soil properties and microbial communities, mitigating harmful rhizosphere microorganisms and protecting against phytopathogen attacks. However, bacterial endophytes typically confer greater benefits to host plants compared to many rhizospheric bacteria, especially under stress conditions (Reinhold-Hurek et al. 2015).

The population density of endophytes varies significantly, primarily influenced by bacterial species and host genotypes, but also by the host's developmental stage, inoculum density, and environmental conditions. Some endophytes are scarce or absent in soil. Nitrogen-fixing endophytes represent a small fraction of the total endophytic bacteria. Increasing their populations in plants could enhance nitrogen fixation (Rosenblueth and Martínez-Romero 2006). Rhizobia dominated the sequences, likely due to the methodology used, constituting around 50% of sequences. Fast-growing γ-Proteobacteria tend to outgrow slower-growing α-Proteobacteria in culture-dependent studies. Endophytic bacteria also inhabit legume nodules, as seen in red clover nodules where *Rhizobium rhizogenes* and *R. leguminosarum bv. trifolii* were found (Sturz et al. 2000). Generally, endophytic bacteria do not form nodules. Benefits of endophytic bacteria to plants include growth stimulation through nitrogen fixation, phytohormone production, biocontrol of root-zone pathogens through antifungal or antibacterial agents, siderophore production, nutrient competition, induction of systemic acquired resistance, and enhanced mineral availability (Sturz et al. 2000; Rosenblueth and Martínez-Romero 2006).

Some bacterial endophytes from healthy plants can inhibit plant growth via metabolite production. These endophytes may become pathogenic under specific conditions like CO_2 accumulation or O_2 depletion or interactions with other microorganisms. Mixed inoculations of growth-inhibiting endophytes can promote plant growth with the inoculation order affecting outcomes (Sturz et al. 2000). Endophytes like *Herbaspirillum rubrisubalbicans* can cause disease in some host but remain symptomless in most others (James et al. 1997). Some endophytes, including *Salmonella* and *Burkholderia cepacia,* are human pathogens, posing health risks through raw produce. This has led to calls for reassessment of *Burkholderia* use in agriculture (Rosenblueth and Martínez-Romero 2006).

Endophytes like *Nocardia and Mycobacterium*, related to human pathogens, have been found in Citrus and Scots pine, respectively. Many effective biocontrol agents, such as *Stenotrophomonas maltophilia and Pantoea agglomerans*, are opportunistic human pathogens (Berg et al. 2005). Their resistance to plant defenses may also confer resistance to human defenses. Close genetic relationships between plant and human pathogens, such as *Klebsiella variicola*, raise concerns about lateral gene transfer and agricultural use (Rosenblueth et al. 2004). Wild plant races and certain varieties create conditions that support nitrogen fixation by endophytes, benefiting crops like rice, sugarcane, and maize. The presence of endophytes can modulate plant genes, affecting nitrogen assimilation, carbon metabolism, growth, and defense. *Arabidopsis thaliana* serves as a model for endophyte interactions, allowing for the testing of defined mutants. Research by Iniguez et al. (2005) on *Medicago truncatula* and *Arabidopsis thaliana* mutants revealed that ethylene, a plant signal molecule, decreases endophytic colonization by *Klebsiella* sp. *strain Kp342* and *Salmonella enterica serovar Typhimurium.* Ethylene-insensitive *Medicago truncatula* mutants showed hypercolonization by Kp342. Adding the ethylene precursor 1-aminocyclopropane-1-carboxylic acid reduced endophytic bacteria in wild-type *Medicago truncatula* and wheat. This reduction was reversed by treating plants with the ethylene inhibitor 1-methylcyclopropane.

Iniguez et al. (2005) found that bacterial extracellular components like flagella and type III secretion systems *(TTSS-SPI1)* from Salmonella pathogenicity island 1 reduce endophytic colonization. *Salmonella typhimurium* mutants lacking these components exhibited higher colonization in *Medicago sativa* and wheat seedlings. In *Arabidopsis thaliana* mutants, a salicylic acid (SA)-independent defense response limited colonization by *Klebsiella* sp. *strain Kp342*, whereas both SA-dependent and SA-independent pathways restricted Salmonella typhimurium colonization. Studies on *Salmonella typhimurium* flagella mutants indicated that flagella are recognized through the SA-independent response, while *TTSS-SPI1* is recognized by the SA-dependent response, inducing the SA-dependent *PR1* gene promoter. *Kp342* lacks flagella and *TTSS-SPI1*, avoiding SA-dependent responses and colonizing plants more effectively. This suggests Kp342 may have lost its flagella during plant association evolution (Mengistu 2020).

Bacterial flagellins are recognized by surface receptors with transmembrane proteins and leucine-rich repeat domains in plants, mammals, and insects. Flagellin acts as a trigger in Arabidopsis thaliana, causing oxidative bursts, callose deposition, ethylene production, and the activation of defense-related genes. Plants use chemo-perception systems to detect bacterial molecules (Mengistu 2020).

In Vitis vinifera, an endophytic Burkholderia strain elicited a local defense response (Pacifico et al. 2019). The impact of plant innate immunity on endophytes, as seen in animal-bacterial interactions, is still uncertain. Antimicrobial peptides from maize and rice and restricted carbon sources in apoplastic fluid can limit endophytic growth. Endophytes express specific genes necessary for entering and colonizing plant tissues, promoting plant growth, competing with pathogens, or producing various substances (Khaskheli et al. 2020; Zhang et al. 2022; Pal et al. 2023). Research on endophytic bacteria in their natural plant habitats, which are more complex than laboratory conditions, is essential. Genomic projects on *Azospirillum* spp. (Sant'Anna et al. 2011), *Herbaspirillum* sp. (Straub et al. 2013), *Gluconacetobacter diazotrophicus* (Giongo et al. 2010), and *Klebsiella* spp. (Runcharoen et al. 2017) will enhance our understanding of their molecular interactions with plants. In vivo expression technology (Rediers et al. 2003), rhizosphere studies, and genome and transcriptome analyses are valuable for examining gene expression during endophytic life (Zuccaro et al. 2011).

10.4 Notable Examples Illustrating Recognition Mechanisms in Plant-Endophyte Interactions

As already discussed above that plant-endophyte interactions involve complex recognition mechanisms that facilitate mutualistic relationships between plants and their endophytic microorganisms. This initial recognition helps the plant differentiate between beneficial endophytes and potential pathogens. The specificity of plant-endophyte recognition mechanisms is also influenced by the plant's genetic makeup and environmental factors. Plants possess various genetic loci that regulate the compatibility with specific endophytes that is called receptors. It also regulates the interactions with environmental conditions like soil composition, climate, and plant health that can affect the establishment and maintenance of these symbiotic relationships. Plants Receptor kinases listed in Table 10.1 play crucial roles in perceiving signals from endophytes, initiating downstream signaling cascades, and regulating various responses in host plants during the establishment of interactions.

A few typical examples of recognition systems that function to establish plant-endophyte connection are given in the text below:

Table 10.1 Receptor kinases: gatekeepers of endophyte recognition

S. N.	Receptor kinases	References
1.	Symbiosis receptor-like kinase (*SYMRK*)	Holsters (2008)
2.	Nod factor perception (NFP) protein	Oldroyd (2013)
3.	LysM receptor kinase (*LYK*)	Bozsoki et al. (2020)
4.	Chitin elicitor receptor kinase (*CERK*)	Bozsoki et al. (2020)
5.	Flagellin sensing 2 (FLS2)	Beck et al. (2014)
6.	Ethylene receptor kinase (ERS)	Camehl et al. (2010)
7.	Brassinosteroid-insensitive 1-associated receptor kinase 1 (*BAK1*)	Yan et al. (2023)
8.	Cytokinin receptor (CRE1/AHK4)	Vadassery et al. (2008)
9.	Phytosulfokine receptor 1 (PSKR1)	Shen and Diener (2013)
10.	XA21-like protein XA21G	Liu et al. (2014)
11.	Wall-associated kinase (*WAK*)	Pinski et al. (2019)
12.	Receptor-like protein kinase 1 (*RPK1*)	Afzal et al. (2008)
13.	Catharanthus roseus receptor-like kinase 1-like (*CrRLK1L*)	Yang et al. (2024)
14.	Leucine-rich repeat receptor-like kinase (*LRR-RLK*)	Saijo et al. (2018)
15.	Somatic embryogenesis receptor kinase (*SERK*)	Li et al. (2022)
16.	BRI1-associated receptor kinase 1 (*BAK1*)	Li et al. (2022)
17.	Phosphatidylinositol 3-kinase-related kinases (*PI3KKs*)	Mauceri et al. (2024)
18.	Receptor-like cytoplasmic kinase (*RLCK*)	Saijo et al. (2018)

1. *Flavonoids in Legume-Rhizobia Symbiosis*:

Legume root-secreted flavonoids and isoflavonoids activate the Nodulation protein D (NodD) transcription factor in compatible rhizobia, prompting the transcription of nod genes crucial for Nod factor production. These factors are detected by receptors on the root cell membrane, initiating the signaling cascade essential for nodule formation and infection, providing a habitat for bacteria and facilitating nitrogen fixation (Clúa et al. 2018). Flavonoids, a varied class of polyphenolic compounds, are particularly highlighted for their signaling functions in plant-microbe interactions compared to simpler phenolic acids like hydroxybenzoic and hydroxycinnamic acids (Mandal et al. 2010).

2. *Phenolic Acids in Endophytic Associations*:

Despite the well-defined roles of flavonoids, the specific signaling functions of phenolic acids beyond flavonoids remain understudied. Phenolic acids, the primary polyphenols synthesized by plants, serve as signaling molecules in initiating legume-rhizobia and arbuscular mycorrhizal symbioses, and contribute to plant defense mechanisms by inducing *nod* and *GmGin1* genes for symbiosis establishment, triggering vir gene expression in *Agrobacterium*, and acting as defense agents against soil-borne pathogens, with implications for rhizospheric microbial ecology (Mandal et al. 2010).

3. *Strigolactones and Arbuscular Mycorrhizal Fungi*:

Strigolactones are a group of metabolites derived from carotenoids and released by plant roots, are instrumental in these processes. Their vital role in mycorrhizal development has been elucidated through studies on mutants deficient in strigolactones. Interestingly, strigolactones were previously identified as stimulants for the germination of parasitic plants Striga and Orobanche. They regulate gene expression across various plant processes, including the phosphate (P) and nitrate (N) signaling pathways, as evidenced by their influence on key regulatory genes such as *PHO2* and *NIGT1/HHO* integrators. Moreover, they are central to the regulation of lateral root development, facilitating both the initiation and elongation phases of lateral root growth (Rochange 2010; Boyno et al. 2023). Plant hormones, strigolactone (SL), precise involvement in the interaction between plants and endophytic fungi remains ambiguous. They are thought to stimulate hyphae branching in mycorrhizal fungi while inhibiting the growth of pathogenic fungi. Rozpądek et al. (2018) while investigating the impact of the synthetic SL analog GR24 on *Mucor* sp. observed a concentration-dependent reduction in colony growth rate along with increased H_2O_2 accumulation and alterations in antioxidant enzyme activity. They also observed that symbiosis with *Mucor* sp. positively influenced *Arabidopsis thaliana* growth, with SL being essential for this beneficial association. Mutants *max1* and *max4* failed to exhibit the beneficial phenotype, showing disrupted SA homeostasis and elevated expression of plant defensin genes. Additionally, Mucor sp. displayed the ability to degrade SL within the plant and downregulate SL biosynthesis genes.

4. *Lipochitooligosaccharides (LCOs)*:

Limpens et al. (2015) present interesting review on how symbiotic nitrogen-fixing rhizobium bacteria and arbuscular mycorrhizal fungi utilize lipochitooligosaccharide (LCO) signals to engage in communication with potential host plants. Upon finding a compatible match, a close relationship is established wherein the microsymbiont gains access to root cells. Plants detect microbial LCO molecules through specific LysM-domain-containing receptor-like kinases. These receptors not only initiate a common signaling pathway shared by both symbioses but also regulate innate immune responses. Recent research indicates that symbiotic LCO receptors share close similarities with chitin innate immune receptors, with some of these receptors performing dual roles in both symbiosis and immunity.

Review by Rosier et al. (2018) presents that flavonoids, released by legume roots, prompt the transcription of rhizobia Nod factors (NF), which are lipochitooligosaccharides (LCO) composed of a chitin core of β1, 4 linked polymers of N-acetylglucosamine (GlcNAc). Unlike chitin, which is an ancient molecule associated with fungi and insect exoskeletons, LCOs feature N-acyl moieties comprising fatty acid chains of varying length, saturation, and substitution groups. These unique side chains likely play a pivotal role in determining specificity between rhizobial strains and their plant host targets, serving as the primary agents driving specificity.

5. *Exopolysaccharides (EPS) in Plant Growth-Promoting Rhizobacteria (PGPR) Interactions*:

The utilization of PGPRs offers dual advantages to plants, primarily enhancing nutrient acquisition and stimulating plant defense mechanisms. Rhizobacterial exopolysaccharides (EPSs) represent essential components of soil organic matter with multifaceted biological roles. These include facilitating biofilm formation and survival, enabling host colonization and bacterial autoaggregation. Additionally, EPSs play crucial roles in host specificity, quorum sensing, and early stages of host plant infection. They also serve as signaling molecules during plant development and offer vital protection against environmental stresses (Bramhachari et al. 2018).

6. *Receptor Kinases in Symbiotic Interactions*:

Plants recruit beneficial symbionts from the diverse microorganisms in the soil, such as arbuscular mycorrhizal fungi (AMF) and nitrogen-fixing bacteria, into their roots. These symbionts require a core signaling pathway for initiating symbiotic interactions, which involves signaling and developmental changes to accommodate them. Understanding of these pathways began with genetic screenings in legume models like *Medicago truncatula* and *Lotus japonicus*, identifying mutants defective in nodulation and arbuscular mycorrhizal symbiosis (AMS). This led to the discovery of a common symbiosis signaling pathway (CSSP), initiated by symbiosis receptor kinases (*SYMRK/DMI2/NORK*). *SYMRK* activates calcium oscillations, which in turn trigger calcium and calmodulin-dependent kinase (*CCaMK/DMI3*) activation, ultimately launching symbiosis signaling. However, CSSP mutants may still allow intraradical colonization and arbuscule development, indicating the existence of CSSP-independent signaling pathways (Chiu and Paszkowski 2020).

7. *G-Protein-Coupled Receptors (GPCRs) in Plant-Fungal Interactions*:

Various bioactive molecules and stimuli, including light, protons, taste stimuli, and hormones, transmit their signals from the extracellular environment to the intracellular milieu through interactions with G-protein-coupled receptors (GPCRs). These receptors are categorized into three superfamilies and play essential roles in transducing a multitude of signals in eukaryotic cells. The diversity of function elicited by neurotransmitters, hormones, etc., is attributed to distinct structural receptor subtypes and the availability of G proteins and intracellular effector systems. GPCRs, with seven transmembrane segments, form heterotrimers comprising α, β, and γ subunits, which facilitate signal transduction across the plasma membrane. Upon activation by ligand binding, GPCRs induce guanine nucleotide exchange on the Gα subunit, leading to the dissociation of the GTP–Gα complex from the β–γ-dimer and initiation of intracellular signaling cascades. The signal is terminated when GTP is hydrolyzed to GDP by the intrinsic GTPase activity of Gα. G protein-mediated signaling is tightly regulated by regulators of G-protein signaling (RGS proteins), which accelerate GTP hydrolysis. In fungi, GPCRs regulate essential processes such as growth, morphogenesis, and pathogenicity. They are classified into various families based on their functions, including pheromone receptors, nutrient sensors, and cAMP receptors. In filamentous fungi like *Aspergillus nidulans*, GPCRs are involved in vegetative growth, sporulation, and mating (El-Defrawy and Hesham 2020).

8. *Calcium Signaling: Orchestrating Endophyte Response Pathways*

Wang et al. (2023) report that endophyte colonization in plants initiates a cascade of events, beginning with calcium influx upon recognition by plant receptors, triggering the activation of calcium-dependent protein kinases (CDPKs). This activation, in turn, regulates symbiosis-related gene expression through calcium-dependent transcription factors, influencing nodulation and mycorrhizal symbiosis. Furthermore, calcium signaling pathways modulate reactive oxygen species (ROS) production and facilitate the establishment and maintenance of symbiotic interactions via calcium oscillations. These interactions also involve calcium-mediated activation of ion channels and transporters facilitating nutrient exchange with endophytes. Moreover, calcium-dependent regulation influences cytoskeletal dynamics, membrane trafficking, and defense responses, orchestrating a complex network of signaling cascades during endophyte colonization (Sinha et al. 2024).

Upon recognition of endophytic signals at the cell membrane, they undergo amplification by second messengers through the corresponding intracellular messenger systems, thus amplifying the signal cascade into a cellular response. These second messengers include Ca^2+, IP3, DAG, cAMP, cGMP, as well as intracellular components like nitric oxide, hydrogen peroxide, arachidonic acid, and cADPR. The intracellular calcium messenger system, inositol phospholipid messenger system, and cyclic nucleotide messenger system are widely acknowledged. In the calcium messenger system, intracellular calcium ions oscillate with specific amplitude and frequency. The inositol phosphate metabolism generates IP3 and DAG via phospholipase C, while cyclic nucleotides cAMP and cGMP are produced through adenylate cyclase and guanylate cyclase, respectively. The protein kinase (PKA) is central in the cyclic nucleotide messenger system, interacting synergistically with the calcium signaling system. This intricate signaling network demonstrates crosstalk between various systems in host plants when stimulated by endophytic bacteria, resulting in a comprehensive signal transduction network within plant cells. Notably, Ca^{2+} plays a pivotal role in physiological changes and serves as a common second messenger in endophyte-plant interactions. Following signal recognition, downstream proteins and transporters, such as *DMI1*, *CNGC15*, and *MCA8*, constitute calcium signal encoders, causing calcium concentration oscillations within the cell. The calcium receptor CaM, located in the nucleus, decodes these oscillations, interacting with *CYCLOPS* to transcribe symbiotic signals via downstream target proteins and transcription factors *NSP1* and *NSP2*, ultimately inducing gene expression and secondary metabolite production, completing the symbiotic interaction pathway (Peiter et al. 2007; Chabaud et al. 2011; Nasir et al. 2021).

Aspect of the Signaling Pathways Involved in Endophyte Recognition and Establishment of Symbiosis

- The genetic basis underlying plant root endosymbioses suggests a unified signaling pathway, termed the common symbiosis pathway (CSP), for both fungal and proteobacterial symbionts. Research has mainly used the model legumes *Lotus japonicus* and *Medicago truncatula*, identifying nine CSP genes to date. In rhizobial infection, plants recognize nodulation (Nod) factors from rhizobia via two lysin motif (LysM) type serine/threonine receptor-like kinases, *Nod factor receptor 1 (NFR1)* and *Nod factor receptor 5 (NFR5)*. This recognition signal is then transmitted through the CSP, resulting in nuclear calcium spiking (Chen et al. 2015; Radhakrishnan et al. 2020; Santoyo 2022).
- Key genes upstream of calcium spiking include the receptor-like kinase *SYMRK/DMI2* (symbiosis receptor-like kinase/does not make infections), nucleoporins *NUP85*, *NUP133*, and *NENA*, and cation channels *CASTOR* and *POLLUX/DMI1*. Downstream genes involve those encoding the calcium and calmodulin-dependent kinase *CCaMK/DMI3* and its phosphorylation partner, the transcription activator *CYCLOPS/IPD3* (interacting protein of *DMI3*) (Peiter et al. 2007; Hamilton et al. 2012; Nasir et al. 2021).
- Interestingly, part of this signaling pathway is also involved in the parasitic interaction with root knot nematodes, which invade the host root intercellularly. In *L. japonicus*, *NFR1, NFR5,* and *SYMRK* may play roles in recognizing nematode signals and causing root hair deformations. Mutants deficient in *NFR1* and *NFR5* showed significantly reduced feeding site development (Weerasinghe et al. 2005).
- Apart from the typical root hair infection, rhizobia can penetrate legume roots between cells, a method known as crack entry which is more common under flooded conditions, particularly at lateral root emergence points. Crack entry invasion strategy resembles how *Azoarcus* sp. colonizes plant tissues intercellularly. Similar to intercellular entry by rhizobia, where the infection of a single cell relies on the receptor SYMRK, in rice-Azoarcus sp. interactions, bacteria form intercellular microcolonies and do not reside inside living cells (Hurek and Reinhold-Hurek 2003; Santoyo 2022).
- Nitrogen fixation by rhizobia does not always require symbiosomes; in nonlegume *Parasponia,* rhizobia enter between cells and fix nitrogen in specialized threads. Components of the common symbiosis pathway (CSP), particularly a receptor related to NFR5, are involved in this process. The CSP can also induce plant enzymes called pectinases in rhizobia-induced nodules, aiding in endophyte colonization. In *Lotus japonicus*, the *LjNPL* gene is activated by Nod factors via the CSP, facilitating infection thread development by breaking down the plant cell wall. Components of the CSP pathway might also be activated by the beneficial microbe *Medicago truncatula* (Chen et al. 2015; Genre and Russo 2016).
- The invasion of endophyte triggers the activation of plant defense networks, leading to the production or increase of related signal molecules, modulation of

ion channels, initiation of signaling responses, and synthesis of defense substances (Lo Presti et al. 2015; Lu et al. 2021).

- As the endophyte invades the plants, its external structures such as fungal hyphae, flagella, and cell walls are specifically recognized by the plants, along with the elicitors released by the endophyte, which are also specifically recognized by plant receptors (Schulz and Boyle 2005; Genre and Russo 2016).
- Endophytic elicitors encompass a range of compounds including oligosaccharides (e.g., chitin, chitosan, cyclodextrin), polysaccharides, polypeptides, proteins, glycoproteins, glycolipids, and unsaturated fatty acids (Zhai et al. 2017).
- Plant receptors that recognize these elicitors can be categorized into cell-surface pattern recognition receptors (PRRs) and intracellular nucleotide-binding leucine-rich repeat proteins (NLRs) (Saijo et al. 2018).
- PRRs, such as receptor-like kinases (RLKs) and receptor-like proteins (RLPs), typically possess an extracellular, single-pass transmembrane, and intracellular domain. These receptors perceive and bind pathogen/microbe-associated molecular patterns (PAMPs/MAMPs), thereby activating plant pattern-triggered immunity (PTI) (Chiu and Paszkowski 2020; Lu et al. 2021).
- Endophytic effectors secreted into plant cells to suppress PTI are recognized by plants' NLRs, which generally initiate a stronger effector-triggered immunity (ETI) response in plants (Lo Presti et al. 2015; Lu et al. 2021). Interactions between the two immune systems, PTI and ETI, involve crosstalk and synergy. Despite distinct mechanisms, PTI and ETI converge on similar downstream responses like Reactive Oxygen Species (ROS) production and MAP kinase activation (Becker et al. 2015; Santoyo 2022; Yan et al. 2023).
- In plant cells, signal molecules exhibit crosstalk and communication, forming diverse signal pathways that intersect and connect in series to create a complex signal transduction network. This network synergistically transduces fungal elicitor and invasion signals (Gao et al. 2012; Pieterse et al. 2012; Zhai et al. 2017).
- Host defense signal transduction pathways encompass a variety of signaling molecules, including hormones like jasmonic acid (JA), salicylic acid (SA), ethylene (ETH), abscisic acid (ABA), nitric oxide (NO), reactive oxygen species (ROS), calcium ions (Ca^{2+}), G-proteins, cyclic adenosine monophosphate, and inositol 1,4,5-triphosphate, which play crucial roles in fungal-plant interactions (Camehl et al. 2010; Chabaud et al. 2011; Santoyo 2022; Choudhary and Upadhyay 2023).
- Moreover, other signal transducer-mediated pathways, such as protein kinase-dependent signaling, lipoxygenase pathway, oxylipin signaling, oxidized lipid signaling, and oxidative burst signaling, contribute significantly (Chiu and Paszkowski 2020; Choudhary and Upadhyay 2023; Mauceri et al. 2024).
- Despite the array of signaling pathways, specific ones are selectively involved in distinct fungus-plant interaction scenarios. Each signal transducer serves either as an upstream or downstream signal, regulating or being regulated by other signal molecules (Yan et al. 2023; Sinha et al. 2024; Mauceri et al. 2024).

10.5 Summary and Conclusion

Fungal endophyte-host recognition remains a dynamic and evolving field of study, revealing intricate mechanisms and interactions that underscore the complexity of symbiotic relationships. Recent advances have highlighted the sophisticated molecular dialogues between endophytes and their host plants, including the role of signaling molecules, receptor proteins, and gene expression patterns. These interactions are increasingly recognized for their potential to enhance plant resilience, growth, and stress tolerance, which holds significant implications for agriculture and biotechnology.

Current trends emphasize the use of omics technologies, such as genomics, transcriptomics, and proteomics, to gain a deeper understanding of the functional dynamics within these symbiotic systems. Integrating these approaches with advanced imaging techniques and computational models is paving the way for more precise insights into the mechanisms governing host recognition and endophyte behavior. Looking to the future, research will likely focus on unraveling the evolutionary adaptations that enable fungal endophytes to successfully colonize diverse host species and environments. Additionally, exploring the potential of manipulating these interactions for agricultural innovation, including crop improvement and sustainable practices, will be a key area of thrust. By bridging fundamental research with applied strategies, the field of fungal endophyte-host recognition is poised to unlock new opportunities for enhancing ecosystem health and agricultural productivity.

References

Afzal AJ, Wood AJ, Lightfoot DA (2008) Plant receptor-like serine threonine kinases: roles in signaling and plant defense. Mol Plant-Microbe Interact 21(5):507–517

Ancheeva E, Daletos G, Proksch P (2020) Bioactive secondary metabolites from endophytic fungi. Current Medicinal Chemistry 27(11):1836–1854

Beck M, Wyrsch I, Strutt J, Wimalasekera R, Webb A, Boller T, Robatzek S (2014) Expression patterns of flagellin sensing 2 map to bacterial entry sites in plant shoots and roots. J Exp Bot 65(22):6487–6498

Becker Y, Eaton CJ, Brasell E, May KJ, Becker M, Hassing B et al (2015) The fungal cell-wall integrity MAPK cascade is crucial for hyphal network formation and maintenance of restrictive growth of Epichloë festucae in symbiosis with Lolium perenne. Mol Plant-Microbe Interact 28(1):69–85

Berg G, Eberl L, Hartmann A (2005) The rhizosphere as a reservoir for opportunistic human pathogenic bacteria. Environ Microbiol 7(11):1673–1685

Bhupenchandra I, Chongtham SK, Devi AG, Dutta P, Sahoo MR, Mohanty S, Kumar S, Choudhary AK, Devi EL, Sinyorita S, Devi SH, Mahanta M, Kumari A, Devi HL, Josmee RK, Pusparani A, Pathaw N, Gupta S, Meena M, Janmeda P, Seth CS, Arora A, Sahoo A, Swapnil P (2024) Unlocking the potential of arbuscular mycorrhizal fungi: Exploring role in plant growth promotion, nutrient uptake mechanisms, biotic stress alleviation, and sustaining agricultural production systems. J Plant Growth Regul 43(9):1–39. https://doi.org/10.1007/s00344-024-11467-9

Boyno G, Rezaee Danesh Y, Demir S, Teniz N, Mulet JM, Porcel R (2023) The complex interplay between Arbuscular mycorrhizal fungi and Strigolactone: mechanisms, Sinergies, applications and future directions. Int J Mol Sci 24(23):16774

Bozsoki Z, Gysel K, Hansen SB, Lironi D, Krönauer C, Feng F et al (2020) Ligand-recognizing motifs in plant LysM receptors are major determinants of specificity. Science 369(6504):663–670

Bramhachari PV, Nagaraju GP, Kariali E (2018) Current perspectives on rhizobacterial-EPS interactions in alleviation of stress responses: novel strategies for sustainable agricultural productivity. In: Role of Rhizospheric microbes in soil: volume 1: stress management and agricultural sustainability, pp 33–55

Camehl I, Sherameti I, Venus Y, Bethke G, Varma A, Lee J, Oelmüller R (2010) Ethylene signalling and ethylene-targeted transcription factors are required to balance beneficial and nonbeneficial traits in the symbiosis between the endophytic fungus *Piriformospora indica* and *Arabidopsis thaliana*. New Phytol 185(4):1062–1073

Chabaud M, Genre A, Sieberer BJ, Faccio A, Fournier J, Novero M et al (2011) Arbuscular mycorrhizal hyphopodia and germinated spore exudates trigger Ca^{2+} spiking in the legume and non-legume root epidermis. New Phytol 189(1):347–355

Chandran H, Meena M, Swapnil P (2021) Plant growth-promoting rhizobacteria as a green alternative for sustainable agriculture. Sustain For 13(19):10986. https://doi.org/10.3390/su131910986

Chen X, Miché L, Sachs S, Wang Q, Buschart A, Yang H et al (2015) Rice responds to endophytic colonization which is independent of the common symbiotic signaling pathway. New Phytol 208(2):531–543

Chiu CH, Paszkowski U (2020) Receptor-like kinases sustain symbiotic scrutiny. Plant Physiol 182(4):1597–1612

Choudhary P (2023) Interaction of efficient Rhizospheric bacteria and responses in the farmer's field. In: Metabolomics, proteomes and gene editing approaches in biofertilizer industry. Springer Nature Singapore, Singapore, pp 101–116

Choudhary P, Upadhyay P (2023) Relevance of the antioxidative mechanism during plant-microbe interaction. In: Plant-microbe interaction-recent advances in molecular and biochemical approaches. Academic, pp 123–140

Clúa J, Roda C, Zanetti ME, Blanco FA (2018) Compatibility between legumes and rhizobia for the establishment of a successful nitrogen-fixing symbiosis. Genes 9(3):125

El-Defrawy MM, Hesham AEL (2020) G-protein-coupled receptors in fungi. In: Fungal biotechnology and bioengineering, pp 37–126

Gao FK, Dai CC, Liu XZ (2010) Mechanisms of fungal endophytes in plant protection against pathogens. Afr J Microbiol Res 4(13):1346–1351

Gao Y, Zhao J, Zu Y, Fu Y, Liang L, Luo M et al (2012) Antioxidant properties, superoxide dismutase and glutathione reductase activities in HepG2 cells with a fungal endophyte producing apigenin from pigeon pea (Cajanus cajan (L.) Millsp.). Food Research International 49(1):147–152

Genre A, Russo G (2016) Does a common pathway transduce symbiotic signals in plant–microbe interactions? Front Plant Sci 7:175287

Giongo A, Tyler HL, Zipperer UN, Triplett EW (2010) Two genome sequences of the same bacterial strain, *Gluconacetobacter diazotrophicus* PAl 5, suggest a new standard in genome sequence submission. Stand Genomic Sci 2:309–317

Hamilton CE, Gundel PE, Helander M, Saikkonen K (2012) Endophytic mediation of reactive oxygen species and antioxidant activity in plants: a review. Fungal Divers 54:1–10

Holsters M (2008) SYMRK, an enigmatic receptor guarding and guiding microbial endosymbioses with plant roots. Proc Natl Acad Sci 105(12):4537–4538

Hurek T, Reinhold-Hurek B (2003) *Azoarcus* sp. strain BH72 as a model for nitrogen-fixing grass endophytes. J Biotechnol 106(2–3):169–178

Iniguez AL, Dong Y, Carter HD, Ahmer BM, Stone JM, Triplett EW (2005) Regulation of enteric endophytic bacterial colonization by plant defenses. Mol Plant-Microbe Interact 18(2):169–178

James EK, Olivares FL, Baldani JI, Döbereiner J. (1997) Herbaspirillum, an endophytic diazotroph colonizing vascular tissue of Sorghum bicolor L. Moench. Journal of Experimental Botany 48(3):785–798
Kadam SB, Pable AA, Barvkar VT (2020) Mycorrhiza induced resistance (MIR): a defence developed through synergistic engagement of phytohormones, metabolites and rhizosphere. Functional Plant Biology 47(10):880–890
Kaul S, Sharma T, Dhar MK (2016) "Omics" tools for better understanding the plant–endophyte interactions. Front Plant Sci 7:955
Khaskheli MA, Wu L, Chen G, Chen L, Hussain S, Song D et al (2020) Isolation and characterization of root-associated bacterial endophytes and their biocontrol potential against major fungal phytopathogens of rice (*Oryza sativa* L.). Pathogens 9(3):172
Kim YI, Chang KJ, Ka KH, Hur H, Hong IP, Shim JO et al (2006) Seed germination of *Gastrodia elata* using symbiotic fungi, *Mycena osmundicola*. Mycobiology 34(2):79–82
Kumar R, Swapnil P, Meena M, Selpair S, Yadav BG (2022) Plant growth-promoting rhizobacteria (PGPR): approaches to alleviate abiotic stresses for enhancement of growth and development of medicinal plants. Sustain For 14(23):15514. https://doi.org/10.3390/su142315514
Li Y, Jiao C, Wei Z, Chai S, Jia H, Gao M et al (2022) Analysis of Grapevine's Somatic Embryogenesis Receptor Kinase (SERK) gene family: VqSERK3/BAK1 overexpression enhances disease resistance. Phytopathology 112(5):1081–1092
Limpens E, van Zeijl A, Geurts R (2015) Lipochitooligosaccharides modulate plant host immunity to enable endosymbioses. Annu Rev Phytopathol 53:311–334
Liu W, Liu J, Triplett L, Leach JE, Wang GL (2014) Novel insights into rice innate immunity against bacterial and fungal pathogens. Annu Rev Phytopathol 52:213–241
Lo Presti L, Lanver D, Schweizer G, Tanaka S, Liang L, Tollot M et al (2015) Fungal effectors and plant susceptibility. Annu Rev Plant Biol 66:513–545
Lu H, Wei T, Lou H, Shu X, Chen Q (2021) A critical review on communication mechanism within plant-endophytic fungi interactions to cope with biotic and abiotic stresses. J Fungi 7(9):719
Mandal SM, Chakraborty D, Dey S (2010) Phenolic acids act as signaling molecules in plant-microbe symbioses. Plant Signal Behav 5(4):359–368
Mauceri A, Puccio G, Faddetta T, Abbate L, Polito G, Caldiero C et al (2024) Integrated omics approach reveals the molecular pathways activated in tomato by *Kocuria rhizophila*, a soil plant growth-promoting bacterium. Plant Physiol Biochem 210:108609
Meena M, Yadav G, Sonigra P, Nagda A, Mehta T, Swapnil P, Harish, and Marwal, A. (2022) Role of elicitors to initiate the induction of systemic resistance in plants to biotic stress. Plant Stress 5:100103. https://doi.org/10.1016/j.stress.2022.100103
Mengistu AA (2020) Endophytes: colonization, behaviour, and their role in defense mechanism. Int J Microbiol 2020:6927219
Mousa WK, Raizada MN (2013) The diversity of anti-microbial secondary metabolites produced by fungal endophytes: an interdisciplinary perspective. Front Microbiol 4:65
Nasir F, Bahadur A, Lin X, Gao Y, Tian C (2021) Novel insights into host receptors and receptor-mediated signaling that regulate arbuscular mycorrhizal symbiosis. J Exp Bot 72(5):1546–1557
Oldroyd GE (2013) Speak, friend, and enter: signalling systems that promote beneficial symbiotic associations in plants. Nat Rev Microbiol 11(4):252–263
Pacifico D, Squartini A, Crucitti D, Barizza E, Lo Schiavo F, Muresu R et al (2019) The role of the endophytic microbiome in the grapevine response to environmental triggers. Front Plant Sci 10:1256
Pal G, Saxena S, Kumar K, Verma A, Kumar D, Shukla P et al (2023) Seed endophytic bacterium *Bacillus velezensis* and its lipopeptides acts as elicitors of defense responses against *Fusarium verticillioides* in maize seedlings. Plant Soil 492(1):109–124
Peiter E, Sun J, Heckmann AB, Venkateshwaran M, Riely BK, Otegui MS et al (2007) The *Medicago truncatula* DMI1 protein modulates cytosolic calcium signaling. Plant Physiol 145(1):192–203
Pieterse CM, Van der Does D, Zamioudis C, Leon-Reyes A, Van Wees SC (2012) Hormonal modulation of plant immunity. Annual Review of Cell and Developmental Biology. 28(1):489–521

Pinski A, Betekhtin A, Hupert-Kocurek K, Mur LA, Hasterok R (2019) Defining the genetic basis of plant–endophytic bacteria interactions. Int J Mol Sci 20(8):1947

Radhakrishnan GV, Keller J, Rich MK, Vernié T, Mbadinga Mbadinga DL, Vigneron N et al (2020) An ancestral signalling pathway is conserved in intracellular symbioses-forming plant lineages. Nat Plants 6(3):280–289

Rediers H, Bonnecarrere V, Rainey PB, Hamonts K, Vanderleyden J, De Mot R (2003) Development and application of a dapB-based in vivo expression technology system to study colonization of rice by the endophytic nitrogen-fixing bacterium *Pseudomonas stutzeri* A15. Appl Environ Microbiol 69(11):6864–6874

Reinhold-Hurek B, Bünger W, Burbano CS, Sabale M, Hurek T (2015) Roots shaping their microbiome: global hotspots for microbial activity. Annu Rev Phytopathol 53:403–424

Rich MK, Schorderet M, Reinhardt D (2014) The role of the cell wall compartment in mutualistic symbioses of plants. Front Plant Sci 5:91711

Rochange S (2010) Strigolactones and their role in arbuscular mycorrhizal symbiosis. In: Arbuscular mycorrhizas: physiology and function, pp 73–90

Rosenblueth M, Martínez-Romero E (2006) Bacterial endophytes and their interactions with hosts. Mol Plant-Microbe Interact 19(8):827–837

Rosenblueth M, Martínez L, Silva J, Martínez-Romero E (2004) Klebsiella variicola, a novel species with clinical and plant-associated isolates. Syst Appl Microbiol 27(1):27–35

Rosier A, Medeiros FH, Bais HP (2018) Defining plant growth promoting rhizobacteria molecular and biochemical networks in beneficial plant-microbe interactions. Plant Soil 428:35–55

Rozpądek P, Domka AM, Nosek M, Ważny R, Jędrzejczyk RJ, Wiciarz M, Turnau K (2018) The role of strigolactone in the cross-talk between *Arabidopsis thaliana* and the endophytic fungus Mucor sp. Front Microbiol 9:307553

Runcharoen C, Moradigaravand D, Blane B, Paksanont S, Thammachote J, Anun S et al (2017) Whole genome sequencing reveals high-resolution epidemiological links between clinical and environmental *Klebsiella pneumoniae*. Genome Med 9:1–10

Saijo Y, Loo EPI, Yasuda S (2018) Pattern recognition receptors and signaling in plant–microbe interactions. Plant J 93(4):592–613

Saikkonen K, Helander M, Faeth SH (2004) Fungal endophytes: hitch-hikers of the green world. In: Plant microbiology. Taylor & Francis, pp 81–101

Sant'Anna FH, Almeida LG, Cecagno R, Reolon LA, Siqueira FM, Machado MR et al (2011) Genomic insights into the versatility of the plant growth-promoting bacterium *Azospirillum amazonense*. BMC Genomics 12:1–14

Santoyo G (2022) How plants recruit their microbiome? New insights into beneficial interactions. J Adv Res 40:45–58

Schulz B, Boyle C (2005) The endophytic continuum. Mycol Res 109(6):661–686

Schulz B, Boyle C (2006) What are endophytes? In: Microbial root endophytes. Springer Berlin Heidelberg, Berlin/Heidelberg, pp 1–13

Seth K, Vyas P, Deora S, Gupta AK, Meena M, Swapnil P, Harish. (2023) Understanding plant-plant growth-promoting rhizobacteria (PGPR) interactions for inducing plant defense. In: Swapnil P, Meena M, Harish, Marwal A, Vijayalakshmi S, Zehra A (eds) Plant-microbe interaction—recent advances in molecular and biochemical approaches, vol 2. Academic/Elsevier, pp 201–226. https://doi.org/10.1016/B978-0-323-91876-3.00010-5

Shen Y, Diener AC (2013) Arabidopsis thaliana resistance to fusarium oxysporum 2 implicates tyrosine-sulfated peptide signaling in susceptibility and resistance to root infection. PLoS Genet 9(5):e1003525

Sinha J, Singh Y, Verma PK (2024) Cytoskeleton remodeling: a central player in plant-fungus interactions. J Exp Bot 75:3269

Straub D, Rothballer M, Hartmann A, Ludewig U (2013) The genome of the endophytic bacterium *H. frisingense* GSF30T identifies diverse strategies in the *Herbaspirillum* genus to interact with plants. Front Microbiol 4:49994

Sturz AV, Christie BR, Nowak J (2000) Bacterial endophytes: potential role in developing sustainable systems of crop production. Crit Rev Plant Sci 19(1):1–30

Vadassery J, Ritter C, Venus Y, Camehl I, Varma A, Shahollari B et al (2008) The role of auxins and cytokinins in the mutualistic interaction between Arabidopsis and *Piriformospora indica*. Mol Plant-Microbe Interact 21(10):1371–1383

Vandenkoornhuyse P, Quaiser A, Duhamel M, Le Van A, Dufresne A (2015) The importance of the microbiome of the plant holobiont. New Phytol 206(4):1196–1206

Wang Y, Li Z, Xu M, Xiao Z, Liu C, Du B et al (2023) Signal molecules regulate the synthesis of secondary metabolites in the interaction between endophytes and medicinal plants. Processes 11(3):849

Wani ZA, Ashraf N, Mohiuddin T, Riyaz-Ul-Hassan S (2015) Plant-endophyte symbiosis, an ecological perspective. Applied Microbiology and Biotechnology 99:2955–2965

Weerasinghe RR, Bird DM, Allen NS (2005) Root-knot nematodes and bacterial Nod factors elicit common signal transduction events in Lotus japonicus. Proceedings of the National Academy of Sciences 102(8):3147–3152

Wilson D (1995) Endophyte: the evolution of a term, and clarification of its use and definition. Oikos 73:274–276

Wu P, Wu Y, Liu CC, Liu LW, Ma FF, Wu XY et al (2016) Identification of arbuscular mycorrhiza (AM)-responsive microRNAs in tomato. Front Plant Sci 7:429

Yadav G, Meena M (2021) Bioprospecting of endophytes in medicinal plants of Thar Desert: an attractive resource for biopharmaceuticals. Biotechnol Rep 30:e00629. https://doi.org/10.1016/j.btre.2021.e00629

Yadav G, Meena M (2022) Biological control of plant diseases by endophytes. In: Shah MP, Deka D (eds) Endophytic association: what, why and how (Volume: developments in applied microbiology and biotechnology). Academic/Elsevier, pp 119–135. https://doi.org/10.1016/B978-0-323-91245-7.00007-9

Yan S, Li J, Zhang Q, Jia S, Zhang Q, Wang R et al (2023) Transcriptional response of wolfberry to infestation with the endophytic *Fusarium nematophilum* strain NQ8GII4, vol 108. Plant Dis, p 1514

Yang SY, Lin WY, Hsiao YM, Chiou TJ (2024) Milestones in understanding transport, sensing, and signaling of the plant nutrient phosphorus. Plant Cell 36:1504

Yuan ZL, Zhang CL, Lin FC (2010) Role of diverse non-systemic fungal endophytes in plant performance and response to stress: progress and approaches. J Plant Growth Regul 29:116–126

Yuan Y, Jin X, Liu J, Zhao X, Zhou J, Wang X et al (2018) The *Gastrodia elata* genome provides insights into plant adaptation to heterotrophy. Nat Commun 9(1):1615

Zaher AM, Moharram AM, Davis R, Panizzi P, Makboul MA, Calderón AI (2015) Characterisation of the metabolites of an antibacterial endophyte Botryodiplodia theobromae Pat. of Dracaena draco L. by LC–MS/MS. Natural Product Research. 29(24):2275–2281

Zhang J, Islam MS, Wang J, Zhao Y, Dong W (2022) Isolation of potato endophytes and screening of *Chaetomium globosum* antimicrobial genes. Int J Mol Sci 23(9):4611

Zhai X, Jia M, Chen L, Zheng CJ, Rahman K, Han T, Qin LP (2017) The regulatory mechanism of fungal elicitor-induced secondary metabolite biosynthesis in medicinal plants. Critical Reviews in Microbiology 43(2):238–261

Zuccaro A, Lahrmann U, Güldener U, Langen G, Pfiffi S, Biedenkopf D et al (2011) Endophytic life strategies decoded by genome and transcriptome analyses of the mutualistic root symbiont *Piriformospora indica*. PLoS Pathog 7(10):e1002290

11 Metatranscriptomics: To Analyze Pattern of Gene Expression and Functional Activities of the Endophytic Community

M. Kavino, Duraisamy Saravanakumar, and Dipti Pandurang Mohite

Abstract

Metatranscriptomics has emerged as a powerful tool for exploring the diversity, dynamics, and functions of endophytic microbial communities within plant ecosystems. This chapter provides an overview of metatranscriptomics and its application in studying endophytes. Metatranscriptomic analysis enables comprehensive characterization of endophytic communities at the transcriptional level, shedding light on their taxonomic classification, gene expression profiles, functional activities, and interactions with host plants. By elucidating the molecular dialogues and signaling pathways involved in host-microbe interactions, metatranscriptomics facilitates understanding of mechanisms underlying plant growth promotion, disease suppression, and colonization by microbes. Furthermore, metatranscriptomics offer insights into the ecological dynamics and evolutionary processes shaping endophytic communities, as well as their potential biotechnological applications in agriculture. Through comparative metatranscriptomic analyses, gene expression patterns can be identified, distinguishing habitats with and without endophytes. This chapter highlights the significance of metatranscriptomics in advancing our understanding of endophytic communities and their roles in plant health, ecosystem functioning, and agricultural sustainability.

M. Kavino (✉) · D. P. Mohite
Department of Fruit Science, Horticultural College and Research Institute, Tamil Nadu Agricultural University, Coimbatore, Tamil Nadu, India
e-mail: kavino.m@tnau.ac.in

D. Saravanakumar
The University of the West Indies, St. Augustine Campus, Trinidad and Tobago, West Indies

M. Meena et al. (eds.), *Multi-omics Approach to Investigate Endophyte Diversity*, Microorganisms for Sustainability 53, https://doi.org/10.1007/978-981-96-4100-0_11

Keywords

Metatranscriptomics · Endophytes · Taxonomical classifications · Functional profiling · Biotechnological applications

11.1 Introduction

Metatranscriptomics is an ultra-modern omics technique that has significantly altered our notion of microbial communities by virtue of its ability to deliver gene expression profiles within complex ecosystems (Zhou et al. 2015). Starting from the much broader field of metagenomics, metatranscriptomics applies transcriptomic analysis to whole microbial communities in order to understand comprehensively their functional activities and metabolic processes (Franzosa et al. 2018). The recent interest in studying endophytic microbial communities residing in plant tissues is due to the important role they play in promoting plant health and growth and ecological functions. Metatranscriptomic analysis elucidates taxonomic composition, gene expression profiles, and functional activities of endophytic microbes which is important in understanding the intricate interactions between microorganisms and their host plants. This will vastly help in agriculture, ecosystem management, and biotechnology. Metatranscriptomics is the current state-of-the-art molecular biology technique that investigates complete RNA transcripts from microbial communities in any given environment. While metagenomics explains genomic abundance for certain microorganisms, metatranscriptomics provides details of gene expression patterns and functional activities of microbial communities during natural functioning (Sunagawa et al. 2013). This is important because metatranscriptomics is able to unlock the complexity of dynamics and interactions taking place in microbial ecosystems. With the active gene expression profile captured from the diverse microbial taxa, these analyses enlighten metabolic, biochemical pathways, and ecological functions executed by these microorganisms in response to the environmental factors (Franzosa et al. 2018). Such techniques allow researchers to infer microbial community adaptation and response to perturbations, including nutrient availability, temperature fluctuation, and interactions with other organisms.

In metatranscriptomics, the context is particularly interesting because endophytic microbial communities are in closer association with their host plants. This way, the transcriptional activities of endophytic microbes inhabiting plant tissues can shed some light on the molecular mechanisms involved in beneficial symbiotic relationships and pathogenic interactions and also in the modulation of plant physiological processes (Dudeja et al. 2021). Furthermore, metatranscriptomic analyses give insights into functional roles of endophytes in nutrient acquisition, stress tolerance, pest and disease resistance, and the production of bioactive compounds potentially applicable in agriculture and biotechnology.

The concept of investigating microbial gene expression in various environmental samples dates back to the early 2000s, when high-throughput sequencing technologies and computational tools for the analysis of biologically complex data made their leap (Shakya et al. 2019). It was realized, that the study of microbial communities that resulted from metagenomics analysis gave valuable information and explored transcriptomes of single species or functional groups and Sanger sequencing (Urich et al. 2008; Gilbert and Hughes 2011), but these approaches need supplementation with information on gene expression in order to estimate the actual microbial functions within plant ecosystem. The invention of NGS technologies, such as Illumina and 454 pyrosequencing, has changed the landscape of metatranscriptomics studies by providing a means for high-throughput sequencing of RNA molecules (Kumar and Yadav 2024). Much of these technologies are enabled to capture the existing diversity of transcripts within complex microbial communities to an unprecedented level of depth and resolution.

In the last decade, enormous amounts of efforts have been taken for the development and optimization of experimental protocols for metatranscriptomic analyses, which involve RNA extraction methods, library preparation techniques, and bioinformatic pipelines processing and interpreting sequencing data. Adding to this, specialized computational tools and algorithms have been developed for taxonomic and functional annotation of metatranscriptomic datasets, which enables meaningful biological inferences to be extracted from such complex transcriptomics data (Bashiardes et al. 2016). This chapter discusses in-depth (i) the use of metatranscriptomics in studying microbial communities of benefit to the plant health and agriculture (ii) types and processes involved in the transcriptomics of microbes.

11.2 Exploring Metatranscriptomics: A Molecular Window into Microbial RNA Transcripts

Metatranscriptomics technology allows scientists to scrutinize the entire assortment of RNA transcripts present in a microbial community, collected from a predetermined environment or any other biological sample. Rather than metagenomics, which looks at only the genomic compilation of microorganisms, due to metatranscriptomics, it is now possible to identify and profile gene expression behavior, in other words, functional activities, patterns of these microbial communities, within their natural habitats (Aransay and Trueba 2016). This will help to broaden our knowledge about functions of the endophytic microbial community in connection with its expressed gene, regulations, and expressed pathways linked to biological and ecological changes (Mauchline et al. 2018). Over the last 10 years, a number of studies have been adopted largely for metatranscriptomics analyses in complex environmental samples to study soil fertility, environmental degradation, impact of biofertilizers, and plant microbe interactions determining suppression or occurrence of diseases (Hayden et al. 2018). Since applications of metatranscriptomics are gaining attention, it has become really very important to understand how we can use this knowledge about microbial genes and their functions in working toward a sustainable environment.

The entire purpose of metatranscriptomics is to isolate and analyze all the RNA molecules, or transcripts, produced by the members within a microbial community present in any given sample. It comprises transcripts from bacteria, archaea, fungi, viruses, or other microorganisms, in some cases even with the addition of transcripts from host organisms like host plants (Cho et al. 2023). The sequencing of these transcripts and their subsequent analysis revealed that the researchers could know the metabolic pathways, biochemical processes, and ecological functions performed by the whole microbial community. Metatranscriptomics has thus emerged in recent years as a powerful approach in the study of endophytic microbial communities in association with the host, in particular with plants. Studying the gene expression profile of endophytic microbes in plant tissues will give an insight into their functional potential in relation to plant health, nutrient cycling, and ecosystem processes. This is pivotal to agriculture, environmental science, since endophytic microbes have central roles in promoting the growth of plants, enhancing or modulating tolerance to stresses, or regulating interactions between plants and microorganisms.

This motion has enabled researchers to manage and guide the potentials of Next Generation Sequencing (NGS) to achieve a broad understanding of functions and prospects of beneficial endophytic microbes in conservational agricultural practices in which profit and productivity can be maximized with the least damage to the environment. The genomic taxonomic diversity analysis by amplicon sequencing of 16S rRNA/18S rRNA or ITS1/ITS4 (Vierheilig et al. 2015; Shakya et al. 2019) sequences linked with microbial communities kicked off NGS about two decades ago. Amplicon sequencing mainly explains the diversity and structure of microbial communities, although several other studies in the past targeted the investigation of microbial taxonomy and phylogeny in different environments (Zuñiga et al. 2017). Another major reason behind the huge popularity of amplicon-based gene sequencing lies in the fact that it has low costs, simple sample preparation protocols and a wide accessibility to bioinformatics tools. Gene amplicon sequencing can provide an unrivalled depth of information related to the nature of the microbiome. However, this technique itself can provide only profiling of community structures; in this case, there is no accurate overview about their functionality and expression of genes and protein of biological and environmental significance (Aguiar-Pulido et al. 2016).

As technology developed, investigators started to work on Whole Genome Sequencing (WGS), which helped in revealing both functional and taxonomic diversity within a given community (Gupta et al. 2018). Shotgun metagenomic sequencing can be used as an informative tool for inferring the taxonomical classes by sequencing whole genomic DNA fragments in a community without any biases expected in amplicon sequencing from this method due to the unnecessity of amplification before sequencing (Zuñiga et al. 2017). However, since WGS is a descriptive analysis, it produces predictive functional annotation profiles (Aransay and Trueba 2016). Besides, shotgun metagenomics cannot distinguish inactive and active members of the microbial community (Azeem et al. 2021). Therefore, it cannot answer a lot of aspects of research pertaining, what or which community contributes to the observed ecosystem activity and which are merely "hitchhiker" in a dormant state (Shakya et al. 2019). This research gap led to the development of metatranscriptomics as a more holistic approach to inferring functional profiles within the microbiome.

11.3 Metatranscriptomics Technologies: Recent Advances

Recent years have seen unprecedented technological improvement in metatranscriptomics, improving the depth, accuracy, and efficiency of transcriptomics analyses. Of various innovations that continue to move this field forward, Single-Cell RNA Sequencing (scRNA-seq) is considered as highly influential in metatranscriptomic analysis. It is applied to identifying genes expressed within individual microbial cells, making up complex communities that have the power of expression profiling, thus making this possible (Zhang et al. 2021). Capture of the transcriptomes of single cells using scRNA-seq could resolve heterogeneity issue within microbial populations and identify rare taxa, showing changes at the cellular level of gene expression patterns. Long-Read Sequencing Technologies like PacBio and Oxford Nanopore sequencing are capable of defining full-length RNA transcripts, thus bypassing the bottleneck of platforms dedicated to short-read sequencing (Poursalavati et al. 2023). Long-read metatranscriptomics enables high-resolution views of alternative splicing events, isoform diversity, RNA modification events, etc., all of which can be harnessed more comprehensively in the representation of transcriptomic complexity within microbial communities.

Microfluidic and nanopore-based sample preparations are some of the advancements in this field developed for the integration of all workflows involved in metatranscriptomic analysis, including RNA extraction and library preparation prior to sequencing (Latorre-Pérez et al. 2020). These technologies reduce the requirement of sample input, processing time, and improve efficiency within sample handing, thus, allowing high-throughput and low-cost metatranscriptomic studies. Another recent advancement for improving the visualization and quantification of specific RNA transcripts within microbial cells directly in their natural environment is Single Molecule Fluorescence In Situ Hybridization or smFISH. This approach can be used for metatranscriptomic for spatial and temporal information about gene expression patterns, cellular localization, and microbial interactions within complex communities (Grieb et al. 2020).

Key milestones in this development of metatranscriptomic techniques are related to sequencing technologies, experimental methodologies, and computational tools. This section highlights major developments that have spurred this technological innovation. High-throughput technologies in the late 2000s included next-generation sequencing technologies for high-throughput sequencing of RNA molecules from complex and derived environmental samples, which revolutionized metatranscriptomics. Those include the platforms of Illumina, 454 pyrosequencing, and Ion Torrent sequencing and they provide entirely new levels of scalability, speed, and cost effectiveness, allowing to actually producing datasets of large size and great depth (Quail et al. 2012). Paralleling developments in the next-generation sequencing platforms were improvements in techniques of RNA isolation and library construction protocols for metatranscriptomic applications. This specialized technique was developed to enhance RNA sequencing by extending rRNA depletion, mRNA enrichment (Kopylova et al. 2012) and strand-specific library preparation to increase sensitivity, specificity, and reproducibility.

11.4 Importance of Metatranscriptomics in Gene Expression and Annotation for Endophytic Microbes

An expanding area of research involves the exploitation of these endophytic microbes for adventive functions like growth promoting bacteria or PGPR that can act as biofertilizer and enhance plant nutrient uptake (Uzoh and Babalola 2018; Meena et al. 2023). The plant growth-promoting rhizobacteria colonize in the different plant parts or in the rhizosphere and directly promote the growth of plants by regulating defense mechanism through production of bioactive molecules or immobilizing nutrients (Gupta et al. 2015). However, how relevant and feasible this will be for strains proposed to be applied worldwide in every climatic situation is still questionable, with an insufficiency in the data on fundamental mechanisms and taxonomic level information (Trivedi et al. 2017). Nevertheless, it is possible for us to further understand and exploit the beneficial microbes and gain knowledge regarding taxonomic and genomic level information through the current state of metatranscriptomics as it stands today. Metatranscriptomics is now being applied in all aspects of molecular research for example, authors Liao et al. 2019 reported that, *Populus trichocarpa* genes for pLTP were upregulated following interaction with *Mortierella elongata*. Furthermore, additional studies have indicated that the affinity of pLTP for lipids represents a basic aspect of their functions. pLTP has also been found to facilitate the fatty acids and translocating phospholipids through cell membranes. Ligand prediction analysis revealed that these *Populus trichocarpa* pLTPs are predicted to bind with saturated fatty acids, such as stearic acid and palmitic acid. Although the biological roles of pLTPs are not clear, many studies suggest that they participate in antimicrobial activity, defensive signaling, cuticle deposition, and cell wall loosening. The knowledge about underlying molecular mechanisms and metabolisms prior to actual application will help in providing suitable techniques and a full grasp of the mechanisms via endophyte metatranscriptomics increases the efficiency and its potential for agricultural progress and research.

There is often complexity in the study of mechanisms involved in plant development in a varied environment, which exist as a complex, dynamic microbial community. This is brought about by the colonization of several species, microorganisms which may exhibit positive or negative, or neutral interactions all in the same organism at the same time. The high-throughput metatranscriptomics development has created a remarkable opportunity to characterize innumerable pathogens, specifically in complex diseases. It has been reported that numerous pathogens causing grapevine trunk have been frequently isolated from a mature grapevine undergoing this disease (Niem et al. 2020). By doing metatranscriptomics analysis, multiple putative expression factors of virulence of these pathogenic species have been identified that would not be retrieved through single-species inoculations (Paolinelli et al. 2022). Moreover, the results showed various pathogenic species responsible for the same diseases.

Another approach like this was investigated to reveal the expression of differentially expressed genes through Affymetrix ATH1 whole-genome Gene Chips microarray for multiple stress responses in the Arabidopsis plants infected with *Botrytis cinera* a necrotrophic fungal pathogen (Sham et al. 2019). This has been shown to upregulate several genes due to *Botrytis* infection—CCR2, CYP71A13, NATA1, SRG1, and EL13-2—are some of the genes responsive to phytohormones like methyl-jasmonate, salicylic acid, abscisic acid, and 1-aminoacyclopropane-1-carbo xylate. This indicates that hormones are the key players in transcriptional activation for *Arabidopsis* defense against *Botrytis* infection. Finally, the involvement of the RAP2.4 gene in plant defense through its modulation of pathogen-derived effectors and endogenous signally molecules are also cited (Yang et al. 2020). Thus, information about the DEGs in response to plant infection will aid in understanding the underlying plant defense mechanisms and utilizing it in the near future for developing disease-suppressive soils as well as genetic transformation of plants with the desired traits.

Hence, this knowledge regarding microbial taxonomic profiles and gene expression patterns against different stresses to the crop will help us in making use of this information for developing a specific and reliable biomarker indicating the type of environmental stress. On this account, an emerging need is realized with respect to their transcriptional diversity, where the application of metatranscriptomics technology is highly recommended. Coupling these with fine meta-omics approaches for the exposition of modifications to the microbiome community and expression of genes, metagenomic and metatranscriptomic analyses find their rightful place in filling the research gaps.

11.5 Unveiling the Complexities of Gene Expression Through Sequencing

Improvements in metagenomic analyses in the sequencing technology have made a quantum leap towards promoting the approaches for study and understanding of gene expression worldwide. The knowledge of the critical roles played by the host gene expression, whether on a cellular or tissue level, this technique has come a long way from the differential display approach (Liang and Pardee 1992) to one of popularly known transcriptome approaches to the use of microarray technology (Pal and Pal 2022). In the recent past, with the emergence of massive parallel sequencing and RNA-seq, new possibilities have opened up in the area of transcriptome analysis which provides insights and dynamic range previously though unthinkable (Wang et al. 2009). Besides the insights eluded from host gene expression profile, transcriptomics positioned the researchers in a better place to investigate the gene expression of intricate bacterial communities in a given environment, including culturable and non-culturable gene expression profiles of microbes. Metatranscriptome analyses data enhances the metagenomics data adding information to the annotation of genes that are transcribed and level of transcription (Franzosa et al. 2014). Active metabolic pathways in bacterial communities can be determined from such functional data and then linked to particular environmental conditions (Bikel et al. 2015). Metatranscriptomics therefore present a greater extent of information compared to metagenomics; it can express information about populations that are transcriptionally active and not simply describe the gene content of bacterial populations, as was observed to be the case in metagenomic analysis.

11.6 Isolation of mRNA for Metatranscriptomics

Usually, a metatranscriptome experiment of the microbiome involves total RNA isolation from bacteria from the location of interest. Accordingly, in eukaryotes, mRNA can be selected by synthesizing cDNA using oligod(T) primers that take advantage of the poly-A tail characterizing mRNA species. Prokaryotic mRNA constitutes only 1–5% of total RNA species; most of these are 16S and 18S rRNAs and tRNAs or ITS1 and ITS4 (Peano et al. 2013), but unlike eukaryotic mRNA, prokaryotic mRNA lacks a poly-A tail, so it cannot be selected for during cDNA synthesis as in the case of eukaryotic mRNA. As such, a number of approaches have been designed and applied for this purpose (Wendisch et al. 2001). An attractive option is the removal of rRNA by probes targeting specific rRNA regions with the help of magnetic beads. It means annealing the probes to target sequences, rRNA, and its removal with the help of a magnet (Sultan et al. 2014). Since this technique contains manipulation of RNA, degradation by contaminating ribonucleases is again a problem faced by this approach. This thereby maintains commonly used laboratory practices for avoiding the introduction of contaminating ribonucleases in protocols so far considered sensitive and the addition of RNase inhibitors to the protocol which has been observed to beproductive strategy. What remains is an enriched population of other mRNAs that are representative of transcriptionally active genes. These RNAs are fractionated, cDNA is synthesized, and adapters are ligated at the ends (following end repair) of cDNA generating a library that is amplified and sequenced massively parallelly. Sequence reads are then mapped to reference genomes and the expressed genes are identified according to the coverage of sequence reads over these regions (Giannoukos et al. 2012).

11.7 Analysis of Metatranscriptomic Sequencing Data

In general, owing to the complexity of their study subjects in metatranscriptome, studies using high throughput sequencing generally generate short reads through Illumina sequencing technology. This has often been applied, particularly in requiring many samples with multiple replicates and deep coverage, like gene expression studies. The reason is that most of the information regarding the samples is priorly unknown, like the microbial composition, its community membership's relative abundance, the size of genomes, and the expression relative to them. Therefore, choosing experimental parameters correctly, like the depth of sequencing for metatranscriptomics, is not an easy task. While long reads can generate full-length or near full-length mRNAs which might help with differentiating among multiple isoforms (Pollard et al. 2018), it also provides longer stretches of sequence for similarity searches. Currently, various technologies available for long read data only have an aiding role thus have been not solely used for metatranscriptome researches.

11.8 Preprocessing

Comparable to most of the other NGS data sets, one of the very first steps in processing RNAseq data is Quality Control (QC) and removal or trimming spurious/erroneous reads to minimize errors. One of many available QC tools (there are literally dozens), for example FastQC, FaQCs (Chen et al. 2018), fastp or Trimmomatic (Bolger et al. 2014) software can be used for short reads which are generated from Illumina sequencers. Another major step to consider in this regard is the actual physical elimination of abundant ribosomal RNA transcripts from the samples because, if not removed, they often make up upward of 90% of all data and do not add any value toward most downstream analyses, including detecting differentially expressed genes or characterizing pathways. These RNA transcripts are usually filtered out with molecular methods before sequencing. However, since they are so abundant in samples, some percentage of the total sequencing is often still rRNA. After sequencing, the rRNAs can be detected for filtering from downstream analyses by tools such as barrnap (Seemann 2014) and SortMeRNA (Kopylova et al. 2012).

11.9 De Novo Assembly

These high-quality, pre-processed reads can be assembled into presumptive transcripts using de novo assemblers. Since most microbiomes otherwise are not captured adequately with reference genomes, de novo assemblers can generate a scaffolds which represents longer genomic segments that can produce a reference set of genes. This will provide the user with the facility for finding homologs more easily, determine the taxonomic origin of a gene, and act as a reference for expression analysis and help in mapping. Metagenomic assemblers, like IDBA-UD (Peng et al. 2012), MEGAHIT (Li et al. 2015), metaQUAST (Goussarov et al. 2024), and metaSPAdes (Nurk et al. 2017), have been developed to handle the complexity of metagenomes that may be similar in sequences in some very conserved regions but differ by orders of magnitude with respect to relative abundances within the microbiome while featuring population (strain-level) variation.

Assemblers like Trans-ABySS (Robertson et al. 2010), BinPacker (Liu et al. 2016), Trinity (Grabherr et al. 2011), Oases (Schulz et al. 2012), IDBA-Tran (Peng et al. 2013), rnaSPAdes (Bushmanova et al. 2019), and SOAPdenovo-Trans (Xie et al. 2014) try to address the problems in transcriptome sequencing and are designed to assemble transcribed data from single organisms. The correlations among some assemblers showed that, in general, the assemblers like Oases, Trinity, and Metavelv improved the number of annotated genes from the resulting contigs; in this case, Trinity outperformed the other assemblers (Celaj et al. 2014).

IDBA-MT (Leung et al. 2013), IDBA-MTP, and Transcript Assembly Graph (Ye and Tang 2016) are some of the de novo assemblers which are created particularly for metatranscriptomes, taking into consideration features of both transcripts and the complexity of microbial communities. IDBA-MT is developed based on IDBA-UD and uses multiple k values in a de Bruijn graph with consideration of features related to mRNAs like uneven sequencing depth and frequent repeat patterns among different mRNAs, hence reducing the mis-assemblies. Likewise, IDBA-MTP was developed from IDBA-MT for assembling lowly expressed mRNAs (Leung et al. 2014). Abiding by this, it uses known protein sequences to guide the assembly. On this regard, it initiates with small values of k and constructs mRNA sequences and then includes those similar to a known set of proteins based on their similarity. TAG is a quite recent assembler also using a de Bruijn graph, but to assemble the corresponding metagenome, which is then used as a reference to map the transcriptome reads and rebuild mRNA sequences by traversing what is called a metagenome assembly graph with mapped transcriptome reads. Because this tool assumes contiguous genes (without splicing), it is quite useless in microbiomes containing eukaryotes. This would also, therefore, imply that there is a depth of the metagenome considered sufficient in the breadth of the community for all expressed genes to map it.

11.10 Taxonomic Profiling

Taxonomic profiling is commonly done with shotgun metagenomic data, a similar suite of tools can be used for reads or contig-based taxonomic assignments to understand what organisms are actively expressing RNA. Another, quite different approach would be to focus on rRNAs only in order to estimate the community's active members, although, as above-mentioned, these are often removed, both in wet-lab protocols and in preprocessing of the raw data. Tools that perform taxonomy classification from reads, such as Trans-ABySS (Lau et al. 2018), GOTTCHA (Freitas et al. 2015), MetaPhlan2 (Truong et al. 2015), can already be applied to metatranscriptomes. Since these tools work with short reads and on the basis of nucleotide matches, the use of these tools is limited to microbiomes study of members that are located close to the existing sequences. These reads, if assembled into longer contigs or even full-length transcripts, can be utilized by the tools (Kim et al. 2016) and Kraken 2 (Wood and Salzberg 2014) and others to possibly identify more members of the sequenced community.

Assignments at the taxonomic level, having reads or predicted coding regions, have a lot of limitations, algorithms necessary to deal with large data volumes or short sequences and the rarity of references in the reference databases. Such issues are compounded together by the fact that most bioinformatics tools make use of only subsets of available genomes or focus on certain organisms. As an example, most of the tool databases do not have eukaryotes. With some new tools and following developments in existing ones, recently there are attempts to add eukaryotic data into their databases, MetaPhlan2 (Truong et al. 2015) and kaiju (Menzel et al. 2016), though their effectiveness for the identification of a eukaryote is still unknown. Moreover, it is often difficult to separate low abundance hits from false positive hits, an intrinsic problem with most microbiome studies. Another confounding factor can be our general lack of knowledge of overall microbial diversity and in any biological system being studied, which can limit the utility of taxonomy classification tools.

11.11 Functional Annotation

One of the primary objectives of metatranscriptomics is to quantify the functional activity of a microbiome. Since the expressed transcripts are proxies to the real phenotype, characterizing the function of transcripts lies at the center of metatranscriptomics tasks. Functional annotation may be performed either with reads or assembled contigs. All toolspecific, database-based read functional profilers, including MetaCLADE (Ugarte et al. 2018), HMM-GRASPx (Zhong et al. 2016) and UProC (Meinicke 2015), require the input of predicted open reading frames from tools such as FragGeneScan (Rho et al. 2010). The latest tools cite only MetaCLADE, which, at the moment, uses a precomputed database of two million probabilistic models derived from 15,000 Pfam domains, thus hundreds of models for any unique domain. Searches against this database yield large numbers of hits per read followed by filtering on the basis of redundancy, probability, and bit-scores (Ugarte et al. 2018).

Alternatively, gene annotation can be assigned directly from assembled contigs. Annotation of assembled transcripts proceeds similarly to the annotation of genomes and metagenomes. This includes gene finding with tools such as Prodigal (Hyatt et al. 2010) and FragGeneScan (Rho et al. 2010), followed by functional assignment in the form of similarity searches using tools like DIAMOND (Buchfink et al. 2015) against some of the functional databases like KEGG (Kanehisa and Goto 2000), NCBI RefSeq (O'Leary et al. 2016) and UniProt (2019). The other tools, pipelines, and platforms include a suite of bioinformatics utilities, such as gene finding and annotation, which combine similarity searches against different databases, for example, Prokka, EDGE Bioinformatics, and MG-RAST. Tools like MinPath or iPath can further map enzymatic functions to known metabolic pathways on the basis of annotations.

11.12 Differential Expression Analyses

Functional profiling and pathway analysis of endophytic microbial communities by metatranscriptomics thus hold great potential for elucidating the metabolic activities, biological functions, and molecular pathways involved in microbial colonization, symbiotic interactions, and ecosystem functioning within plant tissues. Expressed genes can be annotated and metabolic pathways identified that are enriched in endophytic communities allowing researchers to unravel functional roles of microbial taxa and their contribution to plant health, nutrient cycling, stress tolerance, and secondary metabolite production. Together these methods and applications can be used to explore the functional diversity and metabolic potential of endophytic microbial communities through functional profiling of the metagenomic sequences and pathway analysis.

Not only does this give a sense of which members are active and what genes they are expressing at one time point, but it also allows the possibility for the study of differential gene expression with metatranscriptomics in an effort to compare the effects of different conditions and environmental factors on community function or trace the timedependent dynamics of communities. Most of the tools developed for analysis within a single genome can now be used to carry out the metatranscriptomic differential gene expression study. Most of the tools are input-sensitive, with the ease of gene (or transcript) and abundance data per sample input determining the postulated expression under specific conditions or time points. Abundance estimation may be achieved in different ways but typically involves read alignment or mapping of a read to a reference genome, reference assembly, or reference gene set. Commonly, R packages used together with the abundance data for detecting genes that are statistically differentially expressed among large samples of conditions or time points include EdgeR (Robinson et al. 2010), DeSeq2 (Love et al. 2014), and limma (Ritchie et al. 2015). Tools like Generally Applicable Gene-Set/Pathway Analysis (GAGE) might also be used to identify pathways that are enriched in one condition compared with another (Luo et al. 2009). Since replicating metatranscriptomics samples are nontrivial compared to transcriptomic studies with isolate organisms, nonparametric methods that have been implemented in NOISeq (Tarazona et al. 2015) must also be applied here.

Some peculiarity of metatranscriptomic analyses makes differential expression analyses quite problematic because of the huge sequencing diversity of the transcripts, many of which belong to a great variety of organisms. Challenges, such as genes among very closely related organisms and variation in their taxonomic composition, might render an estimation of gene expression-profile incorrect. Lately, a normalization method suggested that there might be a way to reduce the influence of taxonomic diversity by normalizing the count data based on the taxonomic composition, but once again, this method was influenced by representation in taxonomic databases (Klingenberg and Meinicke 2017).

11.13 Analysis of Orthology and Homology

This is an assessment of the evolutionary relationships with respect to the conservation of sequences from annotated genes between different organisms. Orthologous genes are the ones in different species that emerged from a common ancestor and still serve similar functions, while homologous genes are sequence-similar but may have functionally diverged. Analyses of orthology and homology inconsistent with gene conservation or divergence are very informative about the evolutionary history of functional and endophytic microbial communities and help make predictions related to gene function and biological significance. Some of the tools for analysis are OrthoFinder, Orthologous MAtrix (OMA), TransDecoder, MUSCLE, PhyloCSF, PhyML, and KOG (Nichio et al. 2017).

Pathway mapping tools enable visualization of metabolic pathways while network visualization software is used to analyze them. Pathway diagrams model the sequence of steps and enzymatic reactions involved in metabolic pathways and brought forth intermediate metabolites, regulatory points and key enzyme functions. The algorithms of network analysis detect the interconnectedness of metabolic pathways, metabolic crosstalk, and regulatory nodes within metabolic networks, visualizing the level of complexity and organization of metabolic interactions that take place within endophytic microbial communities (Kim et al. 2016).

11.14 Conclusion

In conclusion, metatranscriptomics stands as a powerful tool that has revolutionized our understanding of endophytic communities and their intricate interactions with host plants. By unravelling the dynamic gene expression profiles and functional activities of endophytic microbes, metatranscriptomics has provided unprecedented insights into the ecological dynamics, metabolic capabilities, and biotechnological potential of these microbial communities. From characterizing community composition to elucidating host-microbe interactions and deciphering the molecular mechanisms underlying plant growth promotion and disease suppression, metatranscriptomics has opened new avenues for research and innovation in endophyte studies. Its significance lies in its ability to bridge the gap between genotype and phenotype, providing a comprehensive view of microbial functions and ecological roles within plant ecosystems. As we reflect on the remarkable advancements enabled by metatranscriptomics in endophyte studies, it is evident that there is still much to explore and discover in this burgeoning field. A call to action is warranted for further research and collaboration to unravel the remaining mysteries surrounding endophytic communities and their ecological significance. Collaborative efforts among researchers from diverse disciplines, including microbiology, plant biology, and bioinformatics, are essential for advancing our knowledge and addressing key challenges in endophyte research. Moreover, interdisciplinary approaches integrating experimental and computational methodologies will be instrumental in deciphering complex microbial communities and elucidating their functional roles in plant health, ecosystem dynamics, and biotechnological applications. By fostering collaboration, we can unlock the full potential of metatranscriptomics and endophyte studies for sustainable agriculture, environmental conservation.

References

Aguiar-Pulido V, Huang W, Suarez-Ulloa V, Cickovski T, Mathee K, Narasimhan G (2016) Metagenomics, metatranscriptomics and metabolomics approaches for microbiome analysis: supplementary issue: bioinformatics methods and applications for big metagenomics data. Evol Bioinforma 12:EBO-S36436

Aransay AM, Trueba JLL (eds) (2016) Field guidelines for genetic experimental designs in high-throughput sequencing. Springer

Azeem M, Soundari PG, Ali A, Tahir MI, Imran M, Bashir S, Irfan M, Li G, Zhu YG, Zhang Z (2021) Soil metaphenomics: a step forward in metagenomics. Archives of Agronomy and Soil Science: 1–19

Bashiardes S, Zilberman-Schapira G, Elinav E (2016) Use of metatranscriptomics in microbiome research. Bioinformat Biol Insights 10:BBI-S34610

Bikel S, Valdez-Lara A, Cornejo-Granados F, Rico K, Canizales-Quinteros S, Soberón X, Del Pozo-Yauner L, Ochoa-Leyva A (2015) Combining metagenomics, metatranscriptomics and viromics to explore novel microbial interactions: towards a systems-level understanding of human microbiome. Comput Struct Biotechnol J 13:390–401

Bolger AM, Lohse M, Usadel B (2014) Trimmomatic: a flexible trimmer for Illumina sequence data. Bioinformatics 30(15):2114–2120

Buchfink B, Xie C, Huson DH (2015) Fast and sensitive protein alignment using DIAMOND. Nat Methods 12(1):59–60

Bushmanova E, Antipov D, Lapidus A, Prjibelski AD (2019) rnaSPAdes: a de novo transcriptome assembler and its application to RNA-Seq data. Gigascience 8:giz100

Celaj A, Markle J, Danska J, Parkinson J (2014) Comparison of assembly algorithms for improving rate of metatranscriptomic functional annotation. Microbiome 2:1–13

Chen S, Zhou Y, Chen Y, Gu J (2018) fastp: an ultra-fast all-in-one FASTQ preprocessor. Bioinformatics 34(17):i884–i890

Cho H, Qu Y, Liu C, Tang B, Lyu R, Lin BM, Roach J, Azcarate-Peril MA, Aguiar Ribeiro A, Love MI, Divaris K (2023) Comprehensive evaluation of methods for differential expression analysis of metatranscriptomics data. Brief Bioinform 24(5):bbad279

Dudeja SS, Suneja-Madan P, Paul M, Maheswari R, Kothe E (2021) Bacterial endophytes: molecular interactions with their hosts. J Basic Microbiol 61(6):475–505

Franzosa EA, Morgan XC, Segata N, Waldron L, Reyes J, Earl AM, Giannoukos G, Boylan MR, Ciulla D, Gevers D, Izard J (2014) Relating the metatranscriptome and metagenome of the human gut. Proc Natl Acad Sci 111(22):E2329–E2338

Franzosa EA, McIver LJ, Rahnavard G, Thompson LR, Schirmer M, Weingart G, Lipson KS, Knight R, Caporaso JG, Segata N, Huttenhower C (2018) Species-level functional profiling of metagenomes and metatranscriptomes. Nat Methods 15(11):962–968

Freitas TAK, Li PE, Scholz MB, Chain PS (2015) Accurate read-based metagenome characterization using a hierarchical suite of unique signatures. Nucleic Acids Res 43(10):e69–e69

Giannoukos G, Ciulla DM, Huang K, Haas BJ, Izard J, Levin JZ, Livny J, Earl AM, Gevers D, Ward DV, Nusbaum C (2012) Efficient and robust RNA-seq process for cultured bacteria and complex community transcriptomes. Genome Biol 13:1–13

Gilbert JA, Hughes M (2011) Gene expression profiling: metatranscriptomics. In: High-throughput next generation sequencing: methods and applications, pp 195–205

Goussarov G, Mysara M, Cleenwerck I, Claesen J, Leys N, Vandamme P, Van Houdt R (2024) Benchmarking short-, long-and hybrid-read assemblers for metagenome sequencing of complex microbial communities. Microbiology 170(6):001469

Grabherr MG, Haas BJ, Yassour M, Levin JZ, Thompson DA, Amit I, Adiconis X, Fan L, Raychowdhury R, Zeng Q, Chen Z (2011) Full-length transcriptome assembly from RNA-Seq data without a reference genome. Nat Biotechnol 29(7):644–652

Grieb A, Bowers RM, Oggerin M, Goudeau D, Lee J, Malmstrom RR, Woyke T, Fuchs BM (2020) A pipeline for targeted metagenomics of environmental bacteria. Microbiome 8:1–17

Gupta G, Parihar SS, Ahirwar NK, Snehi SK, Singh V (2015) Plant growth promoting rhizobacteria (PGPR): current and future prospects for development of sustainable agriculture. J MicrobBiochemTechnol 7(2):96–102

Gupta N, Vats S, Bhargava P (2018) Sustainable agriculture: role of metagenomics and metabolomics in exploring the soil microbiota. In: In silico approach for sustainable agriculture, pp 183–199

Hayden HL, Savin KW, Wadeson J, Gupta VV, Mele PM (2018) Comparative metatranscriptomics of wheat rhizosphere microbiomes in disease suppressive and non-suppressive soils for Rhizoctonia solani AG8. Front Microbiol 9:859

Hyatt D, Chen GL, LoCascio PF, Land ML, Larimer FW, Hauser LJ (2010) Prodigal: prokaryotic gene recognition and translation initiation site identification. BMC Bioinformatics 11:119. https://doi.org/10.1186/1471-2105-11-119

Kanehisa M, Goto S (2000) KEGG: kyotoencyclopedia of genes and genomes. Nucleic Acids Res 28(1):27–30

Kim J, Kim MS, Koh AY, Xie Y, Zhan X (2016) FMAP: functional mapping and analysis pipeline for metagenomics and metatranscriptomics studies. BMC Bioinformatics 17:1–8

Klingenberg H, Meinicke P (2017) How to normalize metatranscriptomic count data for differential expression analysis. PeerJ 5:e3859

Kopylova E, Noé L, Touzet H (2012) SortMeRNA: fast and accurate filtering of ribosomal RNAs in metatranscriptomic data. Bioinformatics 28(24):3211–3217

Kumar A, Yadav A (2024) Next generation sequencing in metagenomics and Metatranscriptomics. In: Multi-omics analysis of the human microbiome: from technology to clinical applications. Springer Nature Singapore, Singapore, pp 49–75

Latorre-Pérez A, Villalba-Bermell P, Pascual J, Vilanova C (2020) Assembly methods for nanopore-based metagenomic sequencing: a comparative study. Sci Rep 10(1):13588

Lau MCY, Harris RL, Oh Y, Yi MJ, Behmard A, Onstott TC (2018) Taxonomic and functional compositions impacted by the quality of Metatranscriptomic assemblies. Front Microbiol 9:1235. https://doi.org/10.3389/fmicb.2018.01235

Leung HC, Yiu SM, Parkinson J, Chin FY (2013) IDBA-MT: de novo assembler for metatranscriptomic data generated from next-generation sequencing technology. J Comput Biol 20(7):540–550

Leung HC, Yiu SM, Chin FY (2014) IDBA-MTP: a hybrid metatranscriptomic assembler based on protein information. In: Research in computational molecular biology: 18th annual international conference, RECOMB 2014, Pittsburgh, PA, USA, April 2–5, 2014, proceedings 18. Springer International Publishing, pp 160–172

Li D, Liu CM, Luo R, Sadakane K, Lam TW (2015) MEGAHIT: an ultra-fast single-node solution for large and complex metagenomics assembly via succinct de Bruijn graph. Bioinformatics 31(10):1674–1676

Liang P, Pardee AB (1992) Differential display of eukaryotic messenger RNA by means of the polymerase chain reaction. Science 257(5072):967–971

Liao HL, Bonito G, Rojas JA, Hameed K, Wu S, Schadt CW, Labbé J, Tuskan GA, Martin F, Grigoriev IV, Vilgalys R (2019) Fungal endophytes of Populus trichocarpa alter host phenotype, gene expression and rhizobiome composition. Mol Plant-Microbe Interact 32(7):853–864

Liu J, Li G, Chang Z, Yu T, Liu B, McMullen R, Chen P, Huang X (2016) BinPacker: packing-based de novo transcriptome assembly from RNA-seq data. PLoS Comput Biol 12(2):e1004772

Love MI, Huber W, Anders S (2014) Moderated estimation of fold change and dispersion for RNA-seq data with DESeq2. Genome Biol 15(12):550

Luo W, Friedman MS, Shedden K, Hankenson KD, Woolf PJ (2009) GAGE: 1093 generally applicable gene set enrichment for pathway analysis. BMC Bioinformatics 10:161

Mauchline TH, Hayat R, Clark IM, Hirsch PR (2018) Old meets new: most probable number validation of metagenomic and metatranscriptomic datasets in soil. Lett Appl Microbiol 66(1):14–18

Meena M, Yadav G, Sonigra P, Nagda A, Mehta T, Swapnil P, Harish, Marwal A, Kumar S (2023) Multifarious responses of forest soil microbial community toward climate change. Microb Ecol 86:49–74. https://doi.org/10.1007/s00248-022-02051-3

Menzel P, Ng KL, Krogh A (2016) Fast and sensitive taxonomic classification for metagenomics with Kaiju. Nature communications 7(1):11257

Meinicke P (2015) UProC: tools for ultra-fast protein domain classification. Bioinformatics 31(9):1382–1388

Nichio BT, Marchaukoski JN, Raittz RT (2017) New tools in orthology analysis: a brief review of promising perspectives. Front Genet 8:165

Niem JM, Billones-Baaijens R, Stodart B, Savocchia S (2020) Diversity profiling of grapevine microbial endosphere and antagonistic potential of endophytic pseudomonas against grapevine trunk diseases. Front Microbiol 11:477

Nurk S, Meleshko D, Korobeynikov A, Pevzner PA (2017) metaSPAdes: a new versatile metagenomic assembler. Genome Res 27(5):824–834

O'Leary NA, Wright MW, Brister JR, Ciufo S, Haddad D, McVeigh R, Rajput B, Robbertse B, Smith-White B, Ako-Adjei D, Astashyn A (2016) Reference sequence (RefSeq) database at NCBI: current status, taxonomic expansion and functional annotation. Nucleic Acids Res 44(D1):D733–D745

Pal A, Pal A (2022) DNA microarray. In: Protocols in advanced genomics and allied techniques, pp 221–243

Paolinelli M, Escoriaza G, Cesari C, Garcia-Lampasona S, Hernandez-Martinez R (2022) Characterization of grapevine wood microbiome through a metatranscriptomic approach. Microb Ecol 83(3):658–668

Peano C, Pietrelli A, Consolandi C, Rossi E, Petiti L, Tagliabue L, De Bellis G, Landini P (2013) An efficient rRNA removal method for RNA sequencing in GC-rich bacteria. In: Microbial informatics and experimentation, vol 3, pp 1–11

Peng Y, Leung HC, Yiu SM, Chin FY (2012) IDBA-UD: a de novo assembler for single-cell and metagenomic sequencing data with highly uneven depth. Bioinformatics 28(11):1420–1428

Peng Y, Leung HC, Yiu SM, Lv MJ, Zhu XG, Chin FY (2013) IDBA-Tran: a more robust de novo de Bruijn graph assembler for transcriptomes with uneven expression levels. Bioinformatics 29(13):i326–i334

Pollard MO, Gurdasani D, Mentzer AJ, Porter T, Sandhu MS (2018) Long reads: their purpose and place. Hum Mol Genet 27(R2):R234–R241

Poursalavati A, Javaran VJ, Laforest-Lapointe I, Fall ML (2023) Soil metatranscriptomics: an improved RNA extraction method toward functional analysis using nanopore direct RNA sequencing. Phytobiomes J 7(1):42–54

Quail MA, Smith M, Coupland P, Otto TD, Harris SR, Connor TR, Bertoni A, Swerdlow HP, Gu Y (2012) A tale of three next generation sequencing platforms: comparison of Ion Torrent, Pacific Biosciences and Illumina MiSeq sequencers. BMC Genomics 13:1–13

Rho M, Tang H, Ye Y (2010) FragGeneScan: predicting genes in short and error-prone reads. Nucleic Acids Res 38(20):e191–e191

Ritchie ME, Phipson B, Wu DI, Hu Y, Law CW, Shi W, Smyth GK (2015) Limma powers differential expression analyses for RNA-sequencing and microarray studies. Nucleic Acids Res 43(7):e47–e47

Robertson G, Schein J, Chiu R, Corbett R, Field M, Jackman SD, Mungall K, Lee S, Okada HM, Qian JQ, Griffith M (2010) De novo assembly and analysis of RNA-seq data. Nat Methods 7(11):909–912

Robinson MD, McCarthy DJ, Smyth GK (2010) edgeR: a Bioconductor package for differential expression analysis of digital gene expression data. Bioinformatics 26(1):139–140

Schulz MH, Zerbino DR, Vingron M, Birney E (2012) Oases: robust de novo RNA-seq assembly across the dynamic range of expression levels. Bioinformatics 28(8):1086–1092

Seemann T (2014) Prokka: rapid prokaryotic genome annotation. Bioinformatics 30(14):2068–2069
Shakya M, Lo CC, Chain PS (2019) Advances and challenges in metatranscriptomic analysis. Front Genet 10:904
Sham A, Al-Ashram H, Whitley K, Iratni R, El-Tarabily KA, AbuQamar SF (2019) Metatranscriptomic analysis of multiple environmental stresses identifies RAP2. 4 gene associated with Arabidopsis immunity to Botrytis cinerea. Sci Rep 9(1):17010
Sultan M, Amstislavskiy V, Risch T, Schuette M, Dökel S, Ralser M, Balzereit D, Lehrach H, Yaspo ML (2014) Influence of RNA extraction methods and library selection schemes on RNA-seq data. BMC Genomics 15:1–13
Sunagawa S, Mende DR, Zeller G, Izquierdo-Carrasco F, Berger SA, Kultima JR, Coelho LP, Arumugam M, Tap J, Nielsen HB, Rasmussen S (2013) Metagenomic species profiling using universal phylogenetic marker genes. Nat Methods 10(12):1196–1199
Tarazona S, Furió-Tarí P, Turrà D, Pietro AD, Nueda MJ, Ferrer A, Conesa A (2015) Data quality aware analysis of differential expression in RNA-seq with NOISeq R/Bioc package. Nucleic Acids Res 43(21):e140
Trivedi P, Delgado-Baquerizo M, Trivedi C, Hamonts K, Anderson IC, Singh BK (2017) Keystone microbial taxa regulate the invasion of a fungal pathogen in agro-ecosystems. Soil Biol Biochem 111:10–14
Truong DT, Franzosa EA, Tickle TL, Scholz M, Weingart G, Pasolli E, Tett A, Huttenhower C, Segata N (2015) MetaPhlAn2 for enhanced metagenomic taxonomic profiling. Nat Methods 12(10):902–903
Ugarte A, Vicedomini R, Bernardes J, Carbone A (2018) A multi-source domain annotation pipeline for quantitative metagenomic and metatranscriptomic functional profiling. Microbiome 6:1–27
UniProt Consortium (2019) UniProt: a worldwide hub of protein knowledge. Nucleic Acids Res 47(D1):D506–D515
Urich T, Lanzén A, Qi J, Huson DH, Schleper C, Schuster SC (2008) Simultaneous assessment of soil microbial community structure and function through analysis of the meta-transcriptome. PLoS One 3(6):e2527
Uzoh IM, Babalola OO (2018) Rhizosphere biodiversity as a premise for application in bio-economy. Agric Ecosyst Environ 265:524–534
Vierheilig J, Savio D, Ley RE, Mach RL, Farnleitner AH, Reischer GH (2015) Potential applications of next generation DNA sequencing of 16S rRNA gene amplicons in microbial water quality monitoring. Water Sci Technol 72(11):1962–1972
Wang Z, Gerstein M, Snyder M (2009) RNA-Seq: a revolutionary tool for transcriptomics. Nat Rev Genet 10(1):57–63
Wendisch VF, Zimmer DP, Khodursky A, Peter B, Cozzarelli N, Kustu S (2001) Isolation of Escherichia coli mRNA and comparison of expression using mRNA and total RNA on DNA microarrays. Anal Biochem 290(2):205–213
Wood DE, Salzberg SL (2014) Kraken: Ultrafast metagenomic sequence classification using exact alignments. Genome biology 15:1–2
Xie Y, Wu G, Tang J, Luo R, Patterson J, Liu S, Huang W, He G, Gu S, Li S, Zhou X (2014) SOAPdenovo-Trans: de novo transcriptome assembly with short RNA-Seq reads. Bioinformatics 30(12):1660–1666
Yang SU, Kim H, Kim RJ, Kim J, Suh MC (2020) AP2/DREB transcription factor RAP2. 4 activates cuticular wax biosynthesis in Arabidopsis leaves under drought. Front Plant Sci 11:895
Ye Y, Tang H (2016) Utilizing de Bruijn graph of metagenome assembly for metatranscriptome analysis. Bioinformatics 32(7):1001–1008
Zhang L, Chen F, Zeng Z, Xu M, Sun F, Yang L, Bi X, Lin Y, Gao Y, Hao H, Yi W (2021) Advances in metagenomics and its application in environmental microorganisms. Front Microbiol 12:766364
Zhong C, Edlund A, Yang Y, McLean JS, Yooseph S (2016) Metagenome and metatranscriptome analyses using protein family profiles. PLoS Comput Biol 12(7):e1004991
Zhou J, He Z, Yang Y, Deng Y, Tringe SG, Alvarez-cohen L (2015) High throughput metagenomic technologies for complex microbial community analysis: open and closed formats. MBio 6(1):e02288–e02214
Zuñiga C, Zaramela L, Zengler K (2017) Elucidation of complexity and prediction of interactions in microbial communities. Microb Biotechnol 10(6):1500–1522

Functional Annotation and Characterization of Endophytic Fungal Community

12

Pooja Yadav, Chitra Yadav, Abhishek Joshi, Mukesh Meena, Harish, and Jaya Arora

Abstract

Fungal endophytes are known for their symbiotic relationships with plant tissues such as flowers, fruits, leaves, stems, and roots. Despite their significant ecological roles, complex genomic structure of fungal endophytes presents significant challenges in fully harnessing their potential for various applications. Functional annotation and characterization, particularly through high-throughput omics technologies, have provided critical insights into the genomic features, metabolic pathways, and functional traits of fungal endophytes. Recent discoveries have identified a range of novel genes, pathways, and enzymes that are involved in stress tolerance, plant growth promotion, bioremediation, biodegradation, nutrient cycling, and biocontrol. This chapter offers a comprehensive overview of endophytic fungal communities and highlights recent advancements in functional annotation and characterization of their genome. Furthermore, it explored the multifaceted roles of endophytic fungi and their importance across various sectors, including agriculture, pharmacology, and environmental remediation.

P. Yadav · C. Yadav · A. Joshi · J. Arora (✉)
Laboratory of Biomolecular Technology, Department of Botany, University College of Science, Mohanlal Sukhadia University, Udaipur, Rajasthan, India

M. Meena
Laboratory of Phytopathology and Microbial Biotechnology, Department of Botany, Mohanlal Sukhadia University, Udaipur, Rajasthan, India

Harish
Plant Biotechnology Laboratory, Department of Botany, University College of Science, Mohanlal Sukhadia University, Udaipur, Rajasthan, India

M. Meena et al. (eds.), *Multi-omics Approach to Investigate Endophyte Diversity*, Microorganisms for Sustainability 53, https://doi.org/10.1007/978-981-96-4100-0_12

Keywords

Fungal endophytes · Functional annotation · Genes · Metabolomics · Transcriptome

Abbreviations

ITS	Internal Transcribed Spacer
GO	Gene Ontology
COG	Clusters of Orthologous Groups
KEGG	Kyoto Encyclopedia of Genes & Genomes
BLAST	Basic Local Alignment Search Tool

12.1 Introduction

Endophytes, which encompass both fungi and bacteria, reside within plant tissues, such as stems, leaves, roots, fruits, buds, and inflorescences, without causing any noticeable harm to the plants. The major fungal groups in the endophytic community include Ascomycota, Basidiomycota, and Mucoromycetes (Arora and Ramawat 2017; Burragoni and Jeon 2021). Unlike parasites, endophytic fungi establish mutually beneficial relationships with their host plants, often enduring throughout the plant life cycle (Venieraki et al. 2017). These interactions typically involve the exchange of nutrients, metabolites, and signals, which enhances the overall health and resilience of plants (Patchett and Newman 2021).

Based on their mutualistic or symbiotic interaction with the host plants, fungal endophytes are categorized into two groups. The first group, Balansiaceous or Clavicipitaceous fungi (Class I endophytes), includes free-living symbiotic organisms found in cool-season grasses (Poaceae). These fungi infect host plant ovules and pass their genetic material vertically to progeny through host seeds, colonizing the rhizomes and shoots (Poveda et al. 2021). The second group, Non-Clavicipitaceous or Non-Balansiaceous fungi, includes Class II, III, and IV endophytes. These endophytes are highly diverse and not intimately connected to the host plant, residing in plant tissues in a dormant state without harming the host. They become active when the plant undergoes chemical changes due to damage or environmental stress, entering the plant through roots, shoots, or rhizomes (Rodriguez et al. 2008; Mishra et al. 2021). Class II endophytes colonize through roots, shoots, and rhizomes, undergoing both vertical and horizontal translocation, while Class III and IV endophytes colonize solely through roots and shoots, respectively, undergoing only horizontal translocation (Jha et al. 2023).

In recent years, fungal endophytes have attracted significant attention because of their diverse functional roles and properties. These microorganisms are rich sources of structurally diverse metabolites with antibacterial, antifungal, antioxidant, and antitumor properties, making them valuable for the pharmaceutical industry (Wei et al. 2020; Akram et al. 2023; Yadav and Meena 2022; Bhupenchandra et al. 2024).

In agriculture, fungal endophytes contribute to plant growth and protection by secreting phytohormones and biocontrol agents as well as by enhancing resistance to both abiotic and biotic stresses (Verma et al. 2024a; Yadav and Meena 2021; Chandran et al. 2021). They have the potential to degrade pollutants and detoxify heavy metals, making them valuable for bioremediation efforts (Nandy et al. 2020; He et al. 2023a, b). Their role in nutrient acquisition and ecosystem stability further underscores their importance in promoting sustainable development (García-Latorre et al. 2021; Yadav et al. 2022; Verma et al. 2024b).

Fungal endophytes pose considerable challenges due to their vast diversity and the complex nature of their genomes. Also, the behavior of these fungi can vary significantly depending on the host plant and environmental conditions, making their study and characterization even more difficult (Dos Reis et al. 2022). Therefore, precise isolation and identification of fungal endophytes are critical for understanding their ecological functions and interactions with host plants, as well as their broader environmental impacts. Functional annotation and characterization are particularly important, as they offer detailed insights into the biological, ecological, and functional attributes (Sagita et al. 2021). Recent efforts in the annotation and characterization of fungal endophytes have led to the discovery of several novel genes, pathways, and enzymes with potential applications across various fields (Yadav et al. 2020; Joshi et al. 2023; Yang et al. 2023).

This chapter presents a thorough review of recent advancements in endophytic fungal research. It explores cutting-edge techniques for isolating and characterizing fungal endophytes, while highlighting the importance of functional annotation in deciphering their roles and interactions. This chapter also discusses the relevance of endophytic fungi across the sustainable agriculture, pharmacology, and environmental remediation. .

12.2 Methodological Approaches to Functional Annotation and Characterization of Fungal Endophytes

The initial step of studying fungal endophytes involves sampling them from host plants. This step is crucial, because the diversity and functional attributes of endophytic community are directly influenced by the phytochemical compositions of the host plant. Although several recent studies have focused on the sampling and isolation of fungi from various host plants, there is no universal standard for the optimal number of samples or individual plants required (El-Hasan et al. 2022; Dos Reis et al. 2022). Also, environmental factors play a pivotal role in shaping the distribution of fungal species, both at macro- and micro-scales, even within a single site (Oita et al. 2021). It is generally recommended that sampling strategies be customized to meet the specific objectives of each study.

After sampling, fungal endophytes can be isolated and identified using both culture-dependent and culture-independent approaches (Fig. 12.1). Both methods are widely used to analyze the community structure and diversity of fungal endophytes. Culture-dependent techniques are generally applied to sporulating fungi and involve

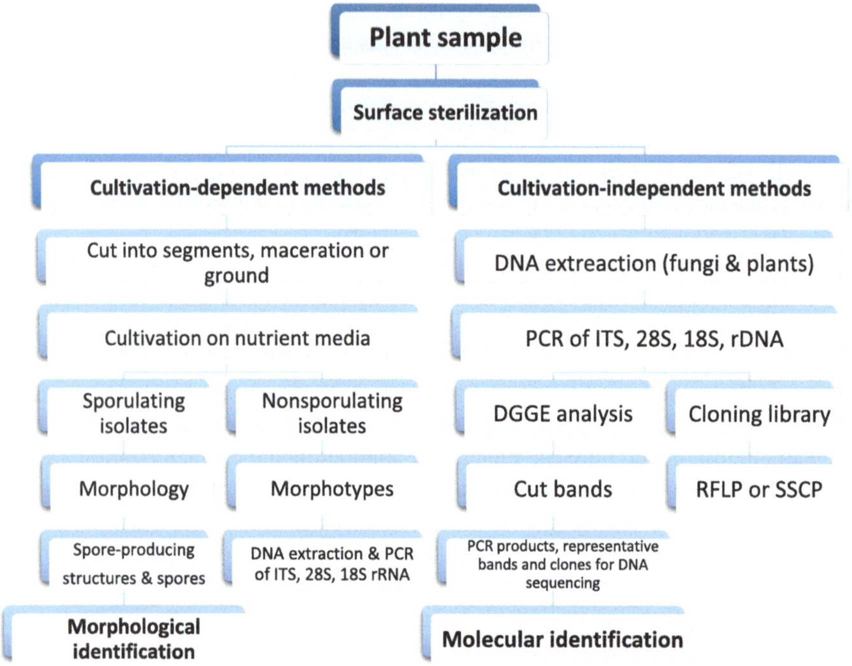

Fig. 12.1 Methodology for isolation and identification of endophytic fungi. (Modified and drawn from Palwe et al. 2021)

several key steps such as surface sterilization of plant tissues and cultivation of endophytes on selective media, followed by purification and identification through morphological characteristics and DNA-based molecular analysis. The success of these steps can be influenced by factors such as the type and age of the plant tissue, the composition of the growth media, and environmental conditions (Sahu et al. 2022; Sohrabi et al. 2024). In recent years, a variety of culture media, such as malt extract agar, malt peptone dextrose agar, potato dextrose agar, starch yeast peptone agar, and yeast peptone dextrose agar, have been used to support the growth of different fungal species, enabling a more comprehensive examination of fungal growth characteristics and their potential applications (Khalil et al. 2021; Tibpromma et al. 2022; André et al. 2024). On the other hand, culture-independent methods are particularly useful for non-sporulating fungi that are difficult to culture using traditional culturing methods. Metagenomics, which involves the extraction of fungal DNA directly from environmental samples, followed by the amplification and sequencing of fungal-specific genes (such as those encoding ribosomal RNA and internal transcribed spacer (ITS) regions), has emerged as a powerful tool for this purpose (Thangavel et al. 2023). These methods allow for direct identification of fungal endophytes from complex microbial communities and provide valuable insights into their structural diversity and ecological attributes (Adeleke et al. 2023).

Functional annotation and characterization are key processes for understanding the biological roles of genes and proteins within a genome. These processes are essential for uncovering the mechanisms by which fungal endophytes interact with their host plants and contribute to beneficial outcomes such as the production of bioactive compounds, plant growth-promoting factors, and enzymes that detoxify or degrade environmental pollutants. Recent advances in omics technologies and bioinformatics have significantly enhanced the functional annotation and characterization of fungal endophytes (Liu-Xu et al. 2022; Verma et al. 2023). High-throughput sequencing techniques, such as next-generation sequencing (NGS), have enabled the sequencing of entire fungal genomes, providing large datasets for functional analysis. RNA sequencing (RNA-seq) allows researchers to identify genes that are actively expressed under specific conditions, offering insights into the functional roles of fungal endophytes in their natural environments (Ren et al. 2022). Mass spectrometry-based approaches such as liquid chromatography-mass spectrometry (LC-MS), gas chromatography-mass spectrometry (GC-MS), tandem mass spectrometry (MS/MS), and two-dimensional gel electrophoresis (2-DE) have been employed to profile and characterize the metabolites and proteins produced by fungal endophytes (Saini et al. 2024).

Bioinformatics tools and databases play a crucial role in functional annotation and characterization of fungal endophytes in recent years. Tools like BLAST (Basic Local Alignment Search Tool), InterProScan, and Pfam allow for the comparison of fungal sequences with known protein families and functional domains. Similarly, databases such as the non-redundant (Nr) protein sequence, Kyoto Encyclopedia of Genes and Genomes (KEGG), Gene Ontology (GO), Clusters of Orthologous Groups (COG), and Carbohydrate-Active Enzymes (CAZy) provide valuable resources for mapping genes to metabolic pathways, structural components, and other functional attributes (Reed et al. 2024; Singh et al. 2024).

12.3 Recent Insights on Functional Annotation and Characterization of Fungal Endophytes

Over the past decade, advancements in annotation and characterization of the genomes of endophytic fungal communities have provided significant insights into their genetic composition and functional capacities. Several novel endophytic strains with unique genetic and metabolic configurations have been discovered from various plant species (Corrêa et al. 2020; Wang et al. 2020). A novel strain, *Dactylonectria alcacerensis* CT-6, was recently isolated from *Corydalis tomentella*. Functional annotation of this species revealed a genome size of 61.8 Mb, containing 45 biosynthetic gene clusters and 783 gene clusters associated with CAZymes (Ming et al. 2023). Similarly, *Talaromyces* sp. DC2 was discovered from the stem of *Catharanthus roseus*. Functional annotation of the *Talaromyces* genome revealed a size of 34.58 Mb, organized into 156 contigs, with a GC content of 46.5%. This genome encodes 11,131 protein-coding genes, of which 10,780 were identified in the non-redundant (NR) database. Furthermore, 8735 genes were identified in the

Pfam database, 8165 in Swiss-Prot, 6682 in GO, 6509 in the KEGG database, and 1230 in CAZy (Quan et al. 2024).

In another study, Huang et al. (2022) discovered a novel strain of *Fusarium oxysporum* TH15 from the roots of *Tetrastigma hemsleyanum*. The genome of this strain is 46.85 Mb in size, organized into 34 contigs, with a GC content of 47.53%. It contains 10,442 genes, including 735 clusters related to CAZymes, 87 clusters encoding glycosyltransferases, 27 for polysaccharide lyases, 114 associated with auxiliary activities, and 348 clusters involved in glycoside hydrolases. Similarly, the functional annotation of *Fusarium multiceps* BPAL1, isolated from the leaves of *Acanthus ilicifolius*, revealed a genome spanning 39.57 Mb, organized into 169 scaffolds with a GC content of 47.62%. This genome encodes 11,675 genes, of which 85.36% (9654 genes) belong to eukaryotic orthologous groups, and 28.41% (3213 genes) are involved in metabolic processes. A total of 33 biosynthetic gene clusters (BGCs) were identified, of which 23 remained uncharacterized (Shah and Chunduri 2023). In a recent study, Wu et al. (2024) isolated and annotated a novel strain of *Fusarium solani* MC503 from the roots of *Macleaya cordata*. The functional annotation revealed that the genome of strain MC503 spans 33.06 Mb, encoding 17,668 protein-coding genes. The main functional categories of these annotated genes include cellular components (6347 genes), molecular functions (8491 genes), biological processes (4843 genes), cellular anatomical entities (5887 genes), catalytic activity (5351 genes), and binding (4456 genes).

By comparing the annotated genomes of endophytic fungi with those of their non-endophytic relatives, researchers have identified the novel genes and pathways involved in symbiosis, metabolism, and other functional properties. The genome of a newly discovered endophytic fungus, *Sarocladium brachiariae* isolated from *Brachiaria brizantha*, was annotated and compared with that the of pathogenic strain *Sarocladium oryzae* (Yang et al. 2019). The analysis revealed that the genome of *S. brachiariae* was 31.86 Mb in size and contained 9903 protein-coding genes. Comparative analysis showed that *S. brachiariae* has 14.9% more CAZymes related to plant cell wall degradation and 33.3% more related to fungal cell wall degradation than *S. oryzae*. Furthermore, synteny analysis of the secondary metabolite gene clusters between the two species revealed that only 37.5% of the clusters exhibited high similarity, suggesting that *S. brachiariae* may produce novel secondary metabolites important for its endophytic nature and antifungal properties. In another recent study, the genome of a novel *Paraphoma chrysanthemicola* DS84, isolated from the roots of *Codonopsis pilosula*, was annotated and compared with those of other *Paraphoma* species (Sun et al. 2023). Sequencing analysis revealed that the genome of *P. chrysanthemicola* spans 44.1 Mb and encodes 37,077 protein-coding genes. The study findings also suggest that the *P. chrysanthemicola* exhibits a mannose preference for monosaccharide utilization and melanin, dimethylcoprogen, and phyllostictine A for their endophytic existence. The key findings from recent functional annotation studies on fungal endophytes are summarized in Table 12.1.

Table 12.1 Functional annotation of fungal endophytes isolated from various host plants

Endophytic species or strains	Host plant (tissue)	Functional annotation of genome	References
Alternaria sp. *SPS-2*	*Echrysantha chrysantha* (leaves)	Genome size: 33.4 Mb; GC content: 51%; N50 scaffold: 2.6 Mb; protein-coding genes: 9789; secondary metabolite biosynthetic gene clusters (BGCs): 22; unknown BGCs: 14; gene clusters (GCs) related to carbohydrate-active enzyme (CAZyme): 644; equisetin, betaenones A–C, alternariol, dimethylcoprogen, and melanin: 8 BGCs	Tao et al. (2022)
Aspergillus sp. Gbtc 2	*Ginkgo biloba* (root)	Genome size: 37,929,866 bp, N50 scaffold: 4,075,796 bp; GC content: 47.45%, predicted protein-coding genes:13307	Wu et al. (2022)
Calcarisporium arbuscula NRRL 3705	*Russulaceae* (fruit-bodies)	Genome size: 45 Mb; predicted genes: 10,001; secretory protein-related GCs: 762; CAZyme-related GCs: 386; P450 enzymes encoding GCs: 177; virulence factor-related GCs: 398; pathogen-host interaction-related GCs: 228; BGCs: 65 with dominance of mycotoxin aurovertins; clusters involved in alternariol biosynthesis: 2	Cheng et al. (2020)
Cladosporium angulosum	*Sesuvium portulacastrum* (root)	Genome size: 29,444,740 bp; number of annotated genes: 6426; identified key GCs: tryptophan synthesis and metabolism, nitrogen metabolism and sterol synthesis pathways	Yang et al. (2023)
Dactylonectria alcacerensis CT-6	*Corydalis tomentella*	Genome size: 61.8 Mb; GC content: 49.86%; BGCs: 45; CAZyme-related GCs: 783	Ming et al. (2023)
Fusarium sp. R1	*Rumex madaio*	Genome size: 51.8 Mb; contigs: 542 with N50 scaffold length of 3.21 Mb; GC content: 50.4%; predicted protein-coding genes: 19,333; 37 BGCs: 37; nonribosomal peptide synthetases (NRPSs) encoding GCs: 17; polyketide synthetases encoding GCs: 13; terpene synthases (Ts) encoding GCs: 3	Liu et al. (2022)
Fusarium sp. VM-40	*Vinca minor* (leaves)	Genome size: 40 Mb; GC content: 47.72%; protein-coding genes: 13,546; BGCs:56; unknown BGCs: 31; CAZyme-encoding GCs: 691	He et al. (2023a, b)

(continued)

Table 12.1 (continued)

Endophytic species or strains	Host plant (tissue)	Functional annotation of genome	References
Fusarium multiceps BPAL1	*Acanthus ilicifolius* (leaves)	Genome size: 39.57 Mb; scaffolds: 169; GC content: 47.62%; BGCs: 33, unknown BGCs: 23; predicted genes: 11675, eukaryotic orthologous group genes: 85.36% (9654); genes involved in metabolism: 28.41% (3213)	Shah and Chunduri (2023)
Fusarium oxysporum strain TH15	*Tetrastigma hemsleyanum* (calabash-shaped root)	Genome size: 46,852,391 bp; contigs:34; GC content: 47.53%; predicted genes:10442; CAZyme-related GC:735; glycosyl transferases encoding GCs: 87; polysaccharide lyases encoding GCs: 27, auxiliary activity-related GCs: 114, glycoside hydrolase-related GCs: 348	Huang et al. (2022)
Fusarium solani strain MC503	*Macleaya cordata* (root)	Genome size: 33,062,279 bp; protein-coding genes: 17,668; major functionally annotated genes: cellular component (6347), molecular function (8491), biological process (4843), cellular anatomical entity (5887), catalytic activity (5351), and binding (4456)	Wu et al. (2024)
Helotiales sp. BL73	*Bergenia pacumbis* (root)	Genome size: 59 Mb; N50 scaffold: 622 kb; GC content: 45.23%; protein coding sequence: 18,437; BGCs: 77; polyketide synthase (PKS) type I encoding GCs:26; PKS type III encoding GCs:3,NRPSs encoding GCs: 25; PKS/NRPS hybrid-related GCs: 6, TS encoding GCs:13; indole-related GCs:4	Oberhofer et al. (2022)
Paraphoma chrysanthemicola	*Codonopsis pilosula* (root)	Genome size: 44.1 Mb; scaffold N50: 36.1 mb; protein-coding genes: 37077; key secondary metabolites: melanin, dimethyl coprogen, phyllostictine A; predominant monosaccharide utilization and glycoside hydrolase modules	Sun et al. (2023)
Talaromyces sp. *DC2*	*Catharanthus roseus* (stem)	Genome size: 34.58 Mb; contigs: 156; GC content: 46.5%; protein coding genes: 11,131; functional genes identified in diverse BLAST databases: non-redundant (NR; 10,780); the protein family (Pfam; 8735); swiss-prot (8165); gene ontology (GO; 6682); Kyoto encyclopedia of genes and genomes (KEGG; 6509); CAZy (1230)	Quan et al. (2024)

12.4 Role of Fungal Endophytes in Various Sector

12.4.1 Fungal Endophytes in Sustainable Agriculture

Fungal endophytes significantly contribute to plant health by secreting soluble and volatile biochemicals that mitigate the harmful effects of various stressors such as drought, salinity, extreme temperatures, and pathogens (Fernando et al. 2021; Rabbee et al. 2024). These metabolites offer protection to plants against functional damage and present a promising alternative to synthetic pesticides and chemical inputs for sustainable agriculture. In a recent study, Santra and Banerjee (2023) identified a volatile-emitting strain of *Diaporthe* sp. CEL3 from *Chloranthus elatior* leaves. The camphor-scented volatiles produced by CEL3 exhibited strong antifungal activity against pathogens, such as *Monilinia fructicola*, a major cause of cherry fruit rot. The other volatiles, including geraniol, mentha-4,8-diene, trans-ocimenol, and trans-verbenol, were found to disrupt fungal cell membranes, leading to protein and intracellular molecule leakage, highlighting their potential as bio-fumigants for controlling postharvest infections. Sornakili et al. (2020) isolated several antagonistic fungal endophytes from rice leaves, which showed significant inhibitory effects on various plant pathogens. These endophytes produce several metabolites, including lactic acid, acetic acid, octaheptanoic acid, and 2,3-butanediol dinitrate, demonstrating their capacity to suppress the growth of fungal and bacterial pathogens. Similarly, Ma et al. (2023) isolated *Fusarium tricinctum* from Kiwi plants (*Actinidia chinensis*), which produces several antibacterial metabolites such as imidazole alkaloids, fusaritricine, fusaritricines, and enniatins. These metabolites have shown potential as biocontrol agents against kiwifruit canker disease, a significant threat to the kiwifruit industry.

Fungal endophytes also enhance plant defenses by activating plant signaling pathways associated with systemic acquired resistance (SAR) and induced systemic resistance (ISR) (Ansari et al. 2020; Lu et al. 2021; Akram et al. 2023; Kumar et al. 2022). These defense mechanisms help plants fend off biotic stressors. For instance, Farhat et al. (2023) isolated fungal strains, including *Chaetomium*, *Curvularia*, *Rhizoctonia*, *Fusarium*, and *Talaromyces*, which enhanced soybean growth and pathogen resistance. This was achieved by increasing the production of salicylic acid (SA), polyphenols, and other resistance markers. In another reports, endophytic infection in *Achnatherum sibiricum* increased resistance to several pathogens, including species from the orders *Erysiphales* and *Curvularia lunata*. This was achieved by activating jasmonic acid (JA) and SA signaling pathways, which are integral to plant immune responses (Shi et al. 2020, 2024).

12.4.2 Fungal Endophytes in Environmental Remediation

Endophytic fungi possess a remarkable ability to metabolize and degrade various environmental pollutants, including hydrocarbons, heavy metals, and organic compounds, making them valuable agents for bioremediation of contaminated soils and water bodies (Kües 2015). Hydrocarbons, including polycyclic aromatic hydrocarbons (PAHs), are widespread pollutants that disperse through the air, soil, and water, causing significant environmental damage. In soil, PAHs bind to soil particles and affect the pore structure and soil mobility. Similarly, in aquatic sediments, PAHs accumulation disrupt the metabolism of aquatic organisms and cause photo-oxidative stress (Kour et al. 2021). A notable example of fungal degradation of PAHs is *Phomopsis liquidambari*, a rice endophyte that degrades phenanthrene both in vitro and in vivo (Fu et al. 2018, 2022). In a recent study, six endophytic fungi, *Aspergillus niger*, *Penicillium* spp., *Penicillium canescens*, *P. atrovenetum*, *Trichocladium opacum*, and *T. lanuginosus*, isolated from the leaves of *Eichhornia crassipes*, were found to utilize various hydrocarbons, including petrol, diesel, kerosene, and spent engine oil, as growth sources (Wahab et al. 2022).

Xenobiotics encompass a wide variety of synthetic chemicals, including pesticides, synthetic azo dyes, antibiotics, polychlorinated biphenyls, and nitroaromatic compounds. These substances adversely affecting the fertility and transparency of soil and water, and harming both terrestrial and marine ecosystems (Wang et al. 2022; Piwowarska and Kiedrzyńska 2022). Over the past years, several endophytic fungal strains capable of degrading xenobiotics, particularly synthetic pesticides and dyes, have been isolated from various plants. These include *Penicillium oxalicum* obtained from *Artemisia annua* (Tian et al. 2018), *Bjerkandera adusta* from *Sinosenecio oldhamianus* (Gao et al. 2020), *Penicillium megasporum* from *Cupressus torulosa* (Agrawal et al. 2021), and *Trichoderma* species from wild Rubiaceae plants (Escudero-Leyva et al. 2022). In a recent study, *Penicillium janthinellum* LM5 isolated from blueberry pulp was reported for degrading the mixture of textile dyes (Yan et al. 2023).

In addition, endophytes from plants growing in polluted environments have also been shown to degrade toxic compounds and reduce heavy metal bioavailability in soils. In some recent reports, endophytic *Penicillium* species isolated from various plants have been reported to degrade heavy metals such as lead (Pb), aluminum (Al), antimony (Sb), cadmium (Cd), and nickel (Ni) (Shi et al. 2017; Bilal et al. 2020; He et al. 2023a, b). Similarly, *Mucor* sp. CM3, isolated from *Eichhornia crassipes*, has been reported to increase cadmium bioremediation (de Almeida et al. 2023). A recent report has highlighted the role of *Cladosporium* sp. and *Rhyzopycnis vagum* in enhancing lead phytoremediation in *Jatropha curcas* and *Reutealis trisperma* on gold mine tailings (Marfuah et al. 2024).

12.4.3 Fungal Endophytes in Pharmacology

Fungal endophytes are an important source of novel bioactive compounds with therapeutic potential. The discovery of paclitaxel (taxol), a powerful anticancer drug, from endophytes associated with Taxus species, demonstrated their immense value in pharmacology (Rana et al. 2020). Since then, numerous structurally and functionally diverse metabolites have been isolated from fungal endophytes, including alkaloids, peptides, flavonoids, phenolics, and camptothecins. These compounds

Table 12.2 Bioactive compounds isolated from endophytic fungi and their bioactivities

Endophytic fungi	Host plant species	Compound	Bioactivity	References
Xylaria feejeensis	*Geophila repens*	6,7-Didehydrointegric acid, 13-carboxyintegric acid	Antibacterial	Rajendran et al. (2023)
Penicillum species	*Taxus wellichiana*	Quercetin	Antibacterial	Adhikari and Pandey (2024)
Paramyrothecium rodidum	*Gynochthotes officinalis*	Pararoine-A	Antitumor	Chen et al. (2024)
Diaporthe species BMX12	*Aquilaria sinesis*	Diaporchalasins A-E	Cytotoxicity and anticancer	Yang et al. (2024)
Curvalaria verruculosa	*Catharanthus roseus*	Vinblastine	Antitumor	Parthasarathy et al. (2020)
Nigrospora zimmermanii	*Catharanthus roseus*	Vincristine	Anticancer	Birat et al. (2022)
Colletotrichum species JS-0367	*Morus alba*	1,3-Dihydroxy-2,8-dimethoxy-6-methylanthraquinone	Anti-inflammatory	Lee et al. (2021)
Shiraia species	*Mosa bamboo*	Hypocrellin-B	Antifungal, anticancer, antiviral	Lu et al. (2024)
Paecilomyces variotii	*Lawsonia alba lam.*	Lawsozaheer, viriditoxin, stigmasta-4,6,8 (14), 22-tetraen-3-one, beta- sitosterol, stigmasterol	Antibacterial	Abbas et al. (2021)
Taxomyces andreanae	*Taxus brevifolia*	Taxol	Anticancer	Cheng et al. (2022)
Burkholderia gladioli	*Lycoris aurea*	Taxoflavin	Antifungal	Li et al. (2019)
Aspergillus terreus	*Catharanthus roseus*	Camptothecin	Antiproliferative	El-Sayed et al. (2024)
Penicillum species HS-11	*Huperzia serrata*	Sorbicatechols E-G	Antibacterial	Ying et al. (2024)
Neurospora terricola HDF-Br-2	*Psedotsuga gaussenii*	Terricoxanthone A-E	Antibacterial and antibiofilm	Chen et al. (2024)
Paecilomyces species JN227071.1	*Moringa olifera*	Cis-13-octadecenoic acid, 1-heptacosanol, 1-Nonadecene, cycloteracosane	Antifungal	Hawar et al. (2023)
Coriolopsis rigida	*Cochlospermum regium*	Tyrosol and p-hydroxyphenylacetamide	Antioxidant	Dantas et al. (2023)

(continued)

Table 12.2 (continued)

Endophytic fungi	Host plant species	Compound	Bioactivity	References
Dimorphiseta acuta	*Gynura japonica*	Emodin	Antibacterial and antioxidant	Riga et al. (2024)
Nigrospora zimmermanii FPD-13	*Poincianella pluviosa*	Nigrosporolide, tyrosol, Decarestrictine-A	Anti-inflammatory	Ferreira et al. (2024)
Aspergillus niger	*Encephalartos whitelockii*	Taxol	Antiproliferative	Gamal et al. (2024)
Fusarium proliferatum	*Dysosma versipellis*	Podophyllotoxin	Anticancer and cytotoxin	Nguyen et al. (2023)
Penicillium polonicum	*Piper nigrum*	Polonilignan	Inhibitor of NO production	Pal et al. (2023)
Colletotrichum iris	*Euphobia lathylris*	Colletotrins A-C	Anticancer	Chen et al. (2023)
Aspergillus fumigatus starin KF436391	*Monarda citriodora*	2-Methlenedecanedioic acid, ergosterol, ditryptophenaline, gliotoxin, pseurotin A brevianamide F, neofipiperazine C, 6-methoxyspirotryprostatin B, Cephalimysin A	Anticancer	Gupta et al. (2024)
Quambalaria cyanescens	*Vitis vinifera*	Resveratrol	Cytotoxic and antioxidant	Meshram et al. (2022)
Dematiopleospora species	*Xanthium spinosum*	Tyrosol, sterigmatocystin, 1-hydroxy-3-methylxanthane, carpesterol, mellein	Antioxidant and phytohormone	Kuchkarova et al. (2024)
Penicillium chrysogenum MT-40	*Huperzia serrata*	Peniciterpenolide A-D, Atlantinae B	Inhibitor of NO production	Jia et al. (2023)
Penicillium sp. *MNP-HS-2*	*Macrozamia communis*	Alternatain E, Mycophenolene-A	Antibacterial and cytotoxic	Elnaggar et al. (2024)
Chaetomium globosum	*Dillenia indica L*	Butanoic acid, 2-methyl-; hexadecanoic acid; 2-propanone, 1-phenyl-; 5-oxo-pyrrolidine-2-carboxylic acid methyl ester; 9,12-octadecadienoic acid	Antioxidant	Kumar and Prasher (2024)

Penicillium sp. *BJR-P2*	*Avicennia marinav*	Penicillol A (**1**), penicillol B (**2**), and citreoviridin H (**3**), and citreoviridin I (**4**)	Anti-inflammatory	Chen et al. (2022)
Cercospora piaropi	*Rosa damascena Mill*	Fulvic acid (F.A.) and anhydrofulvic acid (AFA)	Antimicrobial	Bashir et al. (2023)
Serendipita indicae DSM11827	*Tomato plant*	Viridifloral	Antifungal	Ntana et al. (2021)
Aspergillus sp. *Gbtc-122*	*Ginkgo biloba*	Polyphenols, terpenoids, flavonoides, isoquinoline alkaloids	Pharmaceutical properties	Zou et al. (2021)
Aspergillus flavipes	*Eucommia ulmoides*	3,4-Dihydroxybenzene acetic acid, 3,4-DHBAA methyl ester	Antioxidant	Liu et al. (2021)
Aspergillus aculeatus	*Rosa damascena*	Secalonic acid F-7	Anticancer	Farooq et al. (2020)

exhibit a broad spectrum of pharmacological properties, including antibiotic, antifungal, anti-inflammatory, antiviral, anticancer, and antioxidant activities (Table 12.2).

12.5 Conclusion and Future Perspectives

Recent advancements in high-throughput omics technologies have greatly facilitated functional annotation and characterization of fungal endophytes and revealing numerous novel genes, pathways, and enzymes involved in symbiosis, stress tolerance, growth promotion, and biocontrol. These findings also emphasize the considerable potential of fungal endophytes for application in agriculture, pharmaceuticals, and various industries. However, the vast genetic diversity and complexity of fungal genomes continue to present challenges for precise gene annotation. To address this, the development of more comprehensive fungal databases and advanced bioinformatics tools is essential for enhancing the accuracy of gene function predictions. Moreover, the functional roles of many endophytes, particularly those of non-model plants and underexplored ecosystems, remain poorly understood. Future research efforts should focus on integrating multi-omics approaches to better understand the complexities of fungal biology. Additionally, leveraging machine learning algorithms is pivotal in the interpretation of large datasets, potentially leading to the discovery of novel functional traits that can further expand the biotechnological applications of fungal endophytes.

Acknowledgments PY and CY thankfully acknowledge the CSIR-UGC for funding in the form of JRF vide NTA reference number 211610030312 and 221610048140, respectively.

Declaration The figure is original and drawn by the authors.

References

Abbas Z, Siddiqui BS, Shahzad S, Sattar S, Begum S, Batool A, Choudhary MI (2021) Lawsozaheer, a new chromone produced by an endophytic fungus *Paecilomyces variotii* isolated from *Lawsonia alba* Lam. inhibits the growth of Staphylococcus aureus. Nat Prod Res 35(22):4448–4453

Adeleke BS, Muller D, Babalola OO (2023) A metagenomic lens into endosphere microbial communities, promises, and discoveries. Lett Appl Microbiol 76(2):ovac030

Adhikari P, Pandey A (2024) Phytochemicals, and antioxidants and antibacterial activities of *Taxus wallichiana* Zucc. root associated endophytic fungi. Biologia 79:1895–1907

Agrawal PK, Upadhyay P, Shrivastava R, Sharma S, Garlapati VK (2021) Evaluation of the ability of endophytic fungi from *Cupressus torulosa* to decolorize synthetic textile dyes. J Hazard Toxic Radioact Waste 25(1):06020005

Akram S, Ahmed A, He P, He P, Liu Y, Wu Y, Munir S, He Y (2023) Uniting the role of endophytic fungi against plant pathogens and their interaction. J Fungi 9(1):72

André BLO, Montoya QV, Martiarena MJS, Rodrigues A (2024) Culture-dependent methods revel the diversity of endophytic fungi of *Psidium cattleianum* leaves (Myrtales: Myrtaceae). Braz J Microbiol 55:1–12

Ansari WA, Krishna R, Zeyad MT, Singh S, Yadav A (2020) Endophytic actinomycetes-mediated modulation of defense and systemic resistance confers host plant fitness under biotic stress conditions. In: Microbial versatility in varied environments: microbes in sensitive environments, pp 167–180

Arora J, Ramawat KG (2017) An introduction to endophytes. In: Maheswari DK et al (eds) Endophytes: biology and biotechnology. Springer International Publishing AG, pp 1–23. https://doi.org/10.1007/978-3-319-66541-2_1

Bashir A, Ahmad T, Farooq S, Lone WI, Manzoor MM, Nalli Y, Sultan P, Chaubey A, Ali A, Riyaz-Ul-Hassan S (2023) A secondary metabolite of *Cercospora* sp., associated with *Rosa damascena* Mill., inhibits proliferation, biofilm production, ergosterol synthesis and other virulence factors in *Candida albicans*. Microb Ecol 85(4):1276–1287

Bhupenchandra I, Chongtham SK, Devi AG, Dutta P, Sahoo MR, Mohanty S, Kumar S, Choudhary AK, Devi EL, Sinyorita S, Devi SH, Mahanta M, Kumari A, Devi HL, Josmee RK, Pusparani A, Pathaw N, Gupta S, Meena M, Janmeda P, Seth CS, Arora A, Sahoo A, Swapnil P (2024) Unlocking the potential of arbuscular mycorrhizal fungi: exploring role in plant growth promotion, nutrient uptake mechanisms, biotic stress alleviation, and sustaining agricultural production systems. J Plant Growth Regul 43(9):1–39

Bilal S, Shahzad R, Imran M, Jan R, Kim KM, Lee IJ (2020) Synergistic association of endophytic fungi enhances *Glycine max* L. resilience to combined abiotic stresses: heavy metals, high temperature and drought stress. Ind Crop Prod 143:111931

Birat K, Binsuwaidan R, Siddiqi TO, Mir SR, Alshammari N, Adnan M, Nazir R, Ejaz B, Malik MQ, Dewangan RP, Ashraf SA, Panda BP (2022) Report on Vincristine-Producing Endophytic Fungus Nigrospora zimmermanii from Leaves of Catharanthus roseus. Metabolites 12(11):1119. https://doi.org/10.3390/metabo12111119. PMID: 36422259; PMCID: PMC9697034.

Burragoni SG, Jeon J (2021) Applications of endophytic microbes in agriculture, biotechnology, medicine, and beyond. Microbiol Res 245:126691

Chandran H, Meena M, Swapnil P (2021) Plant growth-promoting rhizobacteria as a green alternative for sustainable agriculture. Sustain For 13(19):10986

Chen C, Ye G, Tang J, Li J, Liu W, Wu L, Long Y (2022) New polyketides from mangrove endophytic fungus *Penicillium* sp. BJR-P2 and their anti-inflammatory activity. Mar Drugs 20(9):583

Chen N, Qiu SY, Lu SM, Chen SW, Wei JR, Yang FW, Geng HC, Zhou M (2023) Colletotrins AC, three new sesquiterpenes from the endophytic fungus *Colletotrichum iris* strain S11. Phytochem Lett 55:34–37

Chen HW, Wu XY, Zhao ZY, Huang ZQ, Lei XS, Yang GX, Li J, Xiong J, Hu JF (2024) Terricoxanthones A–E, unprecedented dihydropyran-containing dimeric xanthones from the endophytic fungus *Neurospora terricola* HDF-Br-2 associated with the vulnerable conifer *Pseudotsuga gaussenii*. Phytochemistry 219:113963

Cheng JT, Cao F, Chen XA, Li YQ, Mao XM (2020) Genomic and transcriptomic survey of an endophytic fungus *Calcarisporium arbuscula* NRRL 3705 and potential overview of its secondary metabolites. BMC Genomics 21:1–3

Cheng T, Kolařík M, Quijada L, Stadler M (2022) A re-assessment of Taxomyces andreanae, the alleged taxol-producing fungus, using comparative genomics. IMA fungus 13(1):17

Corrêa CL, Midorikawa GE, Filho EXF, Noronha EF, Alves GS, Togawa RC, Silva-Junior OB, Costa MMDC, Grynberg P, Miller RN (2020) Transcriptome profiling-based analysis of carbohydrate-active enzymes in *Aspergillus terreus* involved in plant biomass degradation. Front Bioeng Biotechnol 8:564527

Dantas SBS, Moraes GKA, Araujo AR, Chapla VM (2023) Phenolic compounds and bioactive extract produced by endophytic fungus *Coriolopsis rigida*. Nat Prod Res 37(12):2037–2042

de Almeida TT, Tschoeke BAP, Quecine MC, Tezzoto T, Gaziola SA, Azevedo RA, Piotto FA, Orlandelli RC, Dourado MN, Azevedo JL (2023) Mechanisms of Mucor sp. CM3 isolated from the aquatic macrophyte *Eichhornia crassipes* (Mart.) Solms to increase cadmium bioremediation. Environ Sci Pollut Res 30(41):93846–93861

Dos Reis JBA, Lorenzi AS, do Vale HMM (2022) Methods used for the study of endophytic fungi: a review on methodologies and challenges, and associated tips. Arch Microbiol 204(11):675

El-Hasan A, Ngatia G, Link TI, Voegele RT (2022) Isolation, identification, and biocontrol potential of root fungal endophytes associated with solanaceous plants against potato late blight (*Phytophthora infestans*). Plan Theory 11(12):1605

Elnaggar MS, Ibrahim N, Elissawy AM, Anwar A, Ibrahim MA, Ebada SS (2024) Cytotoxic and antimicrobial mycophenolic acid derivatives from an endophytic fungus *Penicillium* sp. MNP–HS–2 associated with *Macrozamia communis*. Phytochemistry 217:113901

El-Sayed AS, ElSayed AI, Wadan KM, El-Saadany SS, Abd El-Hady NA (2024) Camptothecin bioprocessing from *Aspergillus terreus*, an endophyte of *Catharanthus roseus*: antiproliferative activity, topoisomerase inhibition and cell cycle analysis. Microb Cell Factories 23(1):15

Escudero-Leyva E, Alfaro-Vargas P, Muñoz-Arrieta R, Charpentier-Alfaro C, Granados-Montero MDM, Valverde-Madrigal KS, Pérez-Villanueva M, Méndez-Rivera M, Rodríguez-Rodríguez CE, Chaverri P, Mora-Villalobos JA (2022) Tolerance and biological removal of fungicides by *Trichoderma* species isolated from the endosphere of wild Rubiaceae plants. Front Agron 3:772170

Farhat H, Urooj F, Sohail N, Ullah S, Ali MS (2023) Plant disease management with amelioration of systemic resistance in soybean by endophytic fungi associated with GC-MS metabolic profiling of *Chaetomium* sp. Physiol Mol Plant Pathol 126:102049

Farooq S, Qayum A, Nalli Y, Lauro G, Chini MG, Bifulco G, Chaubey A, Singh SK, Riyaz-Ul-Hassan S, Ali A (2020) Discovery of a secalonic acid derivative from *Aspergillus aculeatus*, an endophyte of *Rosa damascena* Mill., triggers apoptosis in MDA-MB-231 triple negative breast cancer cells. ACS Omega 5(38):24296–24310

Fernando K, Reddy P, Spangenberg GC, Rochfort SJ, Guthridge KM (2021) Metabolic potential of Epichloë endophytes for host grass fungal disease resistance. Microorganisms 10(1):64

Ferreira MS, Katchborian-Neto A, Nicácio KDJ, Santos MF, Dias DF, Chagas-Paula DA, Soares MG (2024) Phytochemical investigation of *Nigrospora zimmermanii* isolated from *Poincianella pluviosa* (Sibipiruna): metabolites characterisation and screening for anti-inflammatory activity. Nat Prod Res:1–7. https://doi.org/10.1080/14786419.2024.2320733

Fu W, Xu M, Sun K, Hu L, Cao W, Dai C, Jia Y (2018) Biodegradation of phenanthrene by endophytic fungus *Phomopsis liquidambari* in vitro and in vivo. Chemosphere 203:160–169

Fu WQ, Xu M, Zhang AY, Sun K, Dai CC, Jia Y (2022) Remediation of phenanthrene phytotoxicity by the interaction of rice and endophytic fungus *P. liquidambaris* in practice. Ecotoxicol Environ Saf 235:113415

Gamal A, Fikry E, Tawfeek N, El-Shafae AM, El-Sayed AS, El-Domiaty MM (2024) Production and bioprocessing of Taxol from *Aspergillus niger*, an endophyte of *Encephalartos whitelockii*, with a plausible biosynthetic stability: antiproliferative activity and cell cycle analysis. Microb Cell Factories 23(1):78

Gao T, Qin D, Zuo S, Peng Y, Xu J, Yu B, Song H, Dong J (2020) Decolorization and detoxification of triphenylmethane dyes by isolated endophytic fungus, *Bjerkandera adusta* SWUSI4 under non-nutritive conditions. Bioresour Bioprocess 7:1–12

García-Latorre C, Rodrigo S, Santamaría O (2021) Endophytes as plant nutrient uptake-promoter in plants. In: Maheshwari DK, Dheeman S (eds) Endophytes: Mineral Nutrient Management, Sustainable development and biodiversity, vol 26, vol 3. Springer, Cham, pp 247–265

Gupta N, Singh G, Qayum A, Ovais Dar M, Singh S, Katoch M, Sangwan PL (2024) In vitro and in silico anticancer evaluation of secondary metabolites from an Endophytic fungus *Aspergillus fumigatus* isolated from *Monarda citriodora*. ChemistrySelect 9(13):e202400637

Hawar SN, Taha ZK, Hamied AS, Al-Shmgani HS, Sulaiman GM, Elsilk SE (2023) Antifungal activity of bioactive compounds produced by the endophytic fungus *Paecilomyces* sp. (JN227071. 1) against *Rhizoctonia solani*. Int J Biomater 2023(1):2411555

He T, Li X, Iacovelli R, Hackl T, Haslinger K (2023a) Genomic and metabolomic analysis of the endophytic fungus *Fusarium* sp. VM-40 isolated from the medicinal plant Vinca minor. J Fungi 9(7):704

He Y, Li C, Sun Z, Zhang W, He J, Zhao Y, Xu Z, Zhao W (2023b) Penicillium spp. XK10, fungi with potential to repair cadmium and antimony pollution. Appl Sci 13(3):1228

Huang X, Zeng Z, Chen Z, Yang Y, Pang J, Qian Y, Xiang T (2022) Whole-genome sequence resource of *Fusarium oxysporum* strain TH15, a plant growth promoting endophytic fungus isolated from *Tetrastigma hemsleyanum*. PhytoFrontiers™ 2(3):314–319
Jha P, Kaur T, Chhabra I, Panja A, Paul S, Kumar V, Malik T (2023) Endophytic fungi: hidden treasure chest of antimicrobial metabolites interrelationship of endophytes and metabolites. Front Microbiol 14:1227830
Jia F, Zhang M, Fan J, Wang Y, Ge X, Mi X, Liu X, Wang J, Tu PF, Shi SP (2023) Terpenoids from *Penicillium chrysogenum* MT-40, an endophytic fungus isolated from *Huperzia serrata*. Org Chem Front 10(18):4598–4605
Joshi A, Verma KK, Rajput VD, Arora J (2023) Metabolic engineering of microorganisms in advancing biofuel production. In: Moodley P, Ray RC, Gueguim Kana EB (eds) Advances in Lignocellulosic biofuel production systems. Elsevier Inc., Woodhead Publishing, pp 181–202
Khalil AMA, Hassan SED, Alsharif SM, Eid AM, Ewais EED, Azab E, Gobouri AA, Elkelish A, Fouda A (2021) Isolation and characterization of fungal endophytes isolated from medicinal plant *Ephedra pachyclada* as plant growth-promoting. Biomol Ther 11(2):140
Kour D, Kaur T, Devi R, Yadav A, Singh M, Joshi D, Singh J, Suyal DC, Kumar A, Rajput VD, Yadav AN (2021) Beneficial microbiomes for bioremediation of diverse contaminated environments for environmental sustainability: present status and future challenges. Environ Sci Pollut Res 28:24917–24939
Kuchkarova N, Lei L, Luo S, Toshmatov Z, Han C, Zhou S, Zhang C, Shao H (2024) Plant growth regulatory activity of secondary metabolites produced by the invasive *Xanthium spinosum's* Endophytic *Dematiopleospora* sp. J Plant Growth Regul 43:1–12
Kües U (2015) Fungal enzymes for environmental management. Curr Opin Biotechnol 33:268–278
Kumar V, Prasher IB (2024) Phytochemical analysis and antioxidant activity of Endophytic fungi isolated from *Dillenia indica* Linn. Appl Biochem Biotechnol 196(1):332–349
Kumar R, Swapnil P, Meena M, Selpair S, Yadav BG (2022) Plant growth-promoting rhizobacteria (PGPR): approaches to alleviate abiotic stresses for enhancement of growth and development of medicinal plants. Sustain For 14(23):15514
Lee S, Nguyen QN, Phung HM, Shim SH, Kim D, Hwang GS, Kang KS (2021) Preventive effects of anthraquinones isolated from an endophytic fungus, *Colletotrichum* sp. JS-0367 in tumor necrosis factor-α-stimulated damage of human dermal fibroblasts. Antioxidants 10(2):200
Li X, Li Y, Wang R, Wang Q, Lu L (2019) Toxoflavin produced by *Burkholderia gladioli* from *Lycoris aurea* is a new broad-spectrum fungicide. Appl Environ Microbiol 85(9):e00106–e00119
Liu W, Liu Y, Yang F, Han S, Zhang J, Yang H, Cheng Z, Li Q (2021) Asperflaloids A and B from *Aspergillus flavipes* DZ-3, an Endophytic Fungus of *Eucommia ulmoides* Oliver. Molecules 26(12):3514
Liu Y, Xu M, Tang Y, Shao Y, Wang H, Zhang H (2022) Genome features and AntiSMASH analysis of an endophytic strain *Fusarium* sp. R1. Metabolites 12(6):521
Liu-Xu L, Vicedo B, García-Agustín P, Llorens E (2022) Advances in endophytic fungi research: a data analysis of 25 years of achievements and challenges. J Plant Interact 17(1):244–266
Lu H, Wei T, Lou H, Shu X, Chen Q (2021) A critical review on communication mechanism within plant-endophytic fungi interactions to cope with biotic and abiotic stresses. J Fungi 7(9):719
Lu ZM, Zhang RT, Huang XB, Cao XT, Shen XY, Fan L, Hou CL (2024) Optimisation of hypocrellin production in Shiraia-like fungi via genetic modification involving a transcription factor gene and a putative monooxygenase gene. Mycology:1–10. https://doi.org/10.1080/21501203.2023.2295406
Ma JT, Dong XY, Li ZH, Yan H, He J, Liu JK, Feng T (2023) Antibacterial metabolites from kiwi endophytic fungus *Fusarium tricinctum*, a potential biocontrol strain for kiwi canker disease. J Agric Food Chem 71(20):7679–7688
Marfuah DS, Hamim H, Sulistyaningsih YC, Surono S, Setyaningsih L, Saprudin D (2024) Dark septate endophyte inoculation improved Pb phytoremediation of *Jatropha curcas* and *Reutealis trisperma* on gold mine tailings. Biorem J 28(3):325–342

Meshram V, Kapoor N, Dwibedi V, Srivastava A, Saxena S (2022) Extracellular resveratrol producing endophytic fungus, *Quambalaria cyanescens*. S Afr J Bot 146:409–416
Ming Q, Huang X, He Y, Qin L, Tang Y, Liu Y, Huang Y, Zhang H, Li P (2023) Genome mining and screening for secondary metabolite production in the Endophytic fungus *Dactylonectria alcacerensis* CT-6. Microorganisms 11(4):968
Mishra Y, Sharma L, Dhiman M, Sharma MM (2021) Endophytic fungal diversity of selected medicinal plants and their bio-potential applications. In: Sharma VK, Shah MP, Parmar S, Kumar A (eds) Fungi bio-prospects in sustainable agriculture, environment and nano-technology. Academic/Elsevier Inc, pp 227–283
Nandy S, Das T, Tudu CK, Pandey DK, Dey A, Ray P (2020) Fungal endophytes: futuristic tool in recent research area of phytoremediation. S Afr J Bot 134:285–295
Nguyen GT, Nguyen HTH, Tran HT, Tran HT, Ho AN, Tran QH, Pham NB (2023) Enhanced podophyllotoxin production of endophyte *Fusarium proliferatum* TQN5T by host extract and phenylalanine. Appl Microbiol Biotechnol 107(17):5367–5378
Ntana F, Bhat WW, Johnson SR, Jørgensen HJ, Collinge DB, Jensen B, Hamberger B (2021) A sesquiterpene synthase from the endophytic fungus *Serendipita indica* catalyzes formation of viridiflorol. Biomol Ther 11(6):898
Oberhofer M, Malfent F, Zehl M, Urban E, Wackerlig J, Reznicek G, Vignolle GA, Rückert C, Busche T, Wibberg D, Zotchev SB (2022) Biosynthetic potential of the endophytic fungus *Helotiales* sp. BL73 revealed via compound identification and genome mining. Appl Environ Microbiol 88(6):e02510–e02521
Oita S, Ibáñez A, Lutzoni F, Miadlikowska J, Geml J, Lewis LA, Hom EF, Carbone I, U'Ren JM, Arnold AE (2021) Climate and seasonality drive the richness and composition of tropical fungal endophytes at a landscape scale. Commun Biol 4(1):313
Pal PP, Begum SA, Basha AS, Araya H, Fujimoto Y (2023) A new lignan (Polonilignan) and inhibitors of nitric oxide production from *Penicillium polonicum*, an endophytic fungi of *Piper nigrum*. Chem Biodivers 20(3):e202200840
Palwe SD, Borde MY, Sonawane HB (2021) Endophytic fungi: a source of potential anticancer compounds. Stud Fungi 6(1):188–203
Parthasarathy R, Shanmuganathan R, Pugazhendhi A (2020) Vinblastine production by the endophytic fungus *Curvularia verruculosa* from the leaves of *Catharanthus roseus* and its in vitro cytotoxicity against HeLa cell line. Anal Biochem 593:113530
Patchett A, Newman JA (2021) Comparison of plant metabolites in root exudates of *Lolium perenne* infected with different strains of the fungal endophyte *Epichloë festucae* Var. Lolii. J Fungi 7:148
Piwowarska D, Kiedrzyńska E (2022) Xenobiotics as a contemporary threat to surface waters. Ecohydrol Hydrobiol 22(2):337–354
Poveda J, Eugui D, Abril-Urías P, Velasco P (2021) Endophytic fungi as direct plant growth promoters for sustainable agricultural production. Symbiosis 85:1–19
Quan ND, Nguyen NL, Giang TTH, Ngan NTT, Hien NT, Tung NV, Trang NHT, Lien NTK, Nguyen HH (2024) Genome characteristics of the endophytic fungus *Talaromyces* sp. DC2 isolated from Catharanthus roseus (L.) G. Don. J Fungi 10(5):352
Rabbee MF, Ali MS, Islam MN, Rahman MM, Hasan MM, Baek KH (2024) Endophyte mediated biocontrol mechanisms of phytopathogens in agriculture. Res Microbiol 175:104229
Rajendran S, Robertson LP, Kosgahakumbura L, Fernando C, Göransson U, Wang H, Hettiarachchi C, Gunasekera S (2023) Antibacterial eremophilane sesquiterpenoids from *Xylaria feejeensis*, an endophytic fungi of the medicinal plant *Geophila repens*. Fitoterapia 167:105496
Rana KL, Kour D, Kaur T, Devi R, Negi C, Yadav AN, Yadav N, Singh K, Saxena AK (2020) Endophytic fungi from medicinal plants: biodiversity and biotechnological applications. In: Microbial endophytes. Woodhead Publishing, pp 273–305
Reed CJ, Denise R, Hourihan J, Babor J, Jaroch M, Martinelli M, Hutinet G, de Crécy-Lagard V (2024) Beyond blast: enabling microbiologists to better extract literature, taxonomic distributions and gene neighbourhood information for protein families. Microb Genomics 10(2):001183

Ren X, Shan Y, Li X, Fan J, Li Y, Ma L, Wang L, Li X (2022) Application of RNA sequencing to understand the benefits of endophytes in the salt-alkaline resistance of rice seedlings. Environ Exp Bot 196:104820

Riga R, Wardatillah R, Suryani O, Ryplida B, Suryelita S, Azhar M, Handayani D, Artasasta MA, Benu SM, Putra A (2024) Endophytic fungus from *Gynura japonica*: phytochemical screening, biological activities, and characterisation of its bioactive compound. Nat Prod Res 26:1–9

Rodriguez RJ, Henson J, Van Volkenburgh E, Hoy M, Wright L, Beckwith F et al (2008) Stress tolerance in plants via habitat-adapted symbiosis. ISME J 2:404–416

Sagita R, Quax WJ, Haslinger K (2021) Current state and future directions of genetics and genomics of endophytic fungi for bioprospecting efforts. Front Bioeng Biotechnol 9:649906

Sahu PK, Tilgam J, Mishra S, Hamid S, Gupta A, Jayalakshmi K, Verma SK, Kharwar RN (2022) Surface sterilization for isolation of endophytes: ensuring what (not) to grow. J Basic Microbiol 62(6):647–668

Saini P, Verma A, Tiwari H, Mishra V, Gautam V (2024) Omics-based approaches in studying fungal endophytes and their associated secondary metabolites. In: Singh BP, Abdel-Azeem AM, Gautam V, Singh G, Singh SK (eds) Endophytic fungi. Fungal biology. Springer, Cham, pp 209–227

Santra HK, Banerjee D (2023) Antifungal activity of volatile and non-volatile metabolites of endophytes of *Chloranthus elatior* Sw. Front Plant Sci 14:1156323

Shah SP, Chunduri JR (2023) Genome-wide analysis and in silico screening of secondary metabolite potential of endophytic fungi *Fusarium multiceps* BPAL1 obtained in Mumbai, India. Egypt J Basic Appl Sci 10(1):812–823

Shi Y, Xie H, Cao L, Zhang R, Xu Z, Wang Z, Deng Z (2017) Effects of Cd-and Pb-resistant endophytic fungi on growth and phytoextraction of *Brassica napus* in metal-contaminated soils. Environ Sci Pollut Res 24:417–426

Shi X, Qin T, Liu H, Wu M, Li J, Shi Y, Gao Y, Ren A (2020) Endophytic fungi activated similar defense strategies of *Achnatherum sibiricum* host to different trophic types of pathogens. Front Microbiol 11:1607

Shi X, Zhang J, Qin T, Liu J, Gao Y, Ren A (2024) Endophytes infection increased the disease resistance of host *Achnatherum sibiricum* and non-symbiotic neighbours to pathogenic fungi. Environ Exp Bot 227:105952

Singh S, Kumar K, Rao A, Prajapati VK (2024) Bioinformatics approaches in studying the fungal endophyte-derived bioactive compounds with pharmacological relevance. In: Singh BP, Abdel-Azeem AM, Gautam V, Singh G, Singh SK (eds) Endophytic fungi. Fungal biology. Springer, Cham, pp 191–208

Sohrabi M, Samsampour D, Bagheri A (2024) Molecular identification of fungal endophytes of medicinal plant *Citrullus colocynthis* (L.) schrad as a medicinal plant: role of tissue type and sampling location on the diversity. Mol Biotechnol 66(3):424–431

Sornakili A, Thankappan S, Sridharan AP, Nithya P, Uthandi S (2020) Antagonistic fungal endophytes and their metabolite-mediated interactions against phytopathogens in rice. Physiol Mol Plant Pathol 112:101525

Sun W, Feng M, Zhu N, Leng F, Yang M, Wang Y (2023) Genomic characteristics and comparative genomics analysis of the endophytic fungus *Paraphoma chrysanthemicola* DS-84 isolated from *Codonopsis pilosula* root. J Fungi 9(10):1022

Tao J, Bai X, Zeng M, Li M, Hu Z, Hua Y, Zhang H (2022) Whole-genome sequence analysis of an endophytic fungus *Alternaria* sp. SPS-2 and its biosynthetic potential of bioactive secondary metabolites. Microorganisms 10(9):1789

Thangavel M, Mani I, Surendrababu A, Pandi M (2023) Diversity of endophytic mycobiota through metagenomic approach and bioprospecting the phytoconstituent. Adv Chemicobiol Res 2:1–12

Tian H, Ma YJ, Li WY, Wang JW (2018) Efficient degradation of triclosan by an endophytic fungus *Penicillium oxalicum* B4. Environ Sci Pollut Res 25:8963–8975

Tibpromma S, Karunarathna SC, Bhat JD, Suwannarach N, Stephenson SL, Elgorban AM, Al-Rejaie S, Xu J, Mortimer PE (2022) Using culture-dependent and molecular techniques

to identify endophytic fungi associated with tea leaves (*Camellia* spp.) in Yunnan Province, China. Diversity 14(4):287
Venieraki A, Dimou M, Katinakis P (2017) Endophytic fungi residing in medicinal plants have the ability to produce the same or similar pharmacologically active secondary metabolites as their hosts. Hell Plant Protect J 10:51–66
Verma V, Srivastava A, Garg SK, Singh VP, Arora PK (2023) Incorporating omics-based tools into endophytic fungal research. Biotechnol Notes 5:1–7
Verma KK, Joshi A, Song XP, Liang Q, Xu L, Huang HR, Wu KC, Seth CS, Arora J, Li YR (2024a) Regulatory mechanisms of plant rhizobacteria on plants to the adaptation of adverse agroclimatic variables. Front Plant Sci 15:1377793
Verma KK, Joshi A, Song XP, Singh S, Kumari A, Arora J, Singh SK, Solanki MK, Seth CS, Li YR (2024b) Synergistic interactions of nanoparticles and plant growth promoting rhizobacteria enhancing soil-plant systems: a multigenerational perspective. Front Plant Sci 15:1376214
Wahab AA, Adeyemi FM, Dare AP, Ozabor PT, Badmus MK, Ogunlana OE, Akinde SB (2022) Assessment of total petroleum hydrocarbon degradation by endophytic fungi isolated from water hyacinth (*Eichhornia crassipes*). UNIOSUN J Eng Environ Sci 4(1):164–171
Wang B, Yu H, Jia Y, Dong Q, Steinberg C, Alabouvette C, Edel-Hermann V, Kistler HC, Ye K, Ma LJ, Guo L (2020) Chromosome-scale genome assembly of *Fusarium oxysporum* strain Fo47, a fungal endophyte and biocontrol agent. Mol Plant-Microbe Interact 33(9):1108–1111
Wang X, Sial MU, Bashir MA, Bilal M, Raza QUA, Ali Raza HM, Rehim A, Geng Y (2022) Pesticides xenobiotics in soil ecosystem and their remediation approaches. Sustainability 14(6):3353
Wei J, Chen F, Liu Y, Abudoukerimu A, Zheng Q, Zhang X, Yimiti D (2020) Comparative metabolomics revealed the potential antitumor characteristics of four endophytic fungi of *Brassica rapa* L. ACS Omega 5:5939–5950
Wu X, Zou K, Liu X, Fu S, Zhang S, Duan Z, Zhou J, Liang Y (2022) The novel distribution of intracellular and extracellular flavonoids produced by *Aspergillus* sp. Gbtc 2, an endophytic fungus from *Ginkgo biloba* root. Front Microbiol 13:972294
Wu X, Ibrahim N, Liang Y, Liu X (2024) Screening and genomic analysis of alkaloid-producing endophytic fungus *Fusarium solani* Strain MC503 from *Macleaya cordata*. Microorganisms 12(6):1088
Yadav G, Meena M (2021) Bioprospecting of endophytes in medicinal plants of Thar Desert: An attractive resource for biopharmaceuticals. Biotechnol Rep 30:e00629
Yadav G, Meena M (2022) Biological control of plant diseases by endophytes. In: Shah MP, Deka D (eds) Endophytic association: what, why and how (Volume: Developments in applied microbiology and biotechnology). Academic, Elsevier, pp 119–135
Yadav AN, Kour D, Kaur T, Devi R, Yadav N (2020) Functional annotation of agriculturally important fungi for crop protection: current research and future challenges. In: Yadav A, Mishra S, Kour D, Yadav N, Kumar A (eds) Agriculturally important fungi for sustainable agriculture. Fungal biology. Springer, Cham, pp 347–356
Yadav AN, Kour D, Kaur T, Devi R, Yadav A (2022) Endophytic fungal communities and their biotechnological implications for agro-environmental sustainability. Folia Microbiologica 67(2):203–232
Yan J, Wang P, Wang L, Jin Q, Ali AS, He Y, Wang Y, Sun Y, Li A, Adwy W, Ahmed RH (2023) Bio-decolorization of synthetic dyes by a novel endophytic fungus *Penicillium janthinellum* LM5 from blueberry pulp. Biochem Eng J 195:108909
Yang Y, Liu X, Cai J, Chen Y, Li B, Guo Z, Huang G (2019) Genomic characteristics and comparative genomics analysis of the endophytic fungus *Sarocladium brachiariae*. BMC Genomics 20:1–20
Yang N, Zhang W, Wang D, Cao D, Cao Y, He W, Lin Z, Chen X, Ye G, Chen Z, Chen J (2023) A novel endophytic fungus strain of *Cladosporium*: its identification, genomic analysis, and effects on plant growth. Front Microbiol 14:1287582

Yang L, Wang Q, Ma QY, Xie QY, Gai CJ, Wu YG, Dai HF, Zhao YX (2024) Diaporchalasins A–E, new cytochalasins from the endophytic fungus *Diaporthe* sp. BMX12 Isolated from *Aquilaria sinensis*. Chem Biodivers 21:e202400567

Ying Y, Lei P, Xu Y, Lin Y, Yang N, Han Y, Zhang Z, Shan W, Rao G, Wang J (2024) Secondary metabolites from Penicillium sp. HS-11, a fungal endophyte of *Huperzia serrata*. Fitoterapia 175:105943

Zou K, Liu X, Hu Q, Zhang D, Fu S, Zhang S, Huang H, Lei F, Zhang G, Miao B, Meng D (2021) Root endophytes and *Ginkgo biloba* are likely to share and compensate secondary metabolic processes, and potentially exchange genetic information by LTR-RTs. Front Plant Sci 12:704985

Ecological Symphony: Multi-Omics Study of Environment Dynamics of Endophyte Communities

13

M. Nandhini, G. Roopa, B. S. Kumudini, and H. S. Prakash

Abstract

In natural ecosystems, plants interact with a variety of microorganisms, forming intricate relationships that can influence plant health, growth, and responses to environmental challenges. These interactions include symbiotic, commensal, and pathogenic associations, with microbes such as bacteria, fungi, and actinomycetes playing diverse roles ranging from plant growth promotion to pathogen defense. Among these, endophytes are non-pathogenic microorganisms that reside within plant tissues, providing various benefits including enhanced resistance to environmental stress and pathogens. Research on endophytes focuses on understanding their diversity, ecological roles, and potential biotechnological applications. The diversity of endophyte communities varies by plant species, geographic location, and environmental conditions, and is typically studied using cultivation-dependent and independent techniques. Recent advances in multi-omics approaches, including metagenomics, metatranscriptomics, and metabo-

M. Nandhini (✉)
Department of Lifesciences, School of Sciences, Jain (deemed-to-be University), Bengaluru, Karnataka, India
e-mail: m.nandhini@jainuniversity.ac.in

G. Roopa
Department of Biotechnology, Maharani Science College for Women, Mysuru, Karnataka, India

B. S. Kumudini
Department of Biotechnology and Genetics, School of Sciences, Jain (deemed-to-be University), Bengaluru, Karnataka, India

H. S. Prakash
Department of studies in Biotechnology, University of Mysore, Manasagangotri, Mysuru, Karnataka, India

M. Meena et al. (eds.), *Multi-omics Approach to Investigate Endophyte Diversity*, Microorganisms for Sustainability 53, https://doi.org/10.1007/978-981-96-4100-0_13

lomics, have greatly enhanced our ability to investigate these communities. Additionally, the evolution of endophyte-plant interactions, shaped by vertical and horizontal transmission, reflects the complexity of symbiotic relationships that have developed over time. This chapter explores the dynamics of plant-endophyte interactions, the diversity of endophytes across different plant compartments, and the cutting-edge techniques used to identify and study endophyte communities, highlighting their significance in improving plant health and resilience.

Keywords

Endophyte · Symbiosis · Omics technology · Next-generation sequencing

13.1 Introduction

In the natural ecosystem, plants coexist with a wide variety of microorganisms which form a complex web of interdependent relationships, creating a dynamic interaction that can have both beneficial and harmful effects on the involved organisms. The interactions between plants and these microorganisms are essential in determining plant health, growth, and reactions to environmental factors. Investigating the dynamics of these plant-microbe associations is crucial for thoroughly understanding the complex relationships within the plant ecosystem (Ansari et al. 2024). Bacteria, fungi, and actinomycetes can exist as pathogens, symbionts, including endophytes, plant growth-promoting microorganisms, mycorrhiza, and commensals. The type of relationship between the plant and microbe may be symbiosis, commensalism, mutualism, or saprophytism. The type of lifestyle depends on various strategies like nutritional requirements, environmental conditions, internal host signals, soil microbiome, and microbial interactions (Delaye et al. 2013; Mishra et al. 2021). Endophytes are non-pathogenic microbial communities which exist as endosymbionts in plants and assist them in providing benefits to combat pathogens and extreme environmental conditions. The colonization of endophytes within internal plant tissues has been observed in nearly every plant studied, occurring either in intercellular spaces or intracellularly (Subudhi et al. 2018). Nearly 300,000 plant species on Earth are believed to host one or more endophytes.

Research on endophytic microorganisms primarily focuses on three areas: assessing their diversity, elucidating plant-microbe interactions, and evaluating their biotechnological potentials (Verma et al. 2019). Isolating endophytic microorganisms is a crucial step in exploring their phylogeny, diversity, and interactions with plants. The diversity of endophytes accounts for their evolutionary relatedness, taxonomy, plant hosts, and ecological functions. It is observed that the endophyte community varies in different regions of the plant such as rhizosphere, laimosphere, caulosphere, phyllosphere, anthosphere, carposphere, and spermosphere to the plant endosphere compartment (Ma et al. 2013; Compant et al. 2016). The composition of endophytic communities is influenced by various biotic factors, including the plant species or cultivar, geographical regions, climatic conditions, developmental stage, and interactions with other plant-associated microorganisms and their microenvironment. Despite increasing interest in the ecology, evolution, and applications of endophytes, relatively little is known about their geographic and ecological distributions in most plant communities.

Given that the diversity of endophyte communities is enormous, the techniques required to identify the total microbiome count are still in its infancy stage. Generally the study of endophyte diversity and dynamics is done by either cultivation dependent or cultivation independent methods (Sun and Guo 2012; Wijayawardene et al. 2021). The cultivation dependent method involves the surface sterilization of the plant samples, fragmentation, and plating the plant fragments onto the suitable media for fungal and bacterial isolation (Zinniel et al. 2002; Qadri et al. 2013). The culture independent approaches are carried out through DNA isolation of surface sterilized plant samples and subjecting the DNA for PCR with 16S, 18S, ITS1, ITS2, 5.8S, 28S rDNA followed by DGGE, TGGE, RFLP, and SSCP analysis.

In the past two decades, omics-based approaches and systems biology have emerged to identify endophyte communities in plants. Meta genomics, metatranscriptomics, metaproteomics, and meta-metabolomics can be integrated into systems biology, bioinformatics, and computational tools to identify and study the diversity, dynamics, and interactions between plants and endophytes (Dos Reis et al. 2022). Recent studies on endophyte diversity and dynamics encompass both culture dependent and independent techniques (Wang et al. 2019; Mennicken et al. 2024). The study of total endophyte communities entitles a wide range of microbes which can be explored for their benefits and applications. This chapter discusses the different multi-omics approaches to study the diversity and dynamics of endophyte communities in the plants.

13.2 Evolution of Plant Endophyte Interaction

The study of endophyte and their association with host dates back to 1866 where De Barry coined the term "endophyte" for fungi or bacteria occurring inside plant tissues without causing any apparent symptoms in the host (Wilson 1995). The evolution of endophytes can be through vertical or horizontal transmission. Plant-fungus symbioses are believed to have occurred during the early colonization of land by terrestrial plants. The fungal group Glomeromycota has long been considered the primary candidate for interacting with the first terrestrial plants in the Ordovician era. However, members of the Mucoromycotina are also speculated to have formed symbiotic interactions with these early terrestrial plants (Hardoim et al. 2015a, b). Vertical transmission of the endophytes can be carried out through seed or pollen. Endophytes that are consistently passed down through generations must have a pathway from seeds to reproductive organs, either through xylem vessels or via the shoot apical meristem, which differentiates into reproductive organs (Frank et al. 2017). The association between tall fescue and its endophyte is the best-known example of a vertical transmission and it was found that seed borne *Acremonium* endophyte colonized different species of *Festuca, Lolium, Poa*, and *Stipa* (Clay 1993). Studies of herbarium specimens and living plants from natural populations have revealed that the infection frequency of seed-borne endophytes is often high, reaching 100% in certain species (White Jr and Bultman 1987; Truyens et al. 2015). Endophytes can colonize their host through horizontal transmission via the environment, vertical transmission from parent to offspring, or a combination of both modes. Even stomata act as a transmission route for plants at all life stages, but they may be particularly important for foliar endophytes in trees. The bacterial diversity in seeds and seedlings grown under sterile conditions is usually lower than in plants grown in soil, indicating that most endophytes are acquired from the environment (Hardoim et al. 2012). The evolution of the endophytic lifestyle in fungi is complex, involving multi-species interactions, various levels of causation, and multidirectional flows of influence. It is also shaped by stochastic events, such as abiotic and biotic environmental conditions, which drive the life histories of coevolving fungi and host plants (Saikkonen et al. 2004).

13.3 Dynamics of Plant Endophyte Interaction

Endophytic microbiomes can be classified as either systemic or non-systemic endophytes according to their association with plants. The microbial community consistently associated with a plant genotype was referred to as systemic endophytes also known as conserved microbiota. In contrast, non-systemic endophytes or environment-dependent microbiota, also known as adopted microbiota, have evolved due to various ecological factors (Mishra et al. 2021). Colonization of these endophytes is a very complex process involving the recognition of root exudates, movement toward the plant, adhesion to its surface, and intracellular localization through root penetration. Each of these stages is facilitated by specific signaling molecules, leading to dynamic changes in both the host plant and the endophytes. In the case of non-systemic endophytes, a common symbiotic pathway (CSP) is triggered by symbiotic signals from rhizospheric bacteria or fungi, specifically in the form of lipo-chitooligosaccharides (LCOs). These LCOs are recognized by lysine-motif (LysM) receptors located on the plant cell plasma membrane, which then activate the CSP to manage interactions between the plant and rhizospheric microorganisms. LysM receptor families, found in both legume and non-legume plants, can receive signals from rhizobia (Nod factor signals) and arbuscular mycorrhizal (AM) fungi (Myc-LCO signals) (Kumar et al. 2016). Interaction dynamics involve mutualistic benefits, colonization efficacy, and molecular signaling for abiotic and biotic stress management, enhancing plant resilience through nutrient acquisition, bioactive compound synthesis, and antioxidant enzyme production (Pathak et al. 2022).

Experimental studies indicate that neither a single gene nor a group of genes solely determines endophytic behavior. Genomic studies of the endophyte-plant relationship can provide deeper insights into their adaptations. Comparative genomics has been crucial in uncovering the dynamic interactions and broader perspective of endophyte-plant associations, rather than focusing solely on gene functions (Salwan and Sharma 2023). Comparative genomics and metabolomics studies have shown that non-systemic endophytes possess more genes related to degradation and host invasion, as well as putative novel secretion systems with circadian rhythm regulators. In contrast, systemic endophytes have more biosynthetic genes, including those encoding nitrogenase and ribulose diphosphate carboxylase/oxygenase (RuBisCO), and genes associated with the metabolism of plant stress-related compounds (Zhao et al. 2024).

The colonization of foliar endophytes in *Fraxinus excelsior* is strongly influenced by microclimatic conditions and leaf traits, leading to significant variability in their spatial and temporal distribution within a single tree (Scholtysik et al. 2013). Therefore, regular sampling is crucial to gain a comprehensive understanding of the endophyte community within the plant. With the progress in next-generation sequencing (NGS), techniques like genome sequencing, metagenomics, metatranscriptomics, metabolomics, microarray analysis, and comparative genomics can offer a comprehensive understanding of the endophytic microbial lifestyle.

13.4 Benefits of Plant Endophyte Symbiosis

Endophytic organisms establish a symbiotic relationship with plants, offering various benefits that enhance plant health and productivity (Bacon and White 2016). They offer multifaceted benefits to plants such as enhanced growth and development by enhancing nutrient uptake, produce growth-promoting hormones, and improve overall plant vigor, assist in the solubilization and mobilization of essential nutrients, such as phosphorus and nitrogen, produce phytohormones like auxins, cytokinins, and gibberellins, which stimulate plant growth, improve water use efficiency, and reduce water loss through stomatal regulation (Rodriguez et al. 2008), help plants cope with high salinity levels by modulating ion transport and accumulating osmoprotectants, sequester heavy metals, reducing their toxicity to plants, production of antimicrobial compounds, competition for nutrients and space, and induction of systemic resistance against phytopathogens, improve the nutritional quality of crops by enhancing the uptake and assimilation of minerals and nutrients, and increase biomass of the plants (Richardson et al. 2009).

13.5 Identification of Endophyte Communities

Identifying endophyte communities in plants involves a combination of traditional and modern techniques, including microscopy, culture-based methods, molecular techniques, and advanced sequencing technologies (Table 13.1).

Table 13.1 Different omics technologies used to reveal the plant-endophyte symbiosis

Sl. no	Plant	Omics technology used	Outcomes	References
1	Rice	16S rRNA gene based metagenomics	Predicted that the bacterial community is influenced by both the plant's genotype and its environment and identification of bacteria involved in plant growth promotion and antimicrobial activities	Kunda et al. (2018)
2	Tomato	16S rRNA gene based metagenomics	Revealed that the nematode associated bacterial groups and the key infection processes during nematode pathogenesis in tomato root	Tian et al. (2015a, b)
3	*Ephedra sinica*	Metagenomics and metabolomics	Genera of *Phyllosticta* along with Ephedrine and methylephedrine metabolite were found in stem, whereas *Talaromyces*, *Aporospora*, and *Aspergillus* along with metabolites ephedrannin A and mahuannin A were specific and abundant in roots	Zhang et al. (2020)
4	Maize	Shot gun metagenomics	Impact of fertilizer applications on the functional diversity of endophytic microbiomes	Fadiji et al. (2021)
5	*Arabidopsis* sp.	Metatranscriptomics	Comparison of pattern of expression during infection with pathogenic and endophytic strain of Fusarium	Guo et al. (2021)
6	Atractylodes lancea	Transcriptomics and proteomics	Improved production (biomass and sesquiterpenoids accumulation) and enhanced the metabolic flux (sesquiterpenoids biosynthesis pathway)	Yuan et al. (2019a, b)

(continued)

Table 13.1 (continued)

Sl. no	Plant	Omics technology used	Outcomes	References
7	Maize	Transcriptomics	Identification of bacterial and fungal communities	Tao et al. (2024)
8	Sorghum	Transcriptomics	The activation of genes involved in mineral transport (including nitrate and aquaporin transporters), auxin signaling, sugar assimilation (hexokinase), and disease resistance (thaumatin) was observed in the roots of sorghum inoculated with *Trichoderma afroharzianum* T22	Kabir et al. (2024)
9	Sugarcane	Proteomics	A total of 27,508 proteins were identified, with 378 differentially expressed proteins (DEPs) observed in the treated sugarcane variety compared to the control. Among those proteins, 244 were upregulated and 134 were downregulated	Guo et al. (2022)
10	Perennial ryegrass	Proteomics	Upregulation of tspA gene was found which was responsible for symbiotic relationship with the endophyte	Bassett 2007
11	*Vitis vinifera*	Metabolomics	Chlorins, phenolics, flavonoid and terpenoid glycosides, tannins, dihydropyranones, sesquiterpene lactones, and long-chain unsaturated fatty acids were upregulated in endophyte colonized plants	Stranska et al. (2021)
12	*Mesembryanthemum crystallinum*		The *endophytic* strain led to elevated levels of metabolites associated with the tricarboxylic acid cycle and photosynthesis	Kataoka et al. (2021)

13.5.1 Microscopy

Microscopy, including light microscopy, fluorescence microscopy, and electron microscopy, is often used for the initial visualization of endophytes within plant tissues. Fluorescent dyes and specific stains can highlight microbial cells to observe their localization within plant tissues (Kandel et al. 2017).

13.5.2 Culture-Based Methods

Traditional culture-based methods involve isolating endophytes from surface-sterilized plant tissues on selective media. While this method can identify culturable endophytes, it often underrepresents the full diversity, as many endophytes are not easily cultured under laboratory conditions (Petrini 1991).

13.5.3 Molecular Techniques

Molecular methods, particularly those involving the polymerase chain reaction (PCR), are widely used for identifying endophytes. Techniques such as DNA extraction, amplification of specific genes (e.g., 16S rRNA for bacteria and ITS for fungi), and sequencing are common approaches to identify and classify endophytes (Landa et al. 2013).

13.5.4 Next-Generation Sequencing (NGS)

NGS technologies, including metagenomics and amplicon sequencing, have revolutionized the study of endophyte communities. These methods allow for the comprehensive analysis of entire microbial communities directly from plant tissues without the need for culturing. Techniques such as 16S rRNA gene sequencing for bacteria and ITS sequencing for fungi provide deep insights into the diversity and composition of endophyte communities (Tamošiūnė et al. 2019).

13.5.5 Metagenomics and Metatranscriptomics

Metagenomics involves sequencing the collective genomes of the microbial community, providing insights into the functional potential of endophytes. Metatranscriptomics, on the other hand, focuses on the active gene expression profiles of endophytes, offering a dynamic view of their functional roles in plant hosts (de Medeiros Azevedo et al. 2021).

13.5.6 Biomarker-Based Approaches

Specific biomarkers, such as fatty acid methyl esters (FAME) or protein profiles, can also be used to identify and characterize endophyte communities. These methods are less common but can provide unique insights into the microbial makeup of plant tissues (El-Banna et al. 2021).

13.5.7 Community Profiling Using DNA Fingerprinting

Techniques like Denaturing Gradient Gel Electrophoresis (DGGE) and Terminal Restriction Fragment Length Polymorphism (T-RFLP) can be used for profiling microbial communities. These methods allow for the comparison of endophyte communities across different plants or environmental conditions (van Overbeek et al. 2006).

13.6 Whole Genome Sequencing

Whole genome sequencing (WGS) is a powerful tool for studying endophytes, offering in-depth insights into their genetic makeup, functions, and evolutionary relationships. This approach allows to sequence the entire genome of an endophyte, providing a comprehensive understanding of its capabilities and interactions with plant hosts. The WGS gives a comprehensive genomic characterization which helps to identify all genes present, including those involved in symbiosis, secondary metabolite production, and stress tolerance (Xu et al. 2014). By comparing the genomes of endophytic strains with non-endophytic relatives, WGS helps identify genes that are specifically involved in establishing and maintaining symbiosis with plants. These may include genes related to nutrient exchange, signal transduction, and the suppression of plant immune responses (Hardoim et al. 2015a, b). The field of genomics is propelled by cutting-edge technologies like high-throughput DNA sequencing methods, including Illumina HiSeq, PacBio, and Nanopore sequencing, along with single nucleotide polymorphism (SNP) chips, which facilitate the identification and analysis of genetic variations among individuals. Whole genome sequencing analysis can help identify the genes responsible for growth and development, defense mechanisms, nutrient acquisition, production of secondary metabolites, and various other processes within the organism (Jha et al. 2023).

Endophytes are known for producing a variety of secondary metabolites that can benefit the host plant, such as antibiotics, antifungals, and growth-promoting substances. WGS maps the entire set of biosynthetic gene clusters in an endophyte, facilitating the discovery of novel compounds with potential agricultural or pharmaceutical applications (Kusari et al. 2012). WGS can also be used to study the genetic diversity and evolutionary history of endophyte populations. By sequencing multiple strains or isolates, researchers can investigate how endophytes have adapted to different hosts or environments, revealing patterns of host specialization or co-evolution (Arora and Verma 2017). WGS data can be annotated to predict the functions of genes within an endophyte's genome. This involves using bioinformatics tools to identify genes involved in key processes such as nitrogen fixation, phosphate solubilization, or production of plant hormones like auxins and gibberellins (Loper et al. 2012).

Arnold and Engelbrecht (2007) revealed instances of horizontal gene transfer (HGT) between endophytes and other microorganisms or between endophytes and their plant hosts through WGS. This transfer of genetic material can result in new traits that enhance the endophyte's ability to colonize the host or provide benefits such as increased stress tolerance. WGS data can be used to reconstruct the metabolic pathways of endophytes, providing insights into how these organisms obtain nutrients, synthesize essential compounds, and interact metabolically with their host plants. This can also reveal potential metabolic dependencies between the host and the endophyte (Tian et al. 2015a, b). Whole genome sequencing thus plays a crucial role in advancing our understanding of endophytes, providing insights into their genetic diversity, functional capabilities, and interactions with plant hosts.

13.7 Metagenomics

Profiling the microbial community can be achieved by analyzing genomic samples from the entire population of microbes taken from their natural environment. Metagenomic DNA extracted from resilient environments, like endophytic habitats, has been effectively utilized to determine the composition and diversity of microbiomes in these environments through metagenomic techniques (Maropola et al. 2015). Metagenomics enables the identification of the full spectrum of microbial diversity within a plant's tissues, including bacteria, fungi, archaea, and viruses. By sequencing the total DNA extracted from plant tissues, unculturable endophytes can be revealed that might be missed by culture-based methods. It also helps to identify the species present but also provides information about the functional genes within the endophyte community (Fadiji and Babalola 2020).

Metagenomics has significantly advanced the understanding of endophyte communities, revealing their complex interactions with host plants and their ecological roles. Endophytes, residing within plant tissues, contribute to plant growth and resilience against biotic and abiotic stresses through mechanisms such as nutrient uptake and phytohormone synthesis (Adeleke and Babalola 2022). Recent studies utilizing next-generation sequencing have uncovered diverse endophytic fungal and bacterial communities across various plant species, including mangroves and desert plants (Alsaedi et al. 2022). A study on Acanthus ilicifolius identified 307 operational taxonomic units (OTUs), predominantly from the Ascomycota phylum (Shah et al. 2024). Metagenomics approach can be effectively used to reveal the varied tissue specific endophytes and also helps to understand the dynamics of endophytes in different environmental conditions. In a metagenomic study of endophytic bacteria in *Aloe vera*, NGS of 16S rDNA sequences (V3-V4 regions) produced 5,199,102 reads, revealing *Proteobacteria*, *Firmicutes*, *Actinobacteria*, and *Bacteroidetes* as dominant genera. It was observed that root tissues contained the highest microbial diversity, with 23% unique to roots (Akinsanya et al. 2015). A study by Fadiji et al. (2020) explored the diversity and nutrient pathways of root-associated bacterial endophytes in maize grown with different fertilizers using shotgun metagenomics. A total of 28 phyla were identified, with *Bacteroidetes* dominating in organic fertilizer-treated plants, *Firmicutes* in unfertilized plants, and *Proteobacteria* in inorganic fertilizer-treated plants. The results indicated higher bacterial diversity in organic fertilizer sites, with enhanced nitrogen and phosphorus metabolism, suggesting that organic fertilizers promote sustainable agriculture and support novel endophytic bacteria beneficial for biotechnological applications. Root-associated bacterial endophytes of *Panax ginseng* were analyzed at various ages using shotgun metagenomics. *Proteobacteria* and *Actinobacteria* were predominant, with *Alphaproteobacteria* most abundant in 3- to 5-year-old plants. Three-year-old plants had the highest bacterial diversity and showed significant presence of plant growth-promoting traits, including siderophore and auxin-related functions (Hong et al. 2019).

A study done by Puri et al. 2019 examined the impact of soil and climatic conditions on sweet potato bacterial endophyte communities using locked nucleic acid-PCR clamping and metagenomic analysis. Dominant phyla included *Proteobacteria* (85.0%), *Bacteroidetes* (6.6%), and *Actinobacteria* (6.3%), with *Pseudomonas* being the most prevalent genus. The results showed minimal variation in endophytic bacterial communities across different samples, suggesting that soil and climate had little effect on the endophytic microbiome. 16S rRNA gene-based Illumina NGS was performed to investigate the taxonomic diversity and potential plant growth-promoting (PGP) traits of endophytic bacteria in root, shoot, and callus tissues of two aromatic rice cultivars. Significant variability in bacterial OTUs and genera was observed between root and shoot samples. *Sphingomonas* was the only genus present in all samples. Functional annotation revealed PGP trait-related genes involved in nitrogen fixation, phosphate solubilization, and indole acetic acid production, varying by tissue type and genotype (Krishnamoorthy et al. 2021). Metagenomics offers a powerful tool for exploring the hidden diversity and functionality of endophyte communities. It has revolutionized the study of endophytes, providing a comprehensive, culture-independent approach to exploring their diversity, functions, and interactions with host plants.

13.8 Transcriptomics and Metatranscriptomics

Transcriptomics and metatranscriptomics are critical for understanding the functional dynamics of endophytes within their plant hosts. They help in studying the gene expression profiles of endophytes, revealing how they interact with their hosts, respond to environmental conditions, and contribute to plant health. Transcriptomics helps identify which genes are actively expressed when they colonize a plant. This provides insights into the biological processes endophytes engage in during symbiosis, such as nutrient exchange, stress response, and production of bioactive compounds. Transcriptomic studies can reveal the specific transporters and metabolic pathways endophytes use to assimilate and deliver these nutrients to the host plant. Yi et al. (2017) performed global transcriptome analysis of endophytic and soil isolate of *Bacillus mycoides* which revealed that the endophytic isolate exhibited a more pronounced response compared to the soil-derived isolate, with a greater number of significantly differentially expressed genes identified. Both isolates showed altered expression in a common set of genes, previously associated with plant–microbe interactions, including those involved in organic substance metabolism, oxidative reduction, and transmembrane transport. Interestingly, several genes were specifically upregulated in the endosphere isolate EC18, while being downregulated in the soil isolate SB8. These genes primarily encoded membrane proteins, transcriptional regulators, or were involved in amino acid metabolism and biosynthesis.

Endophytes can enhance plant stress tolerance by expressing genes involved in the production of stress-alleviating compounds. Transcriptomic analysis might show upregulation of genes responsible for antioxidant production during drought conditions (Estrada et al. 2013). Metatranscriptomics extends transcriptomic analysis to entire microbial communities, allowing for the study of all active genes in an endophyte community within a plant. This approach provides a comprehensive view of the functional roles of different endophytes and their interactions within the host environment. By analyzing the collective transcriptome of an endophyte community, metatranscriptomics can uncover the diversity of functions performed by different members of the community, such as those involved in plant hormone modulation, pathogen suppression, or secondary metabolite production. Metatranscriptomics reveals how the gene expression profiles of endophyte communities change in response to environmental stimuli, such as pathogen attack or nutrient availability. This helps in understanding the dynamic interactions between endophytes and their plant hosts.

Guo et al. 2021 reported the plant genes that were preferentially expressed during the endophytic interaction were enriched in gene ontology (GO) terms related to the cell cycle, cell growth, development, response to stimuli, and cellular transport. Furthermore, the genes associated with each enriched GO term exhibited a temporal pattern throughout the infection process. Genes involved in the cell cycle were highly enriched in the early stages of infection but gradually decreased in importance, being replaced by genes related to development, response to stimuli and transport.

Transcriptomic studies can reveal how endophytes perceive and respond to plant signals, leading to the activation of symbiosis-related genes. This includes the expression of genes involved in the synthesis of signaling molecules such as phytohormones or quorum-sensing compounds. Endophytes may modulate the plant's immune system, allowing them to establish a stable association without triggering defense responses (Schlaeppi and Bulgarelli 2015). Transcriptomic analysis can identify endophyte genes that help evade or suppress plant immune responses. Comparing the genomes of endophytic and non-endophytic isolates can reveal traits that are likely essential for establishing endophytic behavior and sustaining plant-microbe interactions. These studies also provide valuable insights into the genetic loci responsible for niche adaptation (Dudeja et al. 2021). Endophytes are known for producing unique secondary metabolites with antimicrobial, antifungal, or plant growth-promoting properties. Transcriptomic and meta-transcriptomics studies can identify the gene clusters responsible for the biosynthesis of these compounds, providing targets for biotechnological exploitation and stress-responsive genes that could be transferred to crop plants to enhance their resilience.

13.9 Proteomics and Metaproteomics

Proteomics and metaproteomics play significant roles in studying endophytes by providing a deeper understanding of the proteins expressed by these microorganisms and their functional interactions with plant hosts. These approaches allow researchers to investigate the active molecular mechanisms through which endophytes contribute to plant health, stress tolerance, and other beneficial traits (Rodriguez et al. 2009). Leetanasaksakul et al. 2024 explored the impact of endophytic fungus *Aspergillus cejpii* on rice through proteomic approach and found that the endophyte generated auxin-responsive proteins to control the IAA levels in plant tissue and promoting overall chlorophyll production through the upregulation of proteins involved in the chlorophyll biosynthesis pathway.

Proteomics and metaproteomics can lead to the discovery of novel proteins with unique functions that are not present in the host plant or other well-studied microbes. Proteomics involves the large-scale study of proteins, including their expression, structure, and functions. For endophytes, proteomics helps identify the proteins that are actively produced when these microorganisms colonize plant tissues. Expression of proteins that help plants tolerate abiotic stresses like drought, salinity, or heavy metals. These proteins might include antioxidants, heat shock proteins, and pathogenesis-related proteins or enzymes involved in osmoprotection. Two proteins, H9X056 and H9VDW5, were uniquely detected in the *Pinus taeda* needles of root endophytic fungus Serendipita indica inoculated seedlings subjected to drought stress. The protein H9VNE7 was upregulated over 11 times compared to non-inoculated seedlings under the same stress. It was also observe that *S. indica* inoculation also increased levels of water deficiency-inducible proteins (e.g., LP3-1, LP3-2, LP3-3, and dehydrins) and ribosomal proteins (e.g., A0A385JF23). Additionally, the inoculation led to significant alterations in biosynthesis and metabolism pathways, including phenylpropanoid biosynthesis and 2-oxocarboxylic acid metabolism (Wu et al. 2024). Zeng et al. (2023) identified the differentially expressed proteins (DEPs) in Pseudomonas alcaliphila-inoculated and control rice plants. A total of 116 DEPs were found, with 78 proteins upregulated and 38 down-regulated between the control and endophyte inoculated plants which included pathways of metabolism, plant-pathogen interactions, and biosynthesis of unsaturated fatty acids.

Metaproteomics extends proteomic analysis to entire microbial communities, allowing for the characterization of all proteins expressed by endophytes within a plant. Metaproteomics can reveal the collective metabolic activities of an endophyte community, such as nutrient cycling, pathogen suppression, or plant growth promotion. This is particularly useful for understanding how different endophytes work together to support plant health (Wilmes and Bond 2006). It can also track changes in protein expression profiles of endophyte communities in response to environmental stimuli, such as pathogen attack or abiotic stress. This helps in understanding the adaptive strategies employed by endophytes in different conditions. Metaproteomics can help identify which proteins are expressed in these tissues, providing insights into the localized effects of endophytes on plant health and also give a glimpse of temporal variations, revealing how endophyte activities are coordinated with plant development (Subudhi et al. 2018).

Findings of Bona et al. 2019 highlighted the significant potential of proteomics for characterizing environmental microbial communities and their functions. Specifically, they demonstrated that bacteria from the *Streptomyces, Bacillus, Bradyrhizobium, Burkholderia,* and *Pseudomonas* genera are highly active in protein expression. Additionally, it was noted that these genera are exclusively involved in phosphorus metabolism and the regulation of primary metabolic processes in the rhizosphere.

13.10 Mitogenomics

Mitogenomics of endophytes involves studying the mitochondrial genomes of fungal species residing within plants. Research on various endophytic fungi species like *Edenia* sp., *Leptosphaerulina* sp., *Curvularia trifolii, Pestalotiopsis fici,* and *Guillardia theta* revealed diverse mitogenome sizes, gene contents, and evolutionary relationships within the Pleosporales lineage (Huang et al. 2022). Notably, the mitogenomes of these endophytic fungi contain essential genes like protein-coding genes, tRNA genes, and rRNA genes, aiding in understanding their genetic makeup and evolutionary history (Yuan et al. 2019a, b). Comparative analyses of mitogenomes have highlighted unique features such as the absence of group I introns in certain species and the presence of introns possibly acquired through lateral gene transfer, contributing to a deeper comprehension of endophyte evolution and diversity at the genomic level.

13.11 Microarray-Based Techniques

Microarray-based techniques have significantly advanced the multi-omics study of endophyte communities, particularly in understanding environmental dynamics. It is conducted to explore the mysteries of endophytism, focusing on gene profiling and expression studies of endophytes, which have helped uncover the potential interactions between the host plant and its associated endophytes. Research indicates that these techniques enable the simultaneous analysis of various biological layers, including genomic, transcriptomic, and metabolomic data, which are crucial for elucidating the interactions between endophytes and their environments (Yamazaki et al. 2021). For instance, Van Nostrand et al. (2010) highlight the importance of integrating metagenomic and metatranscriptomic approaches to capture the functional potential and activity of microbial communities in response to environmental changes. Additionally, Kyselková et al. (2009) emphasize that microarray technologies facilitate the identification of specific endophyte taxa and their functional roles, providing insights into their adaptive strategies under varying environmental conditions. However, limitations such as the complexity of data interpretation and the need for robust bioinformatics tools remain challenges in fully harnessing these techniques for environmental studies. Overall, microarray-based multi-omics approaches are pivotal in advancing our understanding of endophyte community dynamics in diverse ecosystems.

13.12 Metabolomics and Meta-Metabolomics

Metabolomics and meta-metabolomics offer powerful tools to investigate the biochemical and metabolic interactions within endophyte communities. These studies analyze the small molecules (peptides, sugars, fatty acids, organic acids, vitamins, etc.) produced by endophytes and their interactions with host plants and the environment, contributing to a deeper understanding of their ecological roles and the potential benefits they provide to plants (Tahir et al. 2022). Metabolomics focuses on profiling and quantifying metabolites produced by individual endophyte species under different environmental conditions. These metabolites can include hormones, antibiotics, and other bioactive compounds that help plants resist pathogens, improve nutrient uptake, or adapt to environmental stresses such as drought or salinity. Metabolomic analysis can identify unique metabolic signatures in endophytes, providing insights into their biochemical roles in promoting plant health (Chen et al. 2021). Metabolomics is a valuable method for identifying and quantifying the complete range of metabolites within an organism, which can then be linked to its phenotypic characteristics in changing environmental conditions (Chen et al. 2022).

Scherling et al. (2009) used gas chromatography-mass spectrometry to analyze the impact of endophytic bacteria on the metabolites of in vitro-grown poplar shoots. A strain of *Paenibacillus* significantly altered the metabolic composition of poplars, with 11 metabolites showing differences between inoculated and non-inoculated plants. The shifts suggest a mutualistic interaction involving nitrogen fixation by the bacteria, leading to increased asparagine and urea levels, and reduced sugars and tricarboxylic acid cycle organic acids. This highlights the role of Paenibacillus in modifying nitrogen assimilation in poplars. 1H-NMR analysis of *Combretum lanceolatum* after inoculation with its endophytic fungi *Diaporthe phaseolorum* notably influenced the plant's metabolic pathways, enhancing the production of key primary metabolites like threonine, malic acid, and N-acetyl-mannosamine, which served as precursors to specialized metabolites important for plant defense mechanisms (Lacerda et al. 2021). The study by Pellissier et al. (2023) utilized advanced metabolome characterization methods, including

UHPLC-HRMS/MS and molecular networking, to examine the metabolomes of host leaves and 15 endophytic fungi. By constructing a molecular network, metabolites from both sources were identified, uncovering over 24,000 features that highlighted the chemodiversity of endophytes and shared metabolites within the plant-fungal community. Interestingly, only a small fraction of metabolites overlapped between the host leaves and fungal samples, despite significant enrichment in the plant metabolome. This suggests that fungal metabolites produced in vitro are scarcely detectable in plant tissues. The study provides a unique and comprehensive insight into the metabolomic interactions between a plant and its microbiome.

Meta-metabolomics takes a broader approach by analyzing the collective metabolome of entire endophyte communities within a host plant or ecosystem. This approach accounts for the interactions among various endophytic species and their combined impact on the host plant's metabolism. It provides a holistic view of the chemical exchanges and signaling processes that occur between the plant and its associated microbiome (Zulfiqar et al. 2024). Such studies reveal the synergistic or antagonistic relationships within endophyte communities, which may lead to the discovery of novel metabolites or secondary metabolites with applications in agriculture or medicine (Harikrishnan et al. 2021). Meta-metabolomics focuses on both endo- and exometabolites, offering valuable insights for studying priming. To effectively model the intricate interaction networks between endo/epiphytes and their plant hosts, it is essential to integrate both targeted and untargeted approaches with systems biology across various -omic scales (Mhlongo et al. 2018). When integrated with other omics approaches—such as metagenomics, metatranscriptomics, and metaproteomics—metabolomics and meta-metabolomics provide a comprehensive understanding of the environmental dynamics and functional capabilities of endophyte communities.

13.13 Fluxomics

Fluxomics of endophytes refers to the study of metabolic fluxes within endophytic microorganisms. This involves analyzing the rates at which metabolites are produced, consumed, and converted within the metabolic pathways of endophytes. This approach provides insights into the metabolic activities of endophytes, their interactions with host plants, and how they contribute to the overall health and growth of the plant. It also offers valuable insights into the dynamic behavior of metabolic networks and aids in identifying crucial metabolic pathways and regulatory processes (Fan 2024).

13.14 Conclusion

Endophytes are a promising source of novel metabolites, industrial enzymes, and plant stress relievers, yet much of their biology remains unknown. The roles of endophytes are shaped by their interactions with the host plant and other members of the host's microbiome, making it crucial to study these relationships as part of a broader system (Kaul et al. 2016). Multi-omics studies allow researchers to link genetic information to biochemical outcomes, revealing how environmental factors influence endophyte metabolism and how endophytes, in turn, affect the plant's physiology and adaptation to changing conditions. This integrated approach is crucial for understanding the resilience of ecosystems under stress, the sustainability of plant-microbe interactions, and the potential for biotechnological applications, such as enhancing crop productivity, bioremediation, or producing novel natural products.

Advanced techniques like genome sequencing, metagenomics, metatranscriptomics, metproteomics, and metabolomics are essential for uncovering the intricate plant-endophyte relationships, offering insights into their ecological roles within plant microbiomes (López-Mondéjar et al. 2017). The multi-omics approach, while serving as a broad tool for evaluating endophyte diversity, also has its limitations. Firstly, there is a lack of standardized tools and techniques for integrative multi-omics, particularly for small-scale microbiomes like the endohyphal microbiome, which hampers comprehensive investigations of microbial diversity and functionality within fungal hosts (Kelliher et al. 2023). While meta-omics approaches have shown promise in uncovering microbial interactions and functional roles, the complexity of these interactions and the variability in gene expression under different environmental conditions can complicate data interpretation (Adeleke and Babalola 2022; Chen et al. 2022). Lastly, the integration of diverse omics data remains a challenge, as it requires sophisticated analytical frameworks that are still under development. These limitations highlight the need for continued methodological advancements in multi-omics research.

References

Adeleke BS, Babalola OO (2022) Meta-omics of endophytic microbes in agricultural biotechnology. Biocatal Agric Biotechnol 42:102332

Akinsanya MA, Goh JK, Lim SP, Ting ASY (2015) Metagenomics study of endophytic bacteria in Aloe vera using next-generation technology. Genomics Data 6:159–163

Alsaedi ZS, Ashy RA, Shami AY, Majeed MA, Alswat AM, Baz L et al (2022) Metagenomic study of the communities of bacterial endophytes in the desert plant Senna Italica and their role in abiotic stress resistance in the plant. Braz J Biol 82:e267584

Ansari MM, Bisht N, Singh T, Chauhan PS (2024) Symphony of survival: insights into cross-talk mechanisms in plants, bacteria, and fungi for strengthening plant immune responses. Microbiol Res 285:127762

Arnold AE, Engelbrecht BMJ (2007) Fungal endophytes nearly double minimum leaf conductance in seedlings of a tropical tree. J Trop Ecol 23(3):369–372

Arora NK, Verma M (2017) Modified microplate method for rapid and efficient estimation of siderophore produced by bacteria. 3 Biotech 7(5):327

Bacon CW, White JF (2016) Functions, mechanisms and regulation of endophytic and epiphytic microbial communities of plants. Symbiosis 68(1–3):87–98

Bassett SA (2007) Identification of proteins involved in the symbiotic interaction between the fungal endophyte Neotyphodium Lollii with perennial ryegrass (Lolium Perenne)

Bona E, Massa N, Novello G, Boatti L, Cesaro P, Todeschini V et al (2019) Metaproteomic characterization of the Vitis vinifera rhizosphere. FEMS Microbiol Ecol 95(1):fiy204

Chen L, Schwier M, Krumbach J, Kopriva S, Jacoby RP (2021) Metabolomics in plant-microbe interactions in the roots. Adv Bot Res 98:133–161
Chen XL, Sun MC, Chong SL, Si JP, Wu LS (2022) Transcriptomic and metabolomic approaches deepen our knowledge of plant–endophyte interactions. Front Plant Sci 12:700200
Clay K (1993) The ecology and evolution of endophytes. Agric Ecosyst Environ 44(1–4):39–64
Compant S, Saikkonen K, Mitter B, Campisano A, Mercado-Blanco J (2016) Editorial special issue: soil, plants and endophytes. Plant Soil 405:1–11
de Medeiros Azevedo T, Aburjaile FF, Ferreira-Neto JRC, Pandolfi V, Benko-Iseppon AM (2021) The endophytome (plant-associated microbiome): methodological approaches, biological aspects, and biotech applications. World J Microbiol Biotechnol 37(12):206
Delaye L, García-Guzmán G, Heil M (2013) Endophytes versus biotrophic and necrotrophic pathogens—are fungal lifestyles evolutionarily stable traits? Fungal Divers 60(1):125–135
Dos Reis JBA, Lorenzi AS, do Vale HMM (2022) Methods used for the study of endophytic fungi: a review on methodologies and challenges, and associated tips. Arch Microbiol 204(11):675
Dudeja SS, Suneja-Madan P, Paul M, Maheswari R, Kothe E (2021) Bacterial endophytes: molecular interactions with their hosts. J Basic Microbiol 61(6):475–505
El-Banna AN, El-Mahrouk ME, Dewir YH, Farid MA, Abou Elyazid DM, Schumacher HM (2021) Endophytic bacteria in banana in vitro cultures: molecular identification, antibiotic susceptibility, and plant survival. Horticulturae 7(12):526
Estrada P, Davila E, Navarro L, López-Ruiz MA (2013) Transcriptome analysis of the endophyte Pestalotiopsis sp. during its interaction with its host plant Camellia sinensis. J Microbiol Biotechnol 23(5):630–636
Fadiji AE, Babalola OO (2020) Metagenomics methods for the study of plant-associated microbial communities: a review. J Microbiol Methods 170:105860
Fadiji AE, Ayangbenro AS, Babalola OO (2020) Metagenomic profiling of the community structure, diversity, and nutrient pathways of bacterial endophytes in maize plant. Antonie Van Leeuwenhoek 113(11):1559–1571
Fadiji AE, Ayangbenro AS, Babalola OO (2021) Shotgun metagenomics reveals the functional diversity of root-associated endophytic microbiomes in maize plant. Curr Plant Biol 25:100195
Fan K (2024) Metabolomics: insights into plant-pathogen, symbiont, and endophyte interactions unveiling chemical communications. Authorea Preprints
Frank AC, Saldierna Guzmán JP, Shay JE (2017) Transmission of bacterial endophytes. Microorganisms 5(4):70
Guo L, Yu H, Wang B, Vescio K, DeIulio GA, Yang H et al (2021) Metatranscriptomic comparison of endophytic and pathogenic Fusarium–Arabidopsis interactions reveals plant transcriptional plasticity. Mol Plant-Microbe Interact 34(9):1071–1083
Guo DJ, Li DP, Singh RK, Singh P, Verma KK, Sharma A et al (2022) Comparative transcriptome analysis of two sugarcane varieties in response to diazotrophic plant growth promoting endophyte Enterobacter roggenkampii ED5. J Plant Interact 17(1):75–84
Hardoim PR, Hardoim CC, van Overbeek LS, van Elsas JD (2012) Dynamics of seed-borne rice endophytes on early plant growth stages. PLoS One 7(2):e30438
Hardoim PR, van Overbeek LS, Elsas JD (2015a) Dynamics of seed-borne rice endophytes on early plant growth stages. PLoS One 10(5):e0127505
Hardoim PR, Van Overbeek LS, Berg G, Pirttilä AM, Compant S, Campisano A et al (2015b) The hidden world within plants: ecological and evolutionary considerations for defining functioning of microbial endophytes. Microbiol Mol Biol Rev 79(3):293–320
Harikrishnan A, Shanmugavalli R, Oshin K, Veena V, Lakshmi B (2021) Commercial exploitation of microbial communal services to enrich plant microbiome. In: Metagenomics and microbial ecology. CRC Press, pp 181–192
Hong CE, Kim JU, Lee JW, Bang KH, Jo IH (2019) Metagenomic analysis of bacterial endophyte community structure and functions in Panax ginseng at different ages. 3 Biotech 9:1–8
Huang L, Mu Y, Zhang X, Chang K, Zhang J (2022) The mitochondrial genome of the endophyte Edenia gomezpompae CRI Eg3 isolated from sweet potato. Mitochondrial DNA Part B 7(3):454–455

Jha P, Kaur T, Chhabra I, Panja A, Paul S, Kumar V, Malik T (2023) Endophytic fungi: hidden treasure chest of antimicrobial metabolites interrelationship of endophytes and metabolites. Front Microbiol 14:1227830

Kabir AH, Thapa A, Hasan MR, Parvej MR (2024) Local signal from Trichoderma afroharzianum T22 induces host transcriptome and endophytic microbiome leading to growth promotion in sorghum. J Exp Bot 75:erae340

Kandel SL, Joubert PM, Doty SL (2017) Bacterial endophyte colonization and distribution within plants. Microorganisms 5(4):77

Kataoka R, Akashi M, Taniguchi T, Kinose Y, Yaprak AE, Turgay OC (2021) Metabolomics analyses reveal metabolites affected by plant growth-promoting endophytic bacteria in roots of the halophyte Mesembryanthemum crystallinum. Int J Mol Sci 22(21):11813

Kaul S, Sharma T, Dhar K, M. (2016) "Omics" tools for better understanding the plant–endophyte interactions. Front Plant Sci 7:955

Kelliher JM, Robinson AJ, Longley R, Johnson LY, Hanson BT, Morales DP et al (2023) The endohyphal microbiome: current progress and challenges for scaling down integrative multi-omic microbiome research. Microbiome 11(1):192

Krishnamoorthy A, Gupta A, Sar P, Maiti MK (2021) Metagenomics of two gnotobiotically grown aromatic rice cultivars reveals genotype-dependent and tissue-specific colonization of endophytic bacterial communities attributing multiple plant growth promoting traits. World J Microbiol Biotechnol 37(4):59

Kumar V, Baweja M, Singh PK, Shukla P (2016) Recent developments in systems biology and metabolic engineering of plant–microbe interactions. Front Plant Sci 7:1421

Kunda P, Dhal PK, Mukherjee A (2018) Endophytic bacterial community of rice (Oryza sativa L.) from coastal saline zone of West Bengal: 16S rRNA gene based metagenomics approach. Meta Gene 18:79–86

Kusari S, Hertweck C, Spiteller M (2012) Chemical ecology of endophytic fungi: origins of secondary metabolites. Chem Biol 19(7):792–798

Kyselková M, Kopecký J, Ságová-Marečková M, Grundmann GL, Moenne-Loccoz Y (2009) Oligonucleotide microarray methodology for taxonomic and functional monitoring of microbial community composition. Plant Soil Environ 55(9):379–388

Lacerda JW, Siqueira KA, Vasconcelos LG, Bellete BS, Dall'Oglio EL, Sousa Junior PT et al (2021) Metabolomic analysis of Combretum lanceolatum plants interaction with Diaporthe phaseolorum and Trichoderma spirale Endophytic fungi through 1H-NMR. Chem Biodivers 18(10):e2100350

Landa BB, López-Díaz C, Jiménez-Fernández D, Montes-Borrego M, Muñoz-Ledesma FJ, Ortiz-Urquiza A, Quesada-Moraga E (2013) In-planta detection and monitorization of endophytic colonization by a Beauveria bassiana strain using a new-developed nested and quantitative PCR-based assay and confocal laser scanning microscopy. J Invertebr Pathol 114(2):128–138

Leetanasaksakul K, Roytrakul S, Kittisenachai S, Lohmaneeratana K, Jantasuriyarat C, Lueangjaroenkit P (2024) Exploring the impact of Endophytic fungus aspergillus cejpii DMKU-R3G3 on Rice: plant growth promotion and molecular insights through proteomic analysis. Agronomy 14(3):498

Loper JE, Hassan KA, Mavrodi DV, Davis EW, Lim CK, Shaffer BT et al (2012) Comparative genomics of plant-associated pseudomonas spp.: insights into diversity and inheritance of traits involved in multitrophic interactions. PLoS Genet 8(7):e1002784

López-Mondéjar R, Kostovčík M, Lladó S, Carro L, García-Fraile P (2017) Exploring the plant microbiome through multi-omics approaches. In: Probiotics in agroecosystem, pp 233–268

Ma B, Lv X, Warren A, Gong J (2013) Shifts in diversity and community structure of endophytic bacteria and archaea across root, stem and leaf tissues in the common reed, Phragmites australis, along a salinity gradient in a marine tidal wetland of northern China. Antonie Van Leeuwenhoek 104:759–768

Maropola MKA, Ramond JB, Trindade M (2015) Impact of metagenomic DNA extraction procedures on the identifiable endophytic bacterial diversity in Sorghum bicolor (L. Moench). J Microbiol Methods 112:104–117

Mennicken S, Paula CCPD, Vogt-Schilb H, Jersáková J (2024) Diversity of Mycorrhizal fungi in temperate orchid species: comparison of culture-dependent and culture-independent methods. J Fungi 10(2):92

Mhlongo MI, Piater LA, Madala NE, Labuschagne N, Dubery IA (2018) The chemistry of plant–microbe interactions in the rhizosphere and the potential for metabolomics to reveal signaling related to defense priming and induced systemic resistance. Front Plant Sci 9:112

Mishra S, Bhattacharjee A, Sharma S (2021) An ecological insight into the multifaceted world of plant-endophyte association. Crit Rev Plant Sci 40(2):127–146

Pathak P, Rai VK, Can H, Singh SK, Kumar D, Bhardwaj N, Kumar A (2022) Plant-endophyte interaction during biotic stress management. Plan Theory 11(17):2203

Pellissier L, Gaudry A, Vilette S, Lecoultre N, Rutz A, Allard PM et al (2023) Comparative metabolomic study of fungal foliar endophytes and their long-lived host Astrocaryum sciophilum: a model for exploring the chemodiversity of host-microbe interactions. Front Plant Sci 14:1278745

Petrini O (1991) Fungal endophytes of tree leaves. In: Microbial ecology of leaves. Springer, pp 179–197

Puri RR, Adachi F, Omichi M, Saeki Y, Yamamoto A, Hayashi S et al (2019) Metagenomic study of endophytic bacterial community of sweet potato (Ipomoea batatas) cultivated in different soil and climatic conditions. World J Microbiol Biotechnol 35:1–8

Qadri M, Johri S, Shah BA, Khajuria A, Sidiq T, Lattoo SK et al (2013) Identification and bioactive potential of endophytic fungi isolated from selected plants of the Western Himalayas. Springerplus 2:1–14

Richardson AE, Barea JM, McNeill AM, Prigent-Combaret C (2009) Acquisition of phosphorus and nitrogen in the rhizosphere and plant growth promotion by microorganisms. Plant Soil 321(1–2):305–339

Rodriguez RJ, Henson J, Van Volkenburgh E, Hoy M, Wright L, Beckwith F, Kim YO, Redman RS (2008) Stress tolerance in plants via habitat-adapted symbiosis. ISME J 2(4):404–416

Rodriguez RJ, White JF, Arnold AE, Redman RS (2009) Fungal endophytes: diversity and functional roles. New Phytol 182(2):314–330

Saikkonen K, Wäli P, Helander M, Faeth SH (2004) Evolution of endophyte–plant symbioses. Trends Plant Sci 9(6):275–280

Salwan R, Sharma V (2023) Insight of plant–endophytes interactions. In: The chemical dialogue between plants and beneficial microorganisms. Academic, pp 167–178

Scherling C, Ulrich K, Ewald D, Weckwerth W (2009) A metabolic signature of the beneficial interaction of the endophyte Paenibacillus sp. isolate and in vitro–grown poplar plants revealed by metabolomics. Mol Plant-Microbe Interact 22(8):1032–1037

Schlaeppi K, Bulgarelli D (2015) The plant microbiome at work. Mol Plant-Microbe Interact 28(3):212–217

Scholtysik A, Unterseher M, Otto P, Wirth C (2013) Spatio-temporal dynamics of endophyte diversity in the canopy of European ash (Fraxinus excelsior). Mycol Prog 12:291–304

Shah SP, George JT, Chunduri JR (2024) Distribution and diversity studies of endophytic and rhizospheric fungi associated with the mangrove, Acanthus ilicifolius, using amplicon sequencing-based metagenomic approach. Ecol Genet Genomics 31:100261

Stranska M, Lovecka P, Vrchotova B, Uttl L, Bechynska K, Behner A, Hajslova J (2021) Bacterial endophytes from Vitis vinifera L.–metabolomics characterization of plant-endophyte crosstalk. Chem Biodivers 18(12):e2100516

Subudhi E, Sahoo RK, Dey S, Das A, Sahoo K (2018) Unraveling plant-endophyte interactions: an omics insight. In: Endophytes and secondary metabolites, pp 1–19

Sun X, Guo LD (2012) Endophytic fungal diversity: review of traditional and molecular techniques. Mycology 3(1):65–76

Tahir AT, Kang J, Bint-e-Mansoor M, Ayub J, Naureen Z, Hafeez FY (2022) Recent trends in characterization of endophytic microorganisms. In: Biocontrol mechanisms of endophytic microorganisms. Academic, pp 31–53

Tamošiūnė I, Andriūnaitė E, Stanys V, Baniulis D (2019) Exploring diversity of bacterial endophyte communities using advanced sequencing technology. In: Microbiome in plant health and disease: challenges and opportunities, pp 447–481

Tao F, Chen F, Liu H, Chen C, Cheng B, Han G (2024) Insight into the composition and differentiation of endophytic microbial communities in kernels via 368 maize transcriptomes. J Adv Res. https://doi.org/10.1016/j.jare.2024.05.018

Tian BY, Cao Y, Zhang KQ (2015a) Metagenomic insights into communities, functions of endophytes and their associates with infection by root-knot nematode, Meloidogyne incognita, in tomato roots. Sci Rep 5(1):17087

Tian B, Liu Y, Wang Y, Keyhani NO, Wang C (2015b) The entomopathogenic fungus Beauveria bassiana (Hypocreales) and its toxins. Toxins 7(3):2264–2288

Truyens S, Weyens N, Cuypers A, Vangronsveld J (2015) Bacterial seed endophytes: genera, vertical transmission and interaction with plants. Environ Microbiol Rep 7(1):40–50

Van Nostrand JD, He Z, Zhou J (2010) Analysis of microbial communities by functional gene arrays. Springer, pp 109–126

van Overbeek LS, van Vuurde J, van Elsas JD (2006) Application of molecular fingerprinting techniques to explore the diversity of bacterial endophytic communities. In: Microbial root endophytes. Springer Berlin Heidelberg, Berlin/Heidelberg, pp 337–354

Verma SK, Kharwar RN, Gond SK, Kingsley KL, White JF (2019) Exploring endophytic communities of plants: methods for assessing diversity, effects on host development and potential biotechnological applications. In: Seed endophytes: biology and biotechnology, pp 55–82

Wang SS, Liu JM, Sun J, Sun YF, Liu JN, Jia N et al (2019) Diversity of culture-independent bacteria and antimicrobial activity of culturable endophytic bacteria isolated from different Dendrobium stems. Sci Rep 9(1):10389

White JF Jr, Bultman TL (1987) Endophyte-host associations in forage grasses. VIII. Heterothallism in Epichloë typhina. Am J Bot 74(11):1716–1721

Wijayawardene NN, Bahram M, Sánchez-Castro I, Dai DQ, Ariyawansa KG, Jayalal U et al (2021) Current insight into culture-dependent and culture-independent methods in discovering Ascomycetous Taxa. J Fungi 7(9):703

Wilmes P, Bond PL (2006) Metaproteomics: studying functional gene expression in microbial ecosystems. Trends Microbiol 14(2):92–97

Wilson D (1995) Endophyte: the evolution of a term, and clarification of its use and definition. Oikos 73:274–276

Wu C, Yang Y, Wang Y, Zhang W, Sun H (2024) Colonization of root endophytic fungus Serendipita indica improves drought tolerance of Pinus taeda seedlings by regulating metabolome and proteome. Front Microbiol 15:1294833

Xu L, Ravnskov S, Larsen J, Nicolaisen M, Druzhinina IS (2014) The genome of the endophytic fungus Trichoderma virens provides clues for plant growth promotion and biological control. Front Microbiol 5:627

Yamazaki S, Mardani-Korrani H, Kaida R, Ochiai K, Kobayashi M, Nagano AJ et al (2021) Field multi-omics analysis reveals a close association between bacterial communities and mineral properties in the soybean rhizosphere. Sci Rep 11(1):8878

Yi Y, de Jong A, Frenzel E, Kuipers OP (2017) Comparative transcriptomics of Bacillus mycoides strains in response to potato-root exudates reveals different genetic adaptation of endophytic and soil isolates. Front Microbiol 8:1487

Yuan J, Zhang W, Sun K, Tang MJ, Chen PX, Li X, Dai CC (2019a) Comparative transcriptomics and proteomics of Atractylodes lancea in response to endophytic fungus Gilmaniella sp. AL12 reveals regulation in plant metabolism. Front Microbiol 10:1208

Yuan XL, Cao M, Shen GM, Zhang HB, Du YM, Zhang ZF, …, Zhang P (2019b) Comparative mitochondrial genome analysis of the tobacco endophytic fungi Leptosphaerulina chartarum and Curvularia trifolii and the phylogenetic implications. https://doi.org/10.20944/preprints201907.0119.v1

Zeng Q, Man X, Huang Z, Zhuang L, Yang H, Sha Y (2023) Effects of rice blast biocontrol strain Pseudomonas alcaliphila Ej2 on the endophytic microbiome and proteome of rice under salt stress. Front Microbiol 14:1129614

Zhang Q, Xue XZ, Miao SM, Cui JL, Qin XM (2020) Differential relationship of fungal endophytic communities and metabolic profiling in the stems and roots of Ephedra sinica based on metagenomics and metabolomics. Symbiosis 81:115–125

Zhao C, Onyino J, Gao X (2024) Current advances in the functional diversity and mechanisms underlying endophyte–plant interactions. Microorganisms 12(4):779

Zinniel DK, Lambrecht P, Harris NB, Feng Z, Kuczmarski D, Higley P et al (2002) Isolation and characterization of endophytic colonizing bacteria from agronomic crops and prairie plants. Appl Environ Microbiol 68(5):2198–2208

Zulfiqar M, Singh V, Steinbeck C, Sorokina M (2024) Review on computer-assisted biosynthetic capacities elucidation to assess metabolic interactions and communication within microbial communities. Crit Rev Microbiol 50:1–40

The manufacturer's authorised representative in the EU is Springer Nature Customer Service Centre GmbH, Europaplatz 3, 69115 Heidelberg, Germany. If you have any concerns regarding our products, please contact ProductSafety@springernature.com

Printed and bound by CPI Group (UK) Ltd, Croydon, CR0 4YY
01/10/2026
02225923-0010